Energiepolitik und Klimaschutz
Energy Policy and Climate Protection

Reihe herausgegeben von
Lutz Mez, Berlin, Deutschland
Achim Brunnengräber, Berlin, Deutschland

AF202046

Diese Buchreihe beschäftigt sich mit den globalen Verteilungskämpfen um knappe Energieressourcen, mit dem Klimawandel und seinen Auswirkungen sowie mit den globalen, nationalen, regionalen und lokalen Herausforderungen der umkämpften Energiewende. Die Beiträge der Reihe zielen auf eine nachhaltige Energie- und Klimapolitik sowie die wirtschaftlichen Interessen, Machtverhältnisse und Pfadabhängigkeiten, die sich dabei als hohe Hindernisse erweisen. Weitere Themen sind die internationale und europäische Liberalisierung der Energiemärkte, die Klimapolitik der Vereinten Nationen (UN), Anpassungsmaßnahmen an den Klimawandel in den Entwicklungs-, Schwellen- und Industrieländern, Strategien zur Dekarbonisierung sowie der Ausstieg aus der Kernenergie und der Umgang mit den nuklearen Hinterlassenschaften.

Die Reihe bietet ein Forum für empirisch angeleitete, quantitative und international vergleichende Arbeiten, für Untersuchungen von grenzüberschreitenden Transformations-, Mehrebenen- und Governance-Prozessen oder von nationalen „best practice"-Beispielen. Ebenso ist sie offen für theoriegeleitete, qualitative Untersuchungen, die sich mit den grundlegenden Fragen des gesellschaftlichen Wandels in der Energiepolitik, bei der Energiewende und beim Klimaschutz beschäftigen.

This book series focuses on global distribution struggles over scarce energy resources, climate change and its impacts, and the global, national, regional and local challenges associated with contested energy transitions. The contributions to the series explore the opportunities to create sustainable energy and climate policies against the backdrop of the obstacles created by strong economic interests, power relations and path dependencies. The series addresses such matters as the international and European liberalization of energy sectors; sustainability and international climate change policy; climate change adaptation measures in the developing, emerging and industrialized countries; strategies toward decarbonization; the problems of nuclear energy and the nuclear legacy. The series includes theory-led, empirically guided, quantitative and qualitative international comparative work, investigations of cross-border transformations, governance and multi-level processes, and national "best practice"-examples. The goal of the series is to better understand societal-ecological transformations for low carbon energy systems, energy transitions and climate protection.

Reihe herausgegeben von

PD Dr. Lutz Mez
Freie Universität Berlin

PD Dr. Achim Brunnengräber
Freie Universität Berlin

Weitere Bände in der Reihe http://www.springer.com/series/12516

Lena Bendlin

Orchestrating Local Climate Policy in the European Union

Inter-Municipal Coordination and the Covenant of Mayors in Germany and France

 Springer VS

Lena Bendlin
Deutsches Institut für Urbanistik – DIFU
Berlin, Germany

Dissertation Freie Universität Berlin, 2017

D188

Printed with the genereous support of the FAZIT-STIFTUNG and the German Association
of Female Academics (DAB).

ISSN 2626-2827　　　　　ISSN 2626-2835　(electronic)
Energiepolitik und Klimaschutz. Energy Policy and Climate Protection
ISBN 978-3-658-26505-2　　　　ISBN 978-3-658-26506-9　(eBook)
https://doi.org/10.1007/978-3-658-26506-9

This Springer VS imprint is published by the registered company Springer Fachmedien Wiesbaden
GmbH part of Springer Nature
The registered company address is: Abraham-Lincoln-Str. 46, 65189 Wiesbaden, Germany

Acknowledgements

They say it takes a village to raise a child, and it is no coincidence that a doctoral thesis is often referred to as a baby – that, like all children, needs a lot of time and attention to grow up. Now the time has come to let this child go out into the big, wide world of academia, and to thank the many people who helped and supported me during the (also many) years of writing this thesis.

First of all, I would like to express my gratitude to my supervisor Miranda A. Schreurs for her inspiration, confidence, and support. She gave me the freedom to undertake an educational intellectual journey, and gave me a hand when it was time to come to terms with it. More than that, she is an extraordinary mentor. I am much obliged to Philipp Lepenies for readily accepting to serve as my second reviewer despite his manifold duties. For their disposition to bring in their expertise to my dissertation commission, my thanks go to Sabine Kropp, Klaus Jacob, and Jan Beermann.

I extend my thanks to those who made this research possible. My work was supported by the Heinrich Böll Foundation, the German Academic Exchange Service, and the Ernst Reuter Society, and printed with the help of the FAZIT foundation and the German Association of Female Academics whose funding is gratefully acknowledged. The Covenant of Mayors, the Climate Alliance and Energy Cities were kind enough to grant me access to their events and conferences. 53 experts generously shared their knowledge and their precious time for a research interview. My hope is that my results will justify their openness, and inspire their work.

The academic village community backing me comprised many more people. I would like to thank my colleagues at the Environmental Policy Research Center and especially Kirsten Jörgensen for illuminative insights into their work, the members of the PhD colloquium for their feedback and company, and Daphne Stelter for her unweary advice and contagious smile throughout the doctoral procedure. A backbone of this project consisted of my writing group; I owe a debt of gratitude to Corinna Altenburg, Richard Forrest, Fanny Frick, Sebastian

Mehling, Lisa Pettibone, Inken Reimer and Maik Günther for their continuous support and priceless critical comments on my writing. I would also like to thank the many discussants and feedback-givers I met at various conferences for their help in refining my argumentation, in particular to Jörg Kemmerzell, Michèle Knodt, and Anne Tews for the possibility to discuss my results with an expert audience, and for sharing interview data; Henner Busch and Paul Fenton for our inspirational exchange and collaboration; Martin Jänicke and Kerstin Tews for their inspiring thoughts on the subject; Wolfram Lamping for his encouraging feedback; and Kristine Kern for her challenging criticism that helped to sharpen my argument.

I am grateful to my friends and family for their invaluable companionship, and their encouragement in times of doubt. To name but a few, I would like to thank Chrissy and Renate for their help with graphic design, and Anna, Tanya, and Leonie for their support during the final spurt towards submission. Thanks to my brother Robin for his proof that there is an afterlife to graduation, and for cheering me on. A very special thank you goes to my parents Barbara and Joachim who gave me the self-confidence that I could do this, and their trust, too. Above all, I would like to thank my husband Kurt who supported me with un-ending diligence and strength. Together we made it.

Contents

Summary

The proliferation of non-state action has raised hopes of accelerating climate change mitigation activities beyond the pace set by international negotiations. Cities and other local governments from all over the world have engaged in local climate policy. The 2015 Paris agreement recognized their contributions as an inherent component of global climate governance. Voluntary commitment systems aim to promote, coordinate and document this groundswell of local climate action. By sponsoring such initiatives, international organizations try to orchestrate local climate policy, i.e. to steer local governments indirectly through intermediate actors in a non-coercive way. Orchestration has also been applied by the European Commission with a view of implementing the climate and energy targets of the European Union. Unable to intervene directly at independent member state administrations, the Commission invited local authorities to sign the Covenant of Mayors and commit to a step-wise process for implementing local sustainable energy action plans. Networks of local and regional authorities and subnational authorities act as intermediaries on behalf of the Commission within this scheme; they promote Covenant signature among member municipalities and provide support for implementation. With more than 7,000 signatories, the Covenant of Mayors is the largest voluntary commitment scheme of its kind.

Local climate policy coordination within the Covenant in Germany and France is explored in view of contributing to recent orchestration theorizing. This approach focuses on the support provided to intermediaries and local governments and on the benefits for participating actors. Qualitative case studies examine Greater Lyon, Rennes Metropolis, Val d'Ille, Rhine-Neckar Region, and Stuttgart Region, five inter-municipal associations – i.e. groupings of neighboring municipalities – that engaged as Covenant Coordinators. Each case study addresses the research questions why and how they did that, and which modes of local climate governance emerged. In total, I conducted 47 expert interviews with 53 staff and politicians from the local up to the European level.

Cross-case analysis demonstrates that inter-municipal associations use their Covenant commitment as a resource for local leadership, territorial competition, and rescaling politics. But they struggle to sustain their commitment over time because of changes in local politics and a mismatch between Covenant support and local needs. Domestic frameworks as determined by national climate policy and territorial organization remain crucial for local climate policy. Enthusiastic accounts of the Covenant demonstrate participants' incentives not to report on difficulties and failure. Orchestration theorizing should control carefully for instances of window dressing in voluntary commitment systems, and account for domestic influences and the importance of timing. Inter-municipal climate policy coordination action can complement, but not supplement coherent climate governance at superordinate levels.

List of Abbreviations and Acronyms

°C .... Degree Celsius

100ee 100%-Erneuerbare-Energien-Regionen
(100% Renewable Energy Regions)

ADEME .... Agence de l'environnement and de la maîtrise de l'énergie
Environment and Energy Management Agency)

ALEC .... Agence locale de l'énergie et du climat
(Local energy and climate agency)

AMICA ..... EU-project "Adaptation and Mitigation – an Integrated Climate
Policy Approach"

art. .... Article

BEI .... Baseline emissions inventory

BMBF .... Bundesministerium für Bildung und Forschung
(Federal Ministry for Education and Research)

BMU .... Bundesministerium für Umwelt, Naturschutz und
Reaktorsicherheit
(Federal Ministry for Environment, Nature Conservation and
Nuclear Safety)

BMVBS .... Bundesministerium für Verkehr, Bau und Stadtentwicklung
(Federal Ministry for Transport, Construction and Urban
Development)

BMVI .... Bundesministerium für Verkehr und digitale Infrastruktur
(Federal Ministry for Transport and Digital Infrastructure)

BMWi .... Bundesministerium für Wirtschaft und Technologie
(Ministry of Economics and Technology)

C40 .... C40 Cities Climate Leadership Group

CEMR .... Council of European Municipalities and Regions

CO_2 .... Carbon dioxide

CoM .... Covenant of Mayors

CoMO .... Covenant of Mayors Office

COP .... Conference of the Parties

CoR .... Committee of the Regions

DREAL ...... Direction régionale de l'environnement, de l'aménagement et du logement
(Regional Directorate of Environment, Planning and Housing)

EDC .... Wirtschaftsförderung Region Stuttgart GmbH
(Stuttgart Region Economic Development Corporation)

EEA .... European Energy Award

e.g. .... exempli gratia (for example)

EMR .... European metropolitan region

et al. .... et alii (and others)

EU .... European Union

GDP .... Gross domestic product

GHG .... Greenhouse gases

GWh/a ...... Gigawatt hours per year

ICLEI .... Local Governments for Sustainability

i.e. .... id est (that is to say)

IEE .... Intelligent Energy Europe

IEEK .... Integriertes Energie- und Klimakonzept
(Integrated Energy and Climate Concept)

IEKP .... Integriertes Energie- und Klimaprogramm
(Integrated Energy and Climate Program)

ifeu .... Institut für Energie- und Umweltforschung Heidelberg
(Institute for Energy and Environmental Research)

KARS .... Klimaanpassung Region Stuttgart
(Climate Adaptation Stuttgart Region)

KEA .... Klimaschutz- und Energieagentur Baden-Württemberg
(Regional climate and energy agency)

Klik .... Strategisches Fachkonzept Klimaanpassung Ludwigsburg
(Strategic Climate Adaptation Concept)

KLIKS ...... Klimaschutzkonzept Stuttgart
(Climate protection plan Stuttgart)

KLIMAKS Klimaanpassungskonzept Stuttgart (Climate adaptation plan)

KlippS ...... Klimaplanungspass Stuttgart (Climate planning pass Stuttgart)

LPAA .... Lima-Paris Action Agenda

METREX .. Network of European Metropolitan Regions and Areas

MKRO ...... Ministerkonferenz für Raumordnung
(Conference of Planning Ministers)

NGO .... Non-Governmental Organization

NKI .... Nationale Klimaschutz-Initiative
(National Climate Protection Initiative)

NUTS .... Nomenclature des unités territoriales statistiques
(Nomenclature of territorial units for statistics)

PCET .... Plan Climat Energie Territorial
(Territorial Climate and Energy Plan)

RES .... Renewable Energy Sources

SCOT .... Schéma de cohérence territoriale
(Territorial Coherence Scheme)

SEAP .... Sustainable Energy Action Plan

SRCAE ..... Schéma regional climat air énergie
(Regional Climate, Air and Energy Scheme)

t .... ton

UN .... United Nations

UNCED .... United Nations Conference on Environment and Development

UNFCCC .. United Nations Framework Convention on Climate Change

VVS .... Verkehrs- und Tarifverbund Stuttgart
(Regional transport association Stuttgart)

List of Figures

1 Introduction

The Paris agreement recognized local governments as crucial contributors to mitigating climate change (Hale 2016).[1] In the years of stalemate within international climate negotiations before 2015, a groundswell of voluntary local climate action raised hopes for an alternative, decentralized approach to global climate governance (Betsill and Rabe 2009; Bulkeley and Kern 2006; Corfee-Morlot et al. 2009). An exemplary attempt by the European Union (EU) to incentivize and coordinate local climate policy is the Covenant of Mayors, the voluntary climate commitment scheme with the largest number of participating municipalities worldwide. More than 7,000 local governments signed the Covenant pledging to achieve the European climate and energy targets. This book takes note of the proliferation of local climate action and the continued advancement of the Covenant, looking beyond its apparent success in terms of membership numbers (Bansard, Pattberg, and Widerberg 2016; Busch 2015; Hakelberg 2014; Hoffmann 2011; Widerberg, Pattberg, and Kristensen 2016). It adopts an (internal) governance perspective and examines the Covenant as an instance of orchestration, a voluntary and indirect governance arrangement based on the involvement of intermediate actors that support its policy goals (Abbott et al. 2012; 2016). The remainder of this introduction aims to situate this undertaking within climate governance research and provides an outline of its approach and proceedings.

Political and scholarly attention to climate change has long focused on intergovernmental negotiations at the international level within the United Nations Framework Convention on Climate Change (UNFCCC) (Bulkeley and Newell 2010; Giddens 2011; Luterbacher and Sprinz 2001). Meanwhile, local governments all over the globe engaged in voluntary climate action. Some local climate plans, especially in Europe, were developed within the framework of Agenda 21,

---

1 This book focusses primarily – though not exclusively – on climate change mitigation (in the following referred to as climate mitigation). If it refers simply to climate governance and climate policy for readability reasons, this is not meant to negate the importance of climate change adaptation (in the following referred to as climate adaptation) and its interlinkages with climate mitigation (see for example Jordan et al. 2010; Moser 2012; Landauer, Juhola, and Söderholm 2015).

© Springer Fachmedien Wiesbaden GmbH, part of Springer Nature 2020
L. Bendlin, *Orchestrating Local Climate Policy in the European Union*,
Energiepolitik und Klimaschutz. Energy Policy and Climate Protection,
https://doi.org/10.1007/978-3-658-26506-9_1

the action plan on sustainable development adopted at the United Nations Conference on Environment and Development (UNCED) in 1992 in Rio de Janeiro (Gordon 1993). After the US withdrew from the Kyoto protocol, American cities and states stepped in and adopted a series of innovative subnational climate policies (Lutsey and Sperling 2008; Rabe 2004; 2007; Selin and VanDeveer 2007; 2011; Urpelainen 2009). A successor to the Kyoto protocol, a binding international agreement, was a long time coming. The 15th Conference of the Parties (COP 15) in Copenhagen in 2009 was widely perceived as a landmark failure of intergovernmental climate negotiations (Bodansky 2010; Dimitrov 2010), which further spurred hopes that voluntary local action could overcome gridlock at the UNFCCC.

Local governments' contributions were recognized as an integral part of global climate governance during the UN Climate Summit in New York (2014), the COP 20 in Lima (2014) and the COP 21 in Paris (2015) (Abbott 2017; Hale 2016). In support of the future agreement, the Peruvian and French COP Presidencies gathered non-state actors under the Lima-Paris Action Agenda (LPAA). Data on existing initiatives are gathered by the Non-State Actor Zone for Climate Action, an online portal that aggregates information on more than 11,000 voluntary climate commitments made under selected schemes. At the COP 21, the LPAA was not scheduled as a side event, but recognized as an equal forth pillar of negotiations alongside national pledges, the financing package, and the agreement (see Hale 2016; Widerberg 2017).

The Paris agreement explicitly acknowledged and encouraged non-state commitments as an indispensable contribution to global decarbonization and was welcomed by nearly 700 non-state actors in the Paris Pledge for Action (see Oberthür 2016a). In what can be described as a new form of "hybrid multilateralism", non-state actors play important roles as implementers, experts and watchdogs, and are increasingly involved in multilateral and transnational climate action (Bäckstrand et al. 2017). The implementation of the pledge and review procedures established by the Paris agreement depend largely on domestic support for climate action (Keohane and Oppenheimer 2016). Subnational governments' contributions will be crucial (Jörgensen, Jogesh, and Mishra 2015).

Cities contribute to, suffer from, and are engaged in tackling climate change (Betsill and Bulkeley 2006; Bulkeley 2010; 2013; Bulkeley and Betsill 2013; Corfee-Morlot et al. 2009; Hasselmann, Antonio Ruiz de Elvira, and Welp 2009; Hoornweg et al. 2011; Johnson, Toly, and Schroeder 2015; Kousky and Schneider 2003; OECD; OECD 2010; van Staden 2014). This is because large shares of global GHG emissions occur in cities, and because cities are challenged by the need to adapt to climate change. They, moreover, have distinct possibilities to foster socio-technical innovation and economies of scale for mitigating climate change. More than half of the world's population now lives in cities, and urbanization continues (UN-HABITAT 2008).

A quickly growing body of literature has enhanced our understanding of local climate policy with regard to its drivers, means, and potential. Local governments engage in voluntary climate action because they (or engaged individuals within them) aim to contribute to climate mitigation (Alber and Kern; Benz et al. 2015), because they believe they are expected to do so (Heinelt and Lamping 2014), and in order to pursue co-benefits (Geels 2013, 26).[2] Co-benefits are essential for justifying local action for the global collective good (Benz et al. 2015, 328); examples include recognition at the international level, reputational benefits, and local impacts (Alber 2009; Schreurs 2010) such as "cost savings, clean air, regional economic development and job creation, alleviation of energy poverty, accessibility and livability of the city" (Alber 2009, 8). Local governments establish GHG emission reduction targets individually and at international conferences.[3] To document their ambitions, actions, and progress, they engage in voluntary commitment systems including the Carbonn Climate Registry, the Climate Group, and the Covenant of Mayors (Abbott 2017).

Local climate policies address infrastructure, the built environment, land use, mobility, and sometimes carbon sequestration, and tend to include sectors such as electricity generation, cooling and heating, energy efficiency, transportation, waste, and water (Castán Broto and Bulkeley 2013; UN-Habitat 2011). Four major modes of local climate governance have been identified (Alber and Kern; Bulkeley et al. 2011; Bulkeley and Kern 2006; OECD 2010). Firstly, local governments engage in self-governing with regard to their own consumption through procurement, public buildings, and other similar activities. Secondly, they govern by provision as a public service provider and in infrastructure development. For instance, their decisions about waste management procedures and

---

2   While this research focuses on voluntary climate action, it should be noted that some local governments are obliged to develop climate policies, including cities in China and Japan (see Schreurs (2010).

3   Examples include the 2007 Bali World Mayors and Local Governments Climate Protection Agreement aiming for a 60% GHG emission reduction by 2050 by municipal authorities worldwide, and 80% in industrialized countries.

public transport significantly impact citizens' carbon footprints. Thirdly, local governments govern by authority through regulation (e.g. building codes) within the scope of their legal competences. Fourthly, local governments play an enabling role as a facilitator for actions of other territorial actors, such as housing companies, businesses, and citizens.

Local climate policies vary significantly depending on the: 1. local impact (and perception) of climate change (e.g. vulnerability to flooding), 2. available competences in relevant policy fields, political will and capacities, 3. support provided by national programs, and 4. integration in municipal climate networks for learning and innovation (Alber and Kern, 2). Transnational municipal climate networks (TMCNs) support local climate policy through information sharing, capacity-building and implementation activities, and rule setting (Andonova, Betsill, and Bulkeley 2009; Bulkeley and Newell 2010). Some date back to the early 1990s; examples include the Climate Alliance, ICLEI's Cities for Climate Protection Program, and Energy Cities.

The euphoria for local climate policy was soon tempered by accounts of implementation difficulties and growing fragmentation of global climate governance. In order to better understand how local climate policy can best be fostered, this book addresses coordination requirements in local climate governance, i.e. local governments' dependence on superordinate levels of government for ensuring their capacities for climate action, and the coherence of their approaches (1.1). With this purpose in mind, it investigates local inter-municipal climate policy coordination in the EU through the Covenant of Mayors (1.2). This EU-initiative is conceptualized as an instance of orchestration, an indirect mode of governance based on voluntary participation. Within a case study research design, five cases from Germany and France allow for cross-case analysis and a highly policy-relevant comparison (1.3). This introduction concludes with an overview and an outline of the following chapters (1.4).

1.1 The Importance of Coordinating Local Climate Policy

The rise of local climate action has inspired a debate in the literature about the contribution this scale or level of government can make to climate mitigation. Proponents of a polycentric approach argue that the involvement of various levels combined with active oversight by stakeholders allows for overcoming free-rider problems in global climate governance (Andersson and Ostrom 2008;

Ostrom 2005; 2009). From the perspectives of comparative federalism and multi-level governance research, a decentralized approach to global climate govern-ance provides opportunities for experimentation at the local level; successful innovations can then spread beyond the initial 'laboratory' as they diffuse hori-zontally or are taken up by superordinate levels (Betsill and Bulkeley 2006; Betsill and Rabe 2009; Biela, Hennl, and Kaiser 2013; Bulkeley 2013; Bulkeley et al. 2013; Castán Broto and Bulkeley 2013; Geels 2013; Jörgensen, Jogesh, and Mishra 2015; Rabe 2007; Schneidewind and Scheck 2013).[4] The diffusion and performance of emerging polycentric patterns of climate governance remain to be tested (Jordan et al. 2015; Jordan et al. 2018).

For the time being, scholars of global climate governance disagree on whether the resulting fragmented governance architectures entail restricted pro-spects to limit climate change (as opposed to an integrated and more powerful climate regime) (Keohane and Victor 2011) or opportunities in terms of bypass-ing and managing reluctant nation states (Abbott 2014) and addressing co-bene-fits of climate policy (Jänicke 2017b; Jänicke, Schreurs, and Töpfer 2015). Either way, fragmentation entails various coordination requirements both horizontally, i.e. between local governments, and vertically, i.e. with superordinate levels of government (Biermann et al. 2009; Gordon and Johnson 2017; Green, Sterner, and Wagner 2014; van Asselt and Zelli 2014; Zürn and Faude 2013). Otherwise, subnational action risks suffering from "its limited reach, legal obstacles, internal costs, external leakage, and conflicting objectives" (Wiener 2007, 1979).

Indeed, empirical research found several caveats of local climate policy. Local governments' mobilization has hardly advanced UNFCCC negotiations; their influence on the process remains limited (Acuto 2013; Nasiritousi, Hjerpe, and Linnér 2014; Setzer 2015). Municipal climate networks' ability to strengthen local climate action has been questioned as they often are limited to "networks of pioneers for pioneers" (Kern and Bulkeley 2009), not operational by focus (Busch 2015), and lack coverage, ambition, and transparency (Bansard, Pattberg, and Widerberg 2016; Benner, Reinicke, and Witte 2004; Betsill and Bulkeley 2006; Bulkeley and Newell 2010; Gordon 2016; Khan 2010). Voluntary climate action only spreads unevenly (Brody et al. 2008; Bulkeley, Broto, and Edwards

---

4   Superordinate levels of government rank 'higher' in the political system without necessarily disposing of comprehensive discretionary power over 'lower' levels of government, which is why this terminology is preferred to 'higher'.

2012), and local governments often struggle to live up to their aspirations (Alber and Kern; Betsill and Bulkeley 2007; Bulkeley 2010; Wiener 2007). Typically,

> "obstacles include the need for a political champion within the local authority, access to financial resources, jurisdiction over emission-producing activities, technical expertise and dedicated person hours, the weighting of (Brody et al. 2008) climate protection against other local priorities, and political will" (Betsill and Rabe 2009, 211).

Many of these obstacles are determined at the national and, in the EU, European level because "[s]ubnational action [...] is hierarchically structured, embedded in national strategies, regulatory frameworks and incentive systems" (Jörgensen, Jogesh, and Mishra 2015, 241; see also e.g. Collier 1997; Marsden et al. 2013; Rosenzweig et al. 2011). Superordinate levels of government endow local governments with the required competences and resources, with advice and operational assistance for the highly technical task of developing, implementing, and evaluating local climate plans, and with incentives for taking action – or not (Andonova, Hale, and Roger 2017). Without favorable frameworks and coordination between and among levels of government, a coherent and complete landscape of local climate policies with climate plans and policies at the regional, national, and international level are not likely to emerge. For instance, a national government might miss its renewable energy development goals if it neglects mobilizing subnational governments, or a city at the center of an agglomeration might fail to increase its share of sustainable transport modes as long as the surrounding municipalities develop new peri-urban housing areas not serviced by public transport, thereby inciting their inhabitants to commute to the center city by car.

Local governments depend on the architecture of global climate governance for tapping their full potential in terms of mitigating climate change. This book is intended to contribute to an improved understanding of the conditions for local climate policy to spread, and to flourish.

1.2 The Unknown Contribution of Orchestration

Disagreements about the suitable level of ambition, and most effective and desirable policies to mitigate climate change at superordinate levels of government made the case for local climate action in the first place. There is no reason to assume that a mere rescaling of climate change provides a simple solution to this wicked problem (Levin et al. 2012; Sippel and Jenssen 2009). As argued above, a comprehensive and enabling framework for local climate policy has to include multiple levels of government: local, regional, national, and international. All levels contribute to determining the enabling and hindering frameworks for local climate policy; a coherent framework requires adjustments between them. Responsibilities can only be assumed with the help of suitable competencies and resources. Ambition and capacity depend on each other. Local preferences for certain action fields or instruments can blend well with available funding schemes, or not.

In the EU with its multi-level governance system, this problem is all the more pronounced (see chapter 2). Firstly, member states' national climate policies and forms of territorial organization diverge significantly, which complicates the task of forging a coherent framework for local action. Where no consensus can be found, the European Commission must rely on voluntary modes of governance. Secondly, the supranational level, i.e. the European Commission, does not have the capacities to directly steer local climate policy. It lacks the competences and resources to do so, but also the proximity required for knowing about local challenges in sufficient detail and having the legitimacy to intervene operationally. For coordinating local climate policy, the European Commission depends on involving national and subnational governments in a non-hierarchical manner.

One EU initiative that aims to do just that is the Covenant of Mayors (in the following referred to as the Covenant). This voluntary commitment scheme invites local and regional governments to sign a pledge and take action to achieve the European climate and energy targets in a three-step procedure. In view of a joint effort to facilitate municipal climate action, the Covenant involves governmental and other actors from multiple levels. Subnational governments act as so-called Territorial Covenant Coordinators (in the following referred to as Covenant Coordinators) that promote the initiative and provide assistance to participants from their territory.

Covenant Coordinators include one kind of government that has hardly attracted attention in climate governance research: inter-municipal associations. These consist of groups of neighboring municipalities that pool certain responsibilities and resources in a common administration.[5] As citizens' living environments increasingly transcend municipal boundaries, inter-municipal associations have developed in response to growing interdependencies between neighboring municipalities and as a means to fulfill tasks that exceed single municipalities' capacities (Desage 2011; Heinz 2007; Le Galès 2011; Le Saout 2011). Inter-municipal associations, i.e. institutionalized forms of inter-municipal cooperation, have therefore spread particularly in contexts of growing agglomerations or metropolises, but are not limited to urban regions. Operational inter-municipal cooperation consists in the joint provision of public services at lower costs or higher standards; cooperation in coordination refers

> "to the regulation of externalities of local policies and to an allocation of costs and resources that is rational from a supra-local perspective. [...] Coordination tasks also include the joint activities of municipalities to influence other levels of the national or European administration" (Hulst and van Montfort 2007b, 11).

Typically, the competences of inter-municipal associations include urban planning, energy, and transportation. Hence, they provide a natural platform for local climate policy coordination.

This raises the question of whether, how and to what extent inter-municipal associations can provide for self-coordination of local climate governance. Findings from local climate policy literature can only tentatively inform our understanding of inter-municipal climate policy coordination because it often focuses on cities and metropolises or does not differentiate between different kinds of subnational governments (Panara and Varney 2013a; cf. for example Betsill and Rabe 2009; Rosenzweig et al. 2011; Wiener 2007). Inter-municipal climate policy coordination may play an important role in establishing enabling frameworks for local climate policy. Therefore, inter-municipal associations' engagement is at the core of this research.

1.3 Research Design

The Covenant is conceptualized here as an instance of orchestration, i.e. an arrangement based on an indirect and voluntary mode of governance that will be introduced in due detail in chapter 3 (Abbott et al. 2015; Abbott 2015; Hale and

---

5   Inter-municipal associations and their specificities in the German and French context will be introduced in more detail in sections 5.1.1.3 and 5.2.1.3.

Roger 2014). The underlying assumption is that the Covenant (the orchestrator) relies on Covenant Coordinators (intermediaries) because it cannot directly steer municipalities (its targets) to the desired extent. The corresponding O-I-T-model explains actors' participation with the mutual exchange of resources in the pursuit of a joint policy goal. This theoretical framework was developed in iterative revisions of a case study research design (Agranoff and Radin 1991; George and Bennett 2005; Yin 2014).

Furthermore, a comparative perspective serves to highlight the influence of domestic politics on climate governance (Purdon 2015). Two countries, Germany and France, were chosen that are both politically relevant in terms of climate mitigation and their influence on European governance, and scientifically pertinent by providing contrasting national frameworks for local climate policy. In-depth case studies investigate five Covenant Coordinators, i.e. the totality of French and German inter-municipal associations engaged in the Covenant: Rhine-Neckar Region and Stuttgart Region from Germany, and Greater Lyon, Rennes Metropolis, and Val d'Ille from France. 47 expert interviews with 53 staff and politicians, triangulated with other primary and secondary sources, serve as a basis for examining the following research questions:

Why do local governments engage in inter-municipal climate policy coordination?

To answer this question, it is important to differentiate between the motivations of intermediaries and targets, i.e. inter-municipal associations and their member municipalities, and between their motivations to engage in the Covenant on the one hand and in inter-municipal climate policy coordination more broadly on the other. Local climate policy coordination is understood as intentional activities of governments in view of enabling and promoting local climate policy, supporting its performance, and ensuring its coherence both horizontally and vertically. Inter-municipal climate policy coordination refers to local climate policy coordination at the inter-municipal level, i.e. coordination by inter-municipal associations, and participation by municipalities.

How do local governments engage in inter-municipal climate policy coordination?

This includes 1. intermediaries' courses of action in implementing their Covenant commitments, i.e. inter-municipal associations' approaches to local climate policy coordination, 2. targets', i.e. municipalities' responses to the proposed framework for municipal climate policy, and 3. the ways in which local governments rely on the Covenant for local climate policy coordination.

What modes of governance result from inter-municipal climate policy coordination?

Closely related to the means and strategies chosen by inter-municipal associations for coordinating local climate policy, different modes of governance emerge. Modes of governance are conceptualized as sets of rules for collective governing (Arts, Lagendijk, and van Houtum 2009; Bäckstrand et al. 2010; Kooiman 2003).[6] Modes of inter-municipal climate governance might coincide, overlap, or differ from modes of climate governance as described in the literature on urban and subnational climate governance (Alber and Kern; Bulkeley et al. 2011; Bulkeley and Kern 2006; OECD 2010).

Building upon single case study reports, answers to these questions are provided for each case. The cross-case analysis serves to evaluate patterns and variations of inter-municipal climate policy coordination. In particular, it explains the difficulties of inter-municipal associations in implementing their Covenant commitments and weaknesses in the overarching governance arrangements as observed in the case studies.

1.4 Summary and Outline

This book explores the role of inter-municipal associations in the Covenant of Mayors, an EU-initiative conceptualized as an instance of orchestration. It does so in view of inter-municipal associations' potential to respond to the need for coordination in local climate policy that results from the fragmented nature of global climate governance. More specifically, it examines why inter-municipal associations engage in local climate policy coordination, how they do it, and what modes of governance they develop. Expert interviews were central tools for data collection during field research. Qualitative, cross-case analysis builds upon five separate case studies from Germany and France. Local governments are shown to make purposeful use of the Covenant in the pursuit of co-benefits. Deficiencies in Covenant support explain local governments' reservations to the Covenant and point to weaknesses in the governance arrangement and in orchestration theorizing.

---

6   Modes of governance and multi-level governance will be discussed in more detail in chapter 2.

Chapter 2 provides a literature review of local climate policy in the EU with a focus on the framework for local action as determined by supranational policies and the European multi-level governance system. The EU aspires to a leadership role in global climate governance, but has to bridge significant differences between member state priorities. As the multi-level system entails multiple interdependencies between and across levels of government, managing these differences constitutes a considerable challenge. This applies to intergovernmental decision-making processes, but also to implementation where the European Commission lacks crucial capacities for fulfilling its supervising role in a hierarchical manner. Scholars of European multi-level governance emphasized opportunities for stalemates as well as for multi-level reinforcement. Local governments are subject to Europeanization and metropolization within rescaling processes, but also benefit from additional venues for interest representation within the multi-level system. Additionally, they can engage in numerous climate and energy-related initiatives funded under EU programs.

Chapter 3 develops the theoretical framework. In response to the proliferation of voluntary non-state climate action and ensuing coordination requirements as depicted above, international organizations have been found to rely on voluntary and indirect modes of governance. These have been captured within the O-I-T-model. After an introduction to orchestration theorizing, the Covenant is conceptualized as an orchestration arrangement. It reflects upon the Covenant in terms of its key features and its role within EU climate governance, and argues that current conceptualizations fail to fully capture the Covenant in several respects. As a lens for further study so as to improve our understanding of the Covenant, the O-I-T-model is then applied to the Covenant, namely to the secretariat, Covenant Coordinators, and Covenant signatories. The chapter concludes with a preview on expected theoretical contributions.

Chapter 4 reflects upon the research design and methodological approach taken. In the absence of empirically grounded theories that would allow for a variable-centered approach to studying inter-municipal climate policy coordination within the orchestration arrangement of the Covenant, a qualitative case study approach was chosen. The final design was developed in an iterative process informed by literature, pilot studies, and preliminary results. The chapter provides a detailed discussion of the case selection process, and justifies Franco-German comparison. It further describes data collection and evaluation procedures with a special focus on expert interviews, and considers the implications of methodological challenges encountered during the research process.

Chapter 5 presents the empirical findings on inter-municipal climate policy coordination in Germany and France. It starts with an overview of domestic frameworks for local climate policy in the respective country in terms of national policies, territorial organization, and specificities of local climate action. Subsequently, case study reports on Covenant Coordinators from that country provide a comprehensive picture of inter-municipal local climate policy coordination within and beyond the Covenant. Each case study report begins with a brief overview of the inter-municipal associations' characteristics and institutional set-up and then covers the run-up to Covenant signature, its implementation and related inter-municipal activities, and finally assesses the current status of Covenant coordination. A summary of findings highlights the commonness of a reluctant municipal response, disappointments in Covenant support, and weak relationships between the Covenant and its intermediaries as issues deserving further analysis.

Chapter 6 analyzes how and to what extent the Covenant orchestrates local climate policy. It finds that inter-municipal associations that engage as Covenant Coordinators aim for co-benefits with regard to local leadership, territorial competition, and rescaling. Their approaches to local climate policy coordination vary importantly in terms of their primary target groups, but resemble each other in that the Covenant commitment is a resource rather than a guiding framework for their activities. Over time, all Covenant Coordinators under study reduced their level of engagement considerably. This signals limitations of the orchestration arrangement in terms of durability, intensity, and causality.

Chapter 7 concludes by discussing theoretical and policy implications. It finds that the Covenant harbors competing interests that complicate comprehensive local climate policy coordination, let alone systematic experimentation and innovation. National governments remain key players in local climate governance. National policies as well as territorial organization determine the local-level co-benefits of local climate action. The O-I-T-model was successfully applied to the study of subnational governments in the EU, but co-benefits and timing are more decisive drivers of participation than presumed in orchestration theorizing. These results might not apply to Covenant participation from all countries, but can be transferred to resonance groups of cases in the realm of voluntary commitment schemes and local climate governance. Finally, it addresses prospects for future research and derives some recommendations for policy making.

2 Local Governments in European Multi-Level Climate Governance

Why is the study of European multi-level governance of particular relevance with regard to decentralized approaches to global climate governance, and why does it require a multi-level governance approach? An approach building exclusively upon intergovernmental relations theory (Betsill and Bulkeley 2006, 142), or upon intergovernmentalist and supranationalist theories of European integration (see e.g. Bieling and Lerch 2006) that focus on single-level interactions (Scharpf 2010), could not capture the complexities of interactions in European climate governance. This chapter scrutinizes the role of local governments in European climate governance and argues that European climate governance entails the need for local climate policy coordination, and that studying the local level is useful in advancing our understanding of governance more broadly. After a brief historical overview of the EU's role in global climate governance (2.1) two main characteristics of EU multi-level policy making are examined, the proliferation of interdependencies between governance actors (2.2), and the resulting challenges and opportunities with regard to (climate policy) decision making and implementation (2.3). Local governments' position within the EU multi-level system not only requires a differentiated assessment (2.4), it also remains a moving target because of ongoing negotiations about the allocation of competences and repeated rescaling (2.5). The chapter concludes with a discussion of the double role of local governments as addressees and actors in their own right in European multi-level climate governance (2.6) and a brief summary (2.7).

2.1 The European Union in Global Climate Governance

European climate policy is closely linked to the EU's approach to international climate negotiations (Barnes 2010, 41). EU member states account for 10% of global GHG emissions as of 2014; the EU thus ranks third among the top emitting regions worldwide after China (30%) and the USA (15%) (Joint Research Centre

© Springer Fachmedien Wiesbaden GmbH, part of Springer Nature 2020
L. Bendlin, *Orchestrating Local Climate Policy in the European Union*,
Energiepolitik und Klimaschutz. Energy Policy and Climate Protection,
https://doi.org/10.1007/978-3-658-26506-9_2

and PBL Netherlands Environmental Assessment Agency 2015, 4). The EU has also strived for a leadership role in terms of international climate negotiations and domestic policies, but with contested outcomes (Gupta and Grubb 2000; Oberthür and Roche Kelly 2008; Parker and Karlsson 2010; Rayner and Jordan 2013; Schreurs 2016c; Schreurs and Thiberghien 2007; Waldmann 2007; Wurzel and Connelly 2010; Wurzel, Connelly, and Liefferink 2017). This section briefly reviews the EU's commitments to the UNFCCC, and provides an overview of developments in European climate policy during and since the Covenant's creation.

In its beginnings, European-level climate policy developed largely in response to international negotiations (Dupont and Oberthür 2015, 2). In 1991, the European Commission issued its first strategy for limiting CO_2 emissions and increasing energy efficiency, one year before the UNFCCC was established at the UNCED. Ever since, the EU is a party to the UNFCCC, which contributed to its ambitions for being recognized as an actor in its own right in international negotiations (Schreurs and Thiberghien 2007; Waldmann 2007). The European Commission has a crucial rule in brokering agreements between EU member states, also as a precondition for a coherent position (and potentially a leadership role) in international climate negotiations (Barnes 2010).[7] Under the Kyoto protocol adopted in 1997, the EU member states committed to a common target within a so-called 'bubble'.[8] For the first commitment period (2008-2012), the EU-15[9] adopted an overall GHG emissions target of -8% as compared to base-year levels (close to 1990 levels) by 2012. Individual national targets within the EU bubble ranged from -28% (Luxemburg) to +27% (Portugal); Germany committed to a reduction of 21%, France to stabilizing its GHG emissions. Both the EU-15 and the EU-28 overachieved their respective targets (European Commission 2006b). This was due largely to "'gratis' reductions made by the largest states" (Jordan et al. 2012, 57) including the windfall profits of German reunification.

---

7   European climate policy making is subject to considerable disagreement between member states; for a discussion of the interlinkages between internal politics and the EU's role in climate diplomacy, see e.g. Oberthür and Pallemaerts (2010a).

8   For a discussion of the EU's role in UNFCCC negotiations leading to the Kyoto protocol and in its implementation, see e.g. Barker et al. (2001); Damro and Méndez (2003); Gupta and Ringius (2001); Oberthür and Ott (1999); Oberthür and Roche Kelly (2008).

9   This refers to the then 15 member states of the EU: Austria, Belgium, Denmark, Finland, France, Germany, Greece, Ireland, Italy, Luxembourg, the Netherlands, Portugal, Spain, Sweden and the United Kingdom.

European climate policy developed incrementally throughout the 1990s (Dupont and Oberthür 2015, 3), but experienced an increased dynamic beginning around 2005 (Geden and Fischer 2008, 30). The Kyoto protocol came into effect thanks to its ratification by Russia, a bargain which appears to have been struck in exchange for EU support for Russia's accession to the World Trade Organization (WTO). At an informal European Council meeting, heads of states and governments called for a common policy approach beyond a mere common energy market. Subsequently, the EU defined a set of three unilateral climate and energy targets for 2020: to reduce GHG emissions by 20% as compared to 1990 levels,[10] to raise the share of renewable resources in energy consumption to 20% (including a minimum of 10% in the transport sector) and to improve energy efficiency by 20% as compared to business as usual. Often referred to as the 20-20-20 targets, these objectives had been introduced by the European Commission in its green paper 'A European Strategy for Sustainable, Competitive and Secure Energy' in March 2006 and officially presented in January 2007. They were adopted by the European Council in March that same year. Within less than a year, in what has been described as "a momentous development" (Jordan and Rayner 2010, 76), the 20-20-20 targets were translated into the so-called climate and energy package, a set of binding legislation on emissions trading, renewable energy promotion, carbon capture and storage, and emissions from cars that was officially adopted in December 2008 (Oberthür and Pallemaerts 2010b; Skjærseth 2014).

The climate and energy package also had implications for climate diplomacy in the run-up to the COP 15 in Copenhagen (van Schaik 2010). The 2007 decision on the 20-20-20 targets included a so-called conditional target that was not deployed in international negotiations: the option to increase the EU's GHG emission reduction target to 30% by 2020 if other industrialized countries committed to comparable targets, and emerging economies accepted to make an appropriate contribution. The climate and energy package further pursued EU attempts to influence developments at the international level by means of exemplary domestic regulation, and underpinned EU ambitions for a leadership role in international climate negotiations by providing credibility and legitimacy (Bendlin 2013; Kulovesi, Morgera, and Muñoz 2011; Oberthür and Roche Kelly

---

10   Unless specified otherwise, all climate and energy targets mentioned in the following refer to 1990 levels.

2008; Schreurs 2016c; Schreurs and Thiberghien 2007). The Covenant of Mayors has to be understood in this context, i.e. as an additional venue for promoting climate action and substantiating the EU's climate governance objectives (see chapter 3). In 2009, the year following the Covenant's creation, the European Council additionally adopted an 80 to 95% GHG emission reduction target by 2050. At the end of that year, the COP 15 – referred to as 'Hopenhagen' in a media campaign – came in for high-flying hopes, but failed to 'seal the deal' for a binding agreement (Bodansky 2010; Dimitrov 2010). Observers contended this was also linked to limited EU leadership in negotiations due to a lack of internal coherence on the appropriate level of ambition and on diplomatic strategy, and to the EU's inability to forge a coalition with other influential actors (Groen and Niemann 2013; Groen, Niemann, and Oberthür 2012). Hence, the EU retained its 20% reduction target for the second commitment period (2013-2020) of the Kyoto protocol under the Doha amendment adopted in 2012.

As a follow-up on the 2020 framework, the European Council in 2014 adopted a set of climate and energy targets till 2030, including a 40% GHG emission reduction, a 27% share of renewables in energy consumption, and an increase of energy efficiency by 27% as compared to projections. These were criticized for their lack of ambition and binding national targets (Bürgin 2014), but are meant to contribute to longer-term objectives as set out in the Roadmap for moving to a competitive low carbon economy by 2050, the Energy Roadmap 2050 and the Transport white paper proposed by the European Commission in 2011 (Hey 2012). The EU was also the first major economic region to present its intended contribution to a future international climate agreement in the run-up to the COP 21. In negotiating the Paris agreement, the EU finally assumed a leading role again, acting as a "leadiator" through bridge-building and coalition-building strategies (Bäckstrand and Elgström 2013; Oberthür and Groen 2016) without which the Paris agreement would hardly have been possible (Wurzel, Connelly, and Liefferink 2017). Although implementation will of course be decisive to judge its contribution to climate mitigation, the Paris agreement later was heralded as a breakthrough for global climate governance as it finally established a binding post-Kyoto framework for the entire international community based on a combination of mandatory and voluntary elements and the goal of limiting global warming to 1.5–2°C (Dimitrov 2016; Hale 2016; Schellnhuber, Rahmstorf, and Winkelmann 2016; Schreurs 2016c). It entered into force in 2016 after the required threshold of global GHG emission had been achieved with EU ratification.

For the time being, the European Environmental Agency estimates that the EU is on track to meet its 2020 climate and energy targets as renewable energy development and decreased energy consumption balance demographic and economic growth and increased coal-burning for energy production. Some member states struggle to achieve some targets; laggards include Germany that is not on track to meet its energy efficiency goal, and France that risks missing its energy efficiency and renewable energy goals (European Environmental Agency 2015). Also, the European Environmental Agency expects a slow-down of progress and estimates that increased efforts will be required to meet the EU's long-term objectives for 2050. Tangible progress at the supranational and national level is far from self-evident because

> "discussions on developing the cornerstones of the EU's climate policy – the EU ETS, renewables and effort-sharing among the EU member states – have generally progressed at snail's pace" (Dupont and Oberthür 2015, 4).

Open tasks include revising the EU ETS, determining national contributions to the 2030 target, furthering renewable energy promotion, making improvements in energy efficiency, and establishing an internal electricity market (Oberthür 2016b). Opposition to more ambitious climate policies by some member states and the ascension of climate-sceptic populist and far-right parties could make challenging further sharp advancements of EU targets and policies (Schreurs 2016c).

Against this background, the European Commission more than welcomes local governments' contributions towards the EU climate and energy targets. Its options for stimulating such contributions are determined by the functionality of the EU multi-level system.

2.2 Modes of Governance in Multi-Level Systems

Although no consensual definition of governance has become accepted to date, the concept's parallel emergence in different disciplines and strands of literature underpins its practical relevance and usefulness for research (Simoulin 2013, 3–5). Two strands of governance literature are most helpful here (Benz et al. 2007; Kohler-Koch and Larat 2009). A first one consists of multi-level governance with respect to the interlinkages in local climate governance and its potential contribution to the problem-solving capacity of EU climate governance (Bache and Flinders 2004; Conzelmann and Smith 2008; Hooghe and Marks 2001a; Kohler-Koch and Eising 1999; Marks 1993; Tömmel 2008; Weibust and Meadowcroft 2014). A second one consists of (new) modes[11] of governance for policy coordination (Arts, Lagendijk, and van Houtum 2009; Eberlein and Kerwer 2004; Treib, Bähr, and Falkner 2007; Windhoff-Héritier and Rhodes 2011). In this perspective, governing is a collective exercise based on the interplay of inter- and intra-governmental structures and processes (Benz 2007). As a result, governance is made up of multiple interactions of considerable diversity, dynamic, and complexity (Kooiman 2003).

In this study of the interplay of governments as opposed to non-governmental actors, a third strand of governance literature on public-private actor relations (Kohler-Koch and Larat 2009) has not been included. However, the multi-level governance approach highlights the diversity of actors at all levels. Other than governments, there are also agencies, companies, non-governmental organizations (NGOs), academia, etc. involved (Andonova, Betsill, and Bulkeley 2009, 52). This perspective is helpful in identifying potentially relevant actors within case studies in order to verify if and how the respective inter-municipal association involves them in local climate policy coordination.

The multi-level governance approach has been developed in the attempt to better understand European politics and the integration processes within the EU (Bache and Flinders 2004; Conzelmann and Smith 2008; Hooghe and Marks 2001a; Marks 1993; Marks, Hooghe, and Blank 1996; Tömmel 2008; Weibust and Meadowcroft 2014). In particular, multi-level governance theorizing initially aimed to understand "efforts by the Union to tap the full potential of local and regional authorities [...] in contributing to the goals of the Union" (Conzelmann 2008, 13). Hence, it is a natural framework for analysis when examining the interplay of levels in European climate governance. Generally speaking, multi-level governance can be described as politics without a center:

"Instead, variable combinations of governments on multiple layers of authority – European, national, and subnational – form policy networks for collaboration. The

---

11   Modes of governance for policy coordination and orchestration in particular will be discussed in more detail the following chapter that develops the theoretical framework for this research.

relations are characterised by mutual interdependence on each other's resources, not by competition for scarce resources" (Hooghe 1996, 18).

No single level of government can achieve its goals alone in multi-level systems. Both authority and resources have been spread by integration and regionalization dynamics towards the supranational and subnational levels (Hooghe and Marks 2001a, xi). Decisions taken at one level entail impacts on others, and actors have to rely on each other's contributions for problem-solving (Benz 2007, 17–18). An early and groundbreaking definition describes multi-level governance as "a system of continuous negotiations among nested governments at several territorial tiers" (Marks 1993, 392; see also Elgström and Jönsson 2004). As governance actors are most generally collective actors, their behavior during these negotiations depends on internal rules and processes. For instance, negotiations can be hampered when a participating government is about to face new elections and thus is hardly inclined to making unpopular concessions. In this way, politics at a given level of governance – or even within one single collective actor – entail impacts on multi-level governance as a whole. As a result, multi-level governance is characterized by multiple interdependencies between the levels of authority involved.

Multi-level governance structures differ between territories and policy fields. Institutional structures of multi-level systems are determined by: 1. territorial organization, 2. the distribution of competences and resources across and within levels of governance, and 3. rules of mutual control between levels of governance (Benz 2007, 22). Two types of multi-level governance can be distinguished (Hooghe and Marks 2001b; 2003; 2010; Marks and Hooghe 2004). Type I draws upon federalism and corresponds to EU multi-level governance. It consists of multi-task, territorially mutually exclusive jurisdictions in a comparatively stable system with a limited number of levels and units. Type II is more prominent in international relations thinking and consists in large numbers of specialized, territorially overlapping jurisdictions in a more flexible, non-layered system.

Multi-level systems have the common feature that governance actors need to develop strategies for coping with interdependencies. Institutionalized forms of collective governing, i.e. modes of governance in multi-level systems are characterized by the need to ally with others, but differ with regard to strategies of cooperation at work. The resulting interplay of actors has been conceptualized in different ways in multi-level governance theorizing depending on the focus of analysis in terms of political systems, levels of governance, and policy fields (Bache and Flinders 2004; Enderlein, Wälti, and Zürn 2010; Piattoni 2010). Kooiman (2003) builds his typology on different kinds of interactions that lead to self-governance, co-governance, or hierarchical governance. Benz et al. (2007) establish a two-dimensional systematization of governance modes based on the continuum from market mechanisms to hierarchical steering on the one hand and instruments such as competition, networks, negotiations etc. on the other. In order to highlight the originality of participatory, deliberative forms of policy making as opposed to classic approaches such as hierarchical regulation, they are often referred to as 'new' modes of governance (Bäckstrand et al. 2010; Conzelmann 2008; Grande 2012; Knill 2008a; Windhoff-Héritier and Rhodes 2011).

Andonova, Betsill, and Bulkeley (2009, 56) argue that modes of governance determine climate governance to the extent that they matter more than the kind of actors involved (state vs. non-state) and the structures they create. In contrast, Sturm and Bauer (2010, 20–21) have criticized multi-level governance theorizing for its neglect of subnational governments' agency and argued that understanding territorial governance required studying cross-level intergovernmental relations. Before discussing the implications of the multi-level system of European climate governance for local governments, the next section considers multi-level challenges and opportunities for European climate governance more broadly.

2.3 Challenges and Opportunities in European Climate Policy Making

This section reviews multi-level governance theorizing with regard to the EU's governability, i.e. its ability to steer and control societal interactions (Heinelt 2008, 54). It proceeds by subsequently examining collective decision making and policy implementation of the European multi-level system, both in general and with regard to climate policy in particular.

Governance has been described as a way to manage ungovernability by co-operation (Simoulin 2013, 4; see also Scharpf 1997). Typical challenges of multi-level governance, Benz (2007, 18–19) argues, consist of the constitutional organization of levels of authority within the state, the legitimacy of political decisions when territorial units included in decision-making and impacted by their outcome are not congruent, and the resulting coordination requirements, and interdependencies between levels of authority due to constitutional provisions and their dependence on each other's contributions for problem-solving. Unavoidably, this has implications for power relations, also between levels of government (Pasquier and Weisbein 2013, 281).

The diversity of approaches to (multi-level) governance analysis as depicted above entails different stances on the performance of multi-level decision making. Assessments whether or not the European multi-level system is characterized by deadlock or dynamic, capable of problem-solving or not, are legion (see for example Arts, Lagendijk, and van Houtum 2009; Eichener 1997; Elgström and Jönsson 2000; Haahr and Walters 2004; Hueglin 1999; Scharpf 1997; Wallace, Pollack, and Young 2015). Often, such analyses actually concentrate on supranational processes, some integrate national-level implementation (e.g. Knill 2008b) or refer to regionalization, i.e. an assumed rise of the regional level in European politics (e.g. Benz and Eberlein 1998; Hooghe and Marks 2001a).

The literature identifies two major downsides of multi-level governance that are both linked to increased interdependencies:

> "First, opaqueness of decision-making and a consequent lack of political accountability, contributing to the famous democratic deficit. And second, creeping stalemate or deadlock, best captured by the model of the joint decision trap developed by Scharpf" (Benz and Eberlein 1998, 3).

The joint decision trap model of European policy making builds on German federalism theorizing in the attempt to understand blockage in European integration during the 1970s and 1980s. Scharpf (1985; 1988; 2006) argues that European governance consists in a multiple-veto player system. Conceptualizing European governance primarily as intergovernmental negotiations, he asserts that nation states will try to preserve institutional interests when determining decision-making rules, even to the detriment of problem solving.

Nevertheless, scholars – including Scharpf himself – refute the idea of *generalized* stalemate in European multi-level governance and argue that the distinct strategies of the various actors involved in EU governance ensure governmentality (see for example Heinelt 2008; Scharpf 2010). In addition to institutional reform, "intergovernmental relations between the European, national, and subnational levels" have undergone "dynamic restructuring processes" as governments "react and adapt to changes in the emerging multi-level system, both in their internal composition and their external relationships" (Benz and Eberlein 1998, 5). Examples of strategies for escaping from deadlock include brokering by the European Commission and arena shifting (Falkner 2011), but also the division of decisions in functional specialties based on sectors of public policy (Peters 1997). Illustrating their stance with a case study of the EU's Emission Trading System (ETS), Müller and Slominski (2013) argue that time-based strategies allow for distributing costs and benefits of a contested decision over time. In addition, a last resort in case of persisting difficulties to decide jointly consist of differentiated integration (Jensen and Slapin 2015; Kölliker 2001). The use of various forms of subterfuge to reconcile unity with diversity, and competition with cooperation, Windhoff-Héritier (1999) argues, represents the greatest challenge for EU policy making.

The dynamics of (multi-level) governance, including governability, have been shown to differ significantly between policy fields (Arts, Lagendijk, and van Houtum 2009; Heinelt and Knodt 2011), which raises the question how EU multi-level decision making impacts EU climate governance. EU climate and energy governance is characterized by the mixed competences of the European and national levels and discontent among member states about the desirable energy mix. The latter are free to choose between different sources and structures for energy supply while the EU is in charge of the energy market, energy security, promoting energy efficiency, energy savings, and renewables, and promoting the interconnection of energy grids.

The decentralization of decision making provides opportunity structures for European climate policy that resemble those observed in comparative federalism (see for example Bomberg 2012; Brown 2012; Rabe 2007). To name but a few contributions to this strand of EU multi-level climate governance theorizing, Schreurs and Thiberghien (2007) assert that the multi-level governance structure of the EU encouraged a process of mutual reinforcement in favor of EU leadership in international climate negotiations. In policy fields where the European Commission and member states have joint competence and the EU Council decides by qualified majority vote, they argue, the multi-level system creates not only veto points (cf. Tsebelis 2002), but also leadership points. Member states, the European Commission, and the European Parliament compete for political leadership in a process of repeated baton passing. In a similar vein, but with a more market-oriented focus of analysis, Jänicke (2015) argues that the EU multi-level system facilitates the diffusion of low-carbon technologies based on mutually reinforcing market cycles, lead markets, and interactions between vertical and horizontal dynamics.

In the attempt to assess whether multi-level reinforcement is likely to persist in the future, Jordan et al. (2012) draw upon the argument developed by Schreurs and Thiberghien (2007) to explain how EU climate governance manages a series of apparent paradoxes: the tensions between 1. the simultaneous desire for greater international actorness and internal diversity, 2. the desire for policy harmonization and differentiated burden sharing, 3. the desire for ambitious targets and a limited set of policy instrument choices, and 4. high ambition and constrained implementation mechanisms. In a more critical assessment of EU climate governance, Jordan et al. (2012, 57) assert that "the EU meets its policy targets but for reasons that are not directly to do with its policies". They conclude that the EU is likely to have maxed out current modes of governance, and will thus have to confront implementation challenges more directly. After discussing governability in terms of EU decision making processes, this leads us to a second challenge in European multi-level climate governance: policy implementation.

Although EU governability has been discussed primarily in terms of decision-making capacity as discussed above, policy implementation transcends a conceptualization as collective decision making (Benz 2010, 214). Multi-level systems are particularly challenged by policy implementation when different levels of government are responsible for policy formulation and decision-making on the hand and for policy implementation on the other. Such a "structural gap between higher-level policy formulation and its lower-level implementation" (Heidbreder 2015, 369) applies in particular to the EU, as has been established by research on compliance with EU law (Knill 2015; Majone 1996). Non-compliance is linked to the complexities of European policy making rather than country-specific factors (Knill 2015).[12] An important institutional factor for implementation difficulties is linked to the fact that the European Commission continues to lack crucial capacities for policy management in implementation (as opposed to its consolidated role as a policy entrepreneur in decision making) (Laffan 1997). The European Commission is formally in charge of implementation for most EU policies, but mainly restricted to monitoring implementation by national and subnational administrations which are formally independent under the principle of national administrative autonomy. It has no direct grasp on national and subnational administrations, and has only limited capacities to actually exercise control.

---

12   The extent of non-compliance with EU law has been discussed controversially Börzel (2011).

In addition to enforcement strategies for ensuring member state compliance, the European Commission has to rely on alternative strategies (Tallberg 2002). This includes involving member state administrations vertically, and connecting them horizontally (Heidbreder 2014, 37–38). A strategy for flexible vertical integration consists of experimentalist arrangements where member state authorities are free to implement predetermined policy goals as they see fit under the condition they participate in peer reviews of performance (Sabel and Zeitlin 2008; Scott and Trubek 2002). A recent example of horizontal integration strategies consists of horizontal capacity pooling, i.e. the networking of competent bodies from member-state administrations with the help of technical tools provided by the European Commission to organize joint policy execution in a decentralized way (Heidbreder 2015). All in all, Heidbreder (2011, 723–24) finds that the European Commission applies "an elaborate set of instruments that interconnect actors present at different levels" ranging from mandatory rules to voluntary horizontal co-ordination which has led to an "incremental shift from pure indirect administration to strongly interlinked multi-level co-operation" (Heidbreder 2011, 723–24). As a result, she concludes, policy implementation by member-state administrations converges procedurally rather than formally.

This explains the importance of targets and framework directives in European climate and energy governance (Hey and Calliess 2013; Knill 2008a) as depicted above (see section 2.1). They provide a framework for national climate and energy policies but preserve member states' leeway in implementation. This opens up experimentation opportunities, similar to dynamics described in the literature on comparative federalism (see for example Bomberg 2012; Brown 2012; Rabe 2007). For instance, Hey and Calliess (2013) argue that different national approaches to renewable energy promotion facilitate a bottom-up process of Europeanization in energy governance. Leeway for implementation also applies to local governments.

Summing up, the European environmental administration is not the stage for politicization and innovation (Hey 2010a). But the governability in European climate policy is based to a large extent on the competent, incremental development of policies in the multi-level administrative system from the European Commission, the member states, and regions down to the municipal level.

2.4   Local Governments in the European Multi-Level System

The multi-level governance approach has been developed in the first instance "to explain European integration and regional empowerment as an integrated phenomenon" (Hooghe and Marks 2001a, 70). The EU multi-level system has changed domestic policy making significantly, but differently at different levels of government (Risse-Kappen 2001). This book is intended to contribute to the body of literature examining the "nature, opportunities and constraints of Type I multilevel governance for local climate policy" as it focuses on "vertical interactions between different levels of government as a factor shaping the capacity for local climate change governance" (Betsill and Bulkeley 2007, 449). Empirical research on local governance can help to substantiate governance concepts, cool the overheated theoretical debate, and help avoid overly optimistic accounts by pointing out failures and abandoned projects; thereby demonstrating that governance research needs to take account of changing political frameworks and the entailed opportunities, but also limitations (Pasquier and Weisbein 2013). When referring to the local level, it is important to distinguish between the local level as a scale of governance and a layer of government.[13] This book focuses on local governments, i.e. politically institutionalized local actors that have regulatory authority within a given territory.[14] Depending on national provisions, local competences include transport, urban development, housing, education, social services, security, waste, water, energy, consumer protection, green spaces, economic development, culture, environment and sustainability (Le Galès 2011, 359–60).

Just as scholars disagree about the governability of the European multi-level system in terms of decision-making and policy implementation, the effects of European integration on subnational authorities and their role in European governance have been discussed controversially (Benz and Eberlein 1998; George 2004; Hooghe and Marks 2001a; Schaffarzik 2007; Seele 2007; Sturm and Dieringer 2010; Tatham 2016; van Bever et al. 2011). Doubts regarding governability in an increasingly interdependent system of intergovernmental relations

---

13   Often, this distinction is not clearly established in local (climate) governance research. Hence, this literature review includes findings from studies of subnational, local, and urban climate policy in (European) multi-level governance. Focusing on intergovernmental relations, though, it does not include findings on private and civil-society actors from the local level.

14   As opposed to local authorities that carry out superordinate regulation only.

did not prove true; nor did a 'Europe of the Regions' bring about the decline of the nation-state (Sturm and Bauer 2010, 14–17). Member states' preferences for territorial organization continue to differ and to impact the scopes of action for subnational governments along with national policies (Martínez Soria 2007). Even so, the EU multi-level system entails some shared impacts on subnational governments across member states that deserve our attention.

Scholarly assessments of the EU's impact on local governments typically fall into two broad perspectives. In one view, subnational governments have been found to lose influence when decisions formerly taken with their involvement are moved to the supranational level where only nation-states and European institutions might participate (Börzel 2001; Hooghe and Marks 2001a). EU law impacts subnational governments directly and indirectly as they are responsible for implementing superordinate legislation, and have to respect the primacy of EU law in their own policies. Examples include EU directives on the environment and public participation that "had a deep impact on the preparation of municipal town planning schemes" (Frenzel 2013, 114).

Another perspective suggests, subnational governments are not only implementers, but inherent participants in European policy making. The Lisbon Treaty enshrines the EU's obligation to respect regional and local self-government (art. 4, 2), explicitly refers to regional and local governments in the principle of subsidiarity (art. 5,2) and states that regional and local implications of legislative proposals shall be taken into account during prior consultations (art. 2, Protocol on Subsidiarity and Proportionality). The scope of direct subnational participation in European politics increased importantly since the 1980s through a dedicated advisory body in the EU institutional framework, the Committee of the Regions (CoR) (Hooghe and Marks 2001a; Hooghe, Marks, and Schakel 2010). In addition, subnational governments intervene at the European level through their umbrella organization, the Council of European Municipalities and Regions (CEMR), and through general-purpose as well as issue-specific, European-wide and regional networks, and individual offices (Kern 2003). Anticipating future implementation obligations, they aim to influence EU legislation (Henneke 2007b).

Moreover, subnational governments obtain new political opportunities beyond traditional conflicts with the national state or between centers and peripheries, Le Galès (2011) points out. Examples include access to resources and expertise other than those of the national level and challenging national political choices. He concludes that subnational governments will mobilize increasingly to seize these opportunities as European integration proceeds, thereby contributing themselves to shaping European institutions. In a similar vein, Keating (2008; 2011) argues that European local governments' political capacity benefitted from a general trend over the last decades from local administration where territorial units act exclusively as agents of central government, to governance which is coined by institutional pluralism, and finally to local government able to shape territorial development significantly. Partaking in power is not a given, at times it may only a promise made by national or EU institutions to obtain subnational governments' participation in governance processes (Lascoumes and Le Galès 2004, 369).

Subnational governments are not passively accepting empowerment as a "spin-off" of national-supranational conflict; rather, they are "actively claiming a role in EU policy making by exploiting […] problems posed for central governments" (George 2004, 122). Hooghe and Marks (2001a, 77–78) therefore refer to multi-level governance as being partly a by-product of struggles between national and subnational governments. As Benz and Eberlein (1998, 3) point out, this can bring about a congruency of interests and a two-way multi-level game between subnational governments "seeking participation in European policy games, and […] the European Commission looking for partners and support for territorial and other policies". This was also the case in the creation of the Covenant of Mayors (see section 3.2.1).

Subnational governments from different countries and levels are impacted differently by European integration processes. Hooghe and Marks (2001a) find that subnational governments' influence depends on domestic territorial organization, the amount of resources available to them, and their participation in the Council of Ministers or general-purpose and issue-specific networks. Governments from the regional level such as the German *Länder* have been found to make active use of opportunities in European multi-level governance and benefit from institutionalized participation in domestic and European decision making (Benz 1998; Börzel 2001; Jeffery 2001). In comparison, governments at the local level have more limited resources at their disposal for engaging in EU politics.[15] Nevertheless, they participate in European programs and lobby European insti-

---

15   For instance, regional governments have larger representative offices in Brussels and more staff available to deal with European issues.

tutions, individually or through their associations (Hamedinger and Wolffhardt 2010). EU initiatives provide them with "unprecedented access to information, legitimacy, and not least, financial support" (Marshall 2005, 668). In a comparative study on European cities, Wolffhardt et al. (2005) find that cities are motivated to get "involved with 'Europe'" because they perceive the EU as a problem solver, stage, threat, alternative, or duty. EU programs offer additional financial and conceptual resources for problem solving. The European stage provides an additional venue for profiling and identity building. Cities try to avert the threat of unwanted EU legislation, and to overcome domestic frameworks by help of alternative programs at the European level when national policies are not in their interest, or lacking. Finally, they react to duties established by European regulation and transpositions into national law.

A secondary effect of European programs consists of legitimizing them as collective actors in European governance (Le Galès 2011, 72–73). Local governments' legitimacy is of particular importance in multi-level systems where the allocation of competences is under constant negotiation.

2.5 Negotiating Competence Allocation in Multi-level Systems

Although Type I multi-level systems such as the EU are relatively stable in terms of the jurisdictions they comprise (see section 2.2), the allocation of competences at the local level is a constant issue of negotiation (Kersting and Vetter 2003). As a guideline for allocating authority in the EU,[16] the principle of subsidiarity calls for political functions to be carried out by the smallest possible unit. Only if the smaller unit cannot solve a problem should it be centralized at the superordinate level.[17] Decisions over competence allocations are subject to a trade-off, as Benz (2007, 23–29) argues. He contends that small territorial units are better able to carry out tasks which require consideration of regional and local specificities or preferences or which would overburden unitary government because of the amount of information to be processed or conflicting interests within territories. But larger units are better suited when territorial policy entails external effects, when investments require economies of scale, when problems and resources are distributed unevenly among territorial units, and when costs of de-

---

16   The EU shares this principle with many federal states, including Germany.
17   For a discussion of the subsidiarity principles' effectiveness in preserving local self-government, see e.g. Emmerich-Fritsche (2007); Horga (2010).

centralized organization exceed those of unitary organization. The multitude of existing levels, though, makes it difficult to determine the most suitable one(s) and requires extensive coordination. Early examples include the first EU Environmental Action Program (EAP) that emphasized the need to "establish the level best suited to the type of pollution and to the geographical zone to be protected" (Collier 2013, 49). This is not without implications for climate and energy policy. For instance, member states' objections to EU-level intervention have contributed to the failure of early EU climate policy initiatives with regard to energy and fiscal policy such as a carbon tax (Collier 1996).

However, there is no objectively best level to govern a given policy problem. This circumstance is captured in the concept of scale that expresses the *idea* that a given policy problem should best be governed at a certain level. As Hooghe and Marks (2016) point out, functional pressures for public goods provision alone do not determine jurisdictional design but interact with people's preferences for self-government and their understanding of community. Simplistic and functionalist claims for a somewhat mechanical implementation of the subsidiarity principle are thus doomed to fail (Négrier 2011, 195–96). Instead, a governance perspective focusses on "the ways in which the scaling of political authority is in itself a contested process" (Betsill and Bulkeley 2006, 154). Since participation in a particular governance arrangement does not necessarily imply a coincidence of interests, partial or extensive conflicts are possible. Power proportions persist and depend on the distribution of political resources within the governance arrangement in question (Pasquier and Weisbein 2013, 281). But in contrast to hierarchical systems, they are constantly negotiated and vary according to issue, actors involved, and timing.

The reallocation of competences between levels of government has been conceptualized as rescaling (Brenner 2004; 2009). Rescaling processes reallocate competences from the top down as in decentralization and devolution reforms, or from the bottom up as in Europeanization, regionalization or metropolization. As Négrier (2011) points out, they are driven by politics, namely by leadership, as opposed to somewhat mechanical functional, institutional or social influences. Rescaling also changes the preconditions for politics and therefore produces winners and losers, depending on whether or not an actor accepts and can adapt to the new rules of the game. Despite that, rescaling should not be analyzed in terms of a zero-sum game but needs to be understood in more complex terms beyond mere top-down or bottom-up transfers of competences, especially when it comes to intergovernmental relations between subnational levels of government (Faure and Muller 2007, 13–14). Rescaling opens up opportunities for introducing additional resources and creating uncertainties, and motivates new interactions and territorial political exchange (Négrier 2005, 211).

Impacts of Europeanization on local governments have been discussed above. Another form of rescaling of particular relevance for the local level consists of metropolization, understood here as the institutional integration of neighboring municipalities within a metropolitan region in view of joint provision of public services such as planning, economic development, and transport (see for example Meijers, Hoogerbrugge, and Hollander 2014).[18] Metropolitan institutions are studied here in the form of inter-municipal associations, but other forms exist as well. Since different policies and services entail different optimal scales, attempts to identify an ideal level of metropolitan government failed as early as in the 1960s, Le Galès (2011, 383–85) points out; experiments with the metropolization of local government nevertheless spread throughout most of the Western world. He further states that most metropolization attempts failed because of central states' technical and arbitrary approach to reform in neglect of local populations' preferences, local governments' anchorage, and political legitimacy. The construction of the metropolis is neither accomplished nor self-evident; intercommunal leaders have to symbolically establish the extended agglomeration as the metropolitan territory (Ben Mabrouk 2007, 118–19). As a result, metropolitan institutions are often an empty shell designed to obtain windfall benefits from national funding schemes, captured by mayors who exert tight control in favor of their respective municipality, and endowed only selectively with competences, as Négrier (2007, 40) points out. Only in rare cases, he continues, do metropolitan institutions become the key players for urban development. He conceptualizes this process in three subsequent phases: territorial composition, which is most politicized, regime establishment, and public policy implementation (Négrier 2007, 41). Naturally, metropolization entails political conflict between local actors. Often, it is encouraged by national and European programs that incite local actors to forge into a collective actor at the metropolitan level in order to obtain and implement funded projects (Le Galès 2011, 383–85). This also applies to local climate governance.

---

18   It should be noted that competing definitions of metropolization coexist in different strands of research literature. Notably, some scholars refer to metropolization as a process of economic transformation and concentration (see for example Krätke (2007)).

2.6 Implications for Local Climate Governance

Generally speaking, European regulation impacts local government in many ways including public services, economic development, public procurement, environmental policy, and transport policy. In climate policy, examples include power-heat cogeneration, renewable energy, and energy efficiency (Knodt and Tews 2014, 281). As emphasized in section 2.3 with regard to the impacts of EU multi-level decision making on governability, leeway in implementation for subnational governments varies significantly across policy fields. For instance, the European Commission exerts tight control on European urban governance (Pasquier and Pinson 2004, 54–55). In contrast, subnational governments' scopes of action in climate policy have resulted in a diverse landscape of subnational climate action (Alber 2009; Alber and Kern; Knodt and Tews 2014; Löfstedt and Collier 2013). Finck (2014) argues that this has enabled local-level experimentation for European climate policy. Still, unfavorable EU policy frameworks such as a preference for low energy prices and a lack of ambition in defining climate and energy targets also constrain local climate policy (Alber 2009; Collier 1997; Löfstedt and Collier 2013). More pessimistic accounts of local climate action seem to be confirmed by an analysis based on 200 large and medium-sized cities from 11 EU member states. It finds that about one-third has no climate mitigation plan, and that the existing targets and plans would not suffice to meet the EU emission reduction targets by 2050 (Reckien et al. 2014).

But local scopes of action are not only *restricted* by Europeanization. Other impacts of Europeanization on local climate policy include, but are not limited to, growing mobilization of local governments in EU climate policy making, and increased horizontal interactions (Kern and Bulkeley 2009). In this way, the EU multi-level system also entails new opportunity structures for local governments to seize, as Knodt and Tews (2014) point out. In particular, they refer to three kinds of non-regulatory climate policy initiatives by the European Commission that address local governments. Firstly, local governments can benefit from distributive policies. Programs supporting local climate and energy policies include the structural, cohesion, and regional funds and temporal schemes such as the Intelligent Energy Europe (IEE) funding scheme that ran from 2003 to 2013 with a budget of €730 million under the Competitiveness and Innovation Framework Program. IEE has since been succeeded by Horizon 2020. This EU funding scheme for research and innovation comes with a budget of nearly €80 billion from 2014 to 2020. Organized in multiannual work programs, it includes a thematic section on climate action, environment, resource efficiency and raw materials.

As mentioned earlier, distributive policies provide particularly strong incentives for metropolization. Examples include a threshold of investment that exceeds individual municipalities' capacities. Other kinds of non-regulatory local climate policy initiatives can entail incentives for inter-municipal cooperation as well. Secondly, the European Commission aims to facilitate coordination and horizontal networking between local governments in view of diffusing best local climate policy practices. Correspondingly, funding schemes require the formation of international partnerships instead of single-actor applications, and to ensure that results are documented and shared with others. This requires extensive resources for networking, developing proposals, and project implementation that can be more easily provided by a joint inter-municipal undertaking. Other examples of coordination and diffusion initiatives by the European Commission include the European Green Capital Award, and the Covenant of Mayors.

A third kind of local climate policy initiative consists of offering information in the form of green and white papers that provide detailed problem descriptions and proposals for possible solutions, and awareness-raising campaigns and events such as the EU Sustainable Energy Week (EUSEW). Individual municipalities can easily be overstrained, be it only in terms of dedicated person hours, making it difficult to seize these opportunities. This is another incentive for inter-municipal cooperation, but also a risk for uneven participation across European municipalities.

Hence, the picture of EU influences on local climate policy is a mixed one. Further research is necessary to judge what aspects of EU climate and energy policy actually experience a mainstreaming according to the European Commission's priorities and to what extent different kinds of subnational governments benefit from capacity-building and empowerment (cf. Hooghe and Marks 2001a, 81). As local governments' multi-level strategies vary importantly depending on factors such as national climate and energy policies and territorial organization, their assessment requires further case study research on the actual relations and mechanisms at work in different national contexts (Andonova, Betsill, and Bulkeley 2009, 56). Also, these trends need not necessarily imply a loss in influence of national governments and agencies (Hooghe and Marks 2001a, 77–78) (Sturm and Bauer 2010, 11–12), but require careful investigation for sound assessment. Against this backdrop, the Covenant of Mayors is examined here as a European instrument for steering local climate policy, and as an opportunity structure for local agency in European climate governance.

2.7 Summary

This literature review served to refine our understanding of the research problem as introduced in section 1.2. It argued that studying European climate governance allows for covering a highly relevant arena of climate policy making. For this, it further derived from the literature on climate policy and European governance, a multi-level governance approach is most appropriate. In this understanding, local governments such as municipalities and inter-municipal associations are not simple implementers of policies determined at superordinate levels (Sturm and Bauer 2010, 20–21), but full-fledged participants in multi-level governance. European local governments operate in a multi-level system that spans from the local through the regional and the national level up to the supranational level of EU institutions. This is not to negate interdependencies with other levels of government, quite the contrary, "it is important to consider how governance arrangements at one level are shaped by arrangements at other levels" (Betsill and Rabe 2009, 214). In contrast to widespread optimism in climate policy literature as depicted above about the potential of the local level to contribute to climate mitigation, governance research provides some more cautious perspectives on governability. Possible outcomes of inter-municipal climate policy coordination range from enhanced performance to reduced governability of local climate governance. Summing up, European multi-level governance entails potentials for stalemate as well as for successful cooperation. The roles of local governments therein depend on policy fields and territorial organization in the respective member states. Under certain circumstances, they can forge an alliance with the European Commission without including, or even in order to circumvent, the national level. An instrument used by the European Commission for this purpose consists in voluntary commitments.

3 Theorizing Orchestration in European Climate Governance

As demonstrated above, climate governance in the EU happens in a multi-level system. Here, no single level can achieve its goals alone. All actors are interdependent and have to rely on others' resources in a continuous process of negotiation that is not simply hierarchical. This raises questions of governability which have been answered differently by scholars depending on the policy fields and actor constellations they focused on. With regard to decision-making, European multi-level governance has been described as a joint decision-making trap, but also as an opportunity structure for multi-level reinforcements in view of ambitious climate policies. Successful reinforcement requires strategies to overcome disagreements between EU member states about important aspects of climate policy, in particular with regard to the energy mix. In response, European climate governance is based to a significant extent on national targets with leeway for individual implementation. At the same time, the allocation of competences within the multi-level system is constantly negotiated which results in occasional rescaling. From the perspective of local governments, EU climate governance thus entails both superordinate specifications as well as avenues for local emancipation.

Local climate governance confronts the European Commission with coordination requirements that it has to attend to based on soft modes of governance. Although in charge of supervision, it has no direct grasp on member state administrations at the national and subnational level that carry out EU legislation. This challenge is all the more crucial in climate governance where voluntary non-state action plays an important role. In particular, numerous local governments have engaged climate action ahead of national regulation, and made voluntary commitments to individual climate and energy targets such as GHG emission reductions or renewable energy promotion. They have important direct contributions to make towards the EU climate and energy targets, and could become laboratories for experimentation and innovation that encourage or even push national

© Springer Fachmedien Wiesbaden GmbH, part of Springer Nature 2020
L. Bendlin, *Orchestrating Local Climate Policy in the European Union*,
Energiepolitik und Klimaschutz. Energy Policy and Climate Protection,
https://doi.org/10.1007/978-3-658-26506-9_3

governments towards more ambitious climate action. However, not all local governments have engaged yet, and some voluntary commitment systems lack ambition, effectiveness, or accountability. A promising mode of climate governance thus consists in coordinating local governments' voluntary commitments systematically in order to fully realize their potential. Since the European Commission cannot intervene directly at the local level, it has to rely on voluntary cooperation of municipalities, or intervene indirectly through an intermediary. The latter approach has been conceptualized as orchestration. Initially developed within international relations theory, the orchestrator-intermediary-target model explains why and how governance actors at different levels voluntarily pool their resources for jointly addressing target actors in view of shared policy goals. One example consists in the Covenant of Mayors, an EU initiative for local sustainable energy. Launched in 2008, it invites subnational authorities to voluntarily commit to the EU climate and energy targets and includes processes for monitoring and coordination across levels of governance.

The theoretical framework is based on the concept of orchestration, a non-coercive, indirect mode of governance, as a theoretical tool to understand the Covenant. This chapter first derives the theoretical framework from the literature (3.1), and then explicates its application to the study of Covenant Coordinators in local climate governance (3.2).

The first section starts by discussing the role of voluntary non-state action in climate governance in view of its proliferation, promises, and pitfalls (3.1.1). It further argues that superordinate levels of governance are confronted with certain coordination requirements in order to provide a functional framework for voluntary local action (3.1.2). Subsequently, it discusses orchestration as a mode of local climate policy coordination (3.1.3). The second section conceptualizes the Covenant as an orchestration arrangement. It starts by introducing the Covenant, i.e. its functioning, its role in EU climate governance, and the questions it raises with regard to drivers, agency, and impacts (3.2.1). Hereafter, it discusses step by step the respective roles of the participating governance actors, namely of the secretariat as an orchestrator that engages inter-municipal associations as intermediaries in order to address municipalities, the actual target actors (3.2.2). The chapter concludes with an outlook on the expected theoretical contributions (3.3).

3.1 Orchestration as Local Climate Policy Coordination

This section lays the groundwork for the theoretical framework in view of understanding local climate policy coordination in the EU. Different strands of literature have shown that voluntary non-state action plays an increasingly important role in global climate governance, and provided various accounts of how it is motivated, implemented, and governed (3.1.1). The decentralized landscape of non-state action entails growing coordination requirements in order to ensure that policies develop in an effective and coherent manner (3.1.2). This task is a challenge for governors whose inability to act on climate change caused the proliferation of voluntary non-state action in the first place. One mode of governance that relies on voluntary cooperation of intermediate actors motivated by the exchange of resources in the pursuit of a shared policy goal has been conceptualized as orchestration. This theory is introduced in more detail (3.1.3) before being applied to the Covenant in the next section (3.2).

3.1.1 Voluntary Non-state Programs in Climate Governance

The proliferation of multi-level governance approaches and voluntary non-state action in climate policy practice and analysis has been described as a three-steps process from hierarchical steering to alternative bottom-up venues to the challenges of climate governance in a multi-level system (Betsill and Rabe 2009). At an early stage, climate change was considered as an issue to be dealt with hierarchically by governments in international negotiations with a top-down approach. Correspondingly, international climate negotiations under the UNFCCC have long been strongly intergovernmental. Subsequently, a decentralized approach emerged where subnational and local governments constitute a site of governance, if not the primary site. In parallel to the international climate negotiations, a number of voluntary non-state initiatives emerged such as partnerships and commitment systems involving various kinds of non-state actors, including local governments.[19] The mobilization of cities in particular has raised hopes for an alternative venue for effectively tackling climate change (Betsill and Bulkeley 2006; Bulkeley et al. 2013; Bulkeley and Betsill 2005; Corfee-Morlot et al. 2009;

---

19   Here, sub-state actors are explicitly included when referring to non-state action, while state actors are conceived as governmental actors from the national level. It should be noted, though, that some authors differentiate between non-state (i.e. non-governmental) and sub-state actors (i.e., below the national level).

OECD; OECD 2010). Voluntary action by subnational authorities, international organizations, civil society organizations, companies, and partnerships has become an important element of global environmental governance in general and of global climate governance in particular (Andonova, Betsill, and Bulkeley 2009; Bulkeley et al. 2014; Hale and Roger 2014; Hoffmann 2011; Roger, Hale, and Andonova 2016).

In recent years, experiences in decentralized climate governance have shifted attention to the requirements and possibilities to cope with multi-level interdependencies and policy overlap through multi-level governance (Betsill and Rabe 2009). Voluntary non-state climate action has been conceptualized from different perspectives, including as transnational governance (Bulkeley et al. 2014), collaborative environmental management (Koontz and Thomas 2006), multi-stakeholder partnerships (Beisheim and Simon 2015), green clubs (Prakash and Potoski 2012), transnational municipal climate networks (Bansard, Pattberg, and Widerberg 2016; Bulkeley et al. 2003; Cao and Ward 2016; Gordon 2013; Gore 2010; Lee 2013), and voluntary commitment systems (Abbott 2017).

Although the importance of the local level has been emphasized by proponents of a decentralized or polycentric approach to multi-level climate governance, voluntary non-state action does not necessarily consist of bottom-up initiatives. Hale and Roger (2014) estimate that nearly a third of transnational climate governance schemes has been initiated, supported, or steered through other means, by states or intergovernmental organizations. For instance, a growing number of multi-stakeholder partnerships have been established by UN agencies; examples include the UN Global Compact and Sustainable Energy for All (Abbott 2017). Hence, they have to be understood in their multi-level context.

Whether voluntary non-state action emerges from the bottom up or within top-down initiatives, the question arises why non-state actors not only engage in voluntary climate action, but also join programs or schemes that network, document, and steer their undertakings. These forms of institutionalization allow for the integration of voluntary non-state action in the global climate regime and have sometimes gained more scholarly interest than non-state action in itself, but are far from self-evident from the perspective of non-state actors. Why make this extra effort? Benz et al. (2015) demonstrated that local governments' engagement in trans-local climate action can be explained by instrumentalist and normative actor orientations (see figure 3.1). Without being mutually exclusive, an instrumentalist orientation "involves actions aimed at the realisation of properly defined goals and the solution of local policy problems" whereas normative orientations are not "necessarily geared towards the realisation of concrete policies" (Benz et al. 2015, 324–25). Both orientations can adopt an internal or external orientation, i.e. they can relate to local politics in a narrower sense or to local governments' external relations.

	Instrumentalist orientation	Normative orientation
Internal perspective	• Safeguarding local policies or institutions • Funding acquisition	• Diffusing innovation through network learning • Awareness raising
External perspective	• Inter-urban competition • Lobbying at superordinate levels of governments	• Global responsibility

Figure 3.1: Local actors' motivations for trans-local climate action
Source: Based on Benz et al. 2015, 324–25.

Despite the numerical proliferation of voluntary non-state action and programs and their potential contributions to climate governance both "directly and through their influence on states" (Abbott 2017, 758), they should not be mistaken for a "silver bullet" (Beisheim and Simon 2015). In their analysis of global urban climate governance in the post-Paris climate regime, Gordon and Johnson (2017, 695) argue that the claim of non-state-actors – here: cities – for a vital role in global climate governance

> "rests on the production of individual and collective effects; governance engagement not only in and by particular cities but, more importantly, the ability to engender coordinated efforts and activities between cities from across the globe."

But the capacities and impacts of voluntary programs largely differ and do not always live up to their mission (Morgenstern and Pizer 2010). Only a minority of transnational climate governance initiatives has been found to be able to realize tangible mitigation impacts on their own. The vast majority either focus on indirect contributions such as knowledge exchange, or depend on favorable state regulation (Michaelowa and Michaelowa 2016). In the attempt to systematize voluntary non-state action, many commitment systems rely on SMART criteria in order to enhance their ambition and accountability, i.e. on commitments that are specific, measurable, achievable, resource-based and time-bound (Doran

1981; Natural Resources Defense Council 2013). But voluntary commitment systems have been found to

> "apply only modest criteria and vetting procedures, provide commitment-makers with limited support, and have few means to hold them accountable. Perhaps most surprising, given their novelty, they include few mechanisms to promote exploration and testing of new approaches or learning from experience" (Abbott 2016, 3).

This echoes longstanding debates in adjacent fields such as sustainability governance. The use of numerical sustainable development indicators at different levels of governance has been discussed critically for a lack of target achievement and distortions as compared to initial policy objectives (see for example Diefenbacher et al. 2016). However, this perspective overlooks other functions of numerical indicators. In Local Agendas 21, an important value of sustainability indicators consists of mobilizing local actors (Pettibone 2015). Lepenies (2014) raises a similar argument from a Foucauldian governmentality perspective in a case study of the Millennium Development Goals (MDG). Although the MDGs define development, and how to measure it, they primarily serve to legitimize the policies and governance actors involved; target achievement is secondary. Hence, they are 'successful' as a governance tool despite methodological flaws and problems in data collection. In sum, these observations demonstrate that SMART criteria allow for structuring and legitimizing governance initiatives and policies, but ensure neither effectiveness, nor innovation beyond known solutions.

Summing up, climate governance not only involves implementation challenges for policies determined at superordinate levels as described in the previous chapter. Also, the proliferation of non-state action has brought about its own challenges and opportunities for climate governance with regard to effectiveness, coherence, and accountability. Increased coordination requirements result.

3.1.2 Coordination Requirements in Climate Governance

The increased participation of non-state actors – public, private, and from civil society – in climate governance entails growing complexity and interdependencies: "Governing across so many scales and across so many dispersed but overlapping networks presents huge problems of coordination and policy coherence" (Bulkeley and Newell 2010, 106). This represents a typical challenge of multilevel governance (Benz 2007, 18–19) and brings about a growing need for coordination between and within levels of government (Corfee-Morlot et al. 2009; Ohlhorst, Tews, and Schreurs 2014). Here, coordination means to align actors' behavior in non-hierarchical manners in view of coherent policies so as to avoid gaps, overlaps, or negative externalities (Eberlein and Kerwer 2004; Jordan and Schout 2008; Peters 1998).

Two developments in climate governance cause coordination requirements to grow. Firstly, the multi-level architecture of climate governance entails multiple interdependencies across scales and types of governance actors that make local level implementation challenging (see chapter 2). Secondly, the rise of voluntary non-state action in climate governance involves the need to ensure effectiveness (see section 3.1.1). These phenomena bring about coordination requirements for local climate policy in two regards: coordination with superordinate levels of government, and coordination of voluntary commitments with other local governments.

Not only does global climate policy involve actors at multiple levels, also the policy field is particularly prone to requiring coordination. Coordination requirements are known to vary across policy fields (Biela, Hennl, and Kaiser 2013, 167). Climate policy and related policy fields such as transport have clear spatial implications, for example in relying on land use planning, and entail (re-)distribution of resources and opportunities between and within territories. Also, climate planning at superordinate levels of government requires translation to the local context for successful implementation. Therefore, coordination requirements in climate policy can be expected to be relatively to extremely high, depending on the sector of climate action. These requirements go beyond decentral coordination driven by knowledge rather than intentional steering as captured by the literature on governance by diffusion, i.e. the voluntary adoption of policy innovations[20] based on imitation, persuasion, and learning (Holzinger, Jörgens, and Knill 2007; Jörgens 2004). Although international agencies have been found to deliberately promote governance by diffusion, this mode of governance hardly allows for coordination between levels of government (Tews and Jänicke 2015).

As argued above, the rise of voluntary commitment systems in climate governance further increases coordination requirements in order to ensure their performance in terms of ambition, implementation, and accountability. Beisheim

---

20   It is important to note here that the concept of the diffusion of policy innovations does not entail any judgement on the quality of the policy that is diffused, or of the motivation of the policy adopter. The policy is understood as innovative in terms of being *new to the adopter*.

and Simon (2015) call for a meta-governance of multi-stakeholder partnerships in global governance for sustainable development to both enable and ensure successful implementation through the provision of support, and oversight mechanisms. Michaelowa and Michaelowa (2016) demonstrate that coordination by superordinate levels enhances the effectiveness of voluntary non-state climate action, although they find local governments to be more ambitious than other kinds of non-state actors in actually reducing emissions. Chan et al. (2015b) call for further integrating non-state climate action into the UNFCCC framework in order to promote ambitious commitments, systematic experimentation, and accountability, and avoid redundancies. This, Chan, Brandi, and Bauer (2016) argue, would allow for mutual reinforcements between international negotiations and non-state action. In a similar vein, Abbott (2016; 2017) conceives voluntary commitment systems as ideal laboratories for experimentation that could enable learning and foster innovation in climate governance if managed systematically. These effects are likely to be stronger under higher levels of national regulation because reduced efforts in domestic lobbying free non-state actors' resources for transnational activities, which also suggests that non-state and state action complement rather than supplement each other (Andonova, Hale, and Roger 2017; Cao and Ward 2016). In this view, voluntary non-state action is not an alternative to, but an element of multi-level climate governance that needs to be integrated in ways that allow for its innovative potential to stimulate the climate regime as a whole. Linkages between the UNFCCC and voluntary non-state action remain understudied, Betsill et al. (2015) argue, and call for further research, a suggestion taken up in the following with regard to linkages between the supranational and the local level in European contributions to global climate governance (see for example section 3.2.1.2).

In light of increasing coordination requirements and growing expectations towards local climate policy performance, the question arises who could reasonably provide climate policy coordination within the local level. From the perspective of superordinate governors such as UN agencies or the European Commission, an additional challenge for coordinating local climate policy derives from the fact that they lack the competences and resources to intervene directly at the local level. Addressing local governments requires extensive funds, information, and legitimacy, also because the number of target actors at the municipal level is extremely high. Hence, governors need allies that bring in the required resources, and willingly cooperate. One such mode of governance has been conceptualized as orchestration and will be introduced in the next section.

3.1.3   Governing through Orchestration

Multi-level climate governance is marked by a need for coordination between and within levels of government. In the EU, soft modes of governance have been shown to play an important role in overcoming conflict between member states and between the European and national level (Conzelmann and Smith 2008). With regard to local governance, an additional feature of the EU consists in the fact that it includes more than two levels which are not directly linked in territorial and administrative organization. This section introduces the concept of orchestration as a mode of governance available to the European level in order to implement EU climate policy targets at the local level.

In non-hierarchical multi-level systems, the capabilities of governance actors – in the following referred to as governors – for hierarchical regulation do not suffice to meet their goals. Thus, they will likely rely on alternative modes of governance. For purposes of theorizing, Abbott, Genschel, Snidal, and Zangl (Abbott et al. 2012) argue, modes of governance can be categorized based on their direct or indirect approach and the use of hard vs. soft power (see figure 3.2). Does a governor directly address the target actors of its policy through mandatory regulation as in hierarchy, or by engaging in collaboration at eye height? If a governor relies on an intermediary, is the relation to the intermediary characterized by discretionary power? Such delegation to outside parties, both an indirect and hard mode of governance, has been conceptualized in principal-agent theory (Hawkins et al. 2006). As these categories fail to capture instances of governance that are indirect and soft at the same time, the authors introduce the concept of *orchestration* (Abbott et al. 2012, 2).

	Direct	Indirect
Hard	Hierarchy	Delegation
Soft	Collaboration	Orchestration

Figure 3.2: Four modes of governance
Source: Abbott et al. 2012, 5.

Orchestration consists in "the mobilization of an intermediary by an orchestrator on a voluntary basis in pursuit of a joint governance goal" (Abbott et al. 2016, 1) as conceptualized in the *O-I-T-model* (see figure 3.3). Abbott et al. first developed this model in the search for improved public support for private sustainability governance (Abbott 2012; Abbott and Bernstein 2015; Abbott and Snidal 2009; 2010). While the initial series of articles has been criticized for remaining largely conceptual and normative (Schleifer 2013), the authors have since delivered a comprehensive presentation of the O-I-T-model in an edited volume on international institutions as orchestrators (Abbott et al. 2015). It presents a series of case studies from the realm of international relations and different policy fields ranging from financial regulation to climate adaptation. Intermediaries covered include the EU, WTO, WHO, and several UN programs. To date, other scholars have adopted the model for applications to the study of various policy fields including sustainable development (Klingebiel and Paulo 2015), biodiversity (Jörgens et al. 2016), global human rights governance (Pegram 2015), financial regulation (Viola 2015), maritime shipping (Listera, Poulsen, and Ponte 2015), transnational climate governance (Chan et al. 2015a), and EU biofuel governance (Schleifer 2013).

Orchestrator → Intermediary → Target

Figure 3.3: O-I-T-Model of indirect governance through orchestration
Source: Abbott et al. 2012.

By definition, orchestration is an indirect mode of governance. A governor that acts as an orchestrator addresses the target actors of its policy indirectly by bringing "third parties into the governance arrangement to act as intermediaries between itself and the targets" (Abbott et al. 2012, 2). Since intermediaries have to perceive participation as beneficial to their own goals, orchestration is also a soft mode of governance based on "soft inducements rather than hierarchical controls" (Abbott 2016, 3). This is "because an orchestrator has no firm control over the activities of intermediaries but must mobilize and facilitate their voluntary cooperation in a joint governance effort" (Abbott et al. 2012, 3).

The O-I-T-model assumes that this cooperation between an orchestrator and an intermediary in view of addressing target actors of a given policy builds on shared policy goals, and complementary capabilities:

> "The orchestrator enlists the intermediary to contribute the missing capabilities; in return, the orchestrator uses its (limited) capabilities to enlist, facilitate and support the intermediary's governance activities. Orchestrator and intermediary are mutually dependent, each unable to achieve its goals without the other. The division of governance responsibilities between orchestrator and intermediary, and the resulting gains from specialization, are at the core of the O-I-T model" (Abbott et al. 2012, 11–12).

This assumes a goal-seeking behavior of orchestrator and intermediary. Goals may be material or ideational, self-seeking or altruistic. Although the O-I-T-model assumes neither perfect information nor perfect rationality, and concedes that actors might be unequally successful in pursuing goals, it implies a relationship that builds on shared or correlated goals. The overlap need not to be complete, but suffices to overcome conflicts of interests, e.g. over particular policy instruments (Abbott et al. 2012, 11–12).

Why exactly would a governor choose to rely on intermediaries, and why would the latter accept to participate? From the perspective of the governor, orchestration is an alternative to direct governance modes when lacking the regulatory competence, operational capacity or legitimating authority to unilaterally achieve its objectives. The governor thus relies on intermediaries' complementary governance capabilities such as local information technical expertise, enforcement capacity, material resources, legitimacy and direct access to targets (Abbott et al. 2012). By means of orchestration, a governor

> "with sufficient legitimacy, focality, and resources can enlist like-minded actors and organizations, deploying material support, reputational incentives, information and guidance, and mobilizing pressure and assistance from third parties, to catalyze, support, and steer the behavior of enlisted actors, enhancing their impact on their ultimate targets or beneficiaries" (Abbott 2017, 739).

Certain features increase a government's likeliness to rely on orchestration (Abbott et al. 2016, 6–7). These include a preference for goal achievement rather than hard control (i.e. democratic rather than authoritarian regimes) and limited capacity for hard control (e.g. international organizations). Also, governments are more likely to choose orchestration over other modes of governance when they have to circumvent veto players, when they prioritize blame avoidance in case of failure over credit claiming in case of success, and when an issue is not salient enough to accept the (higher) costs of delegation.

A governor who intends to resort to orchestration needs to find voluntary collaborators, i.e. intermediaries that "share the [governor]'s basic governance goals and value its material and ideational support" (Abbott et al. 2012, 3–4). Material and intellectual resources of an orchestrator are key to attract intermediaries.

> "Even the ability to pay very basic transaction costs – meetings, staff support, research – can be decisive for some initiatives. Others, importantly, depend on capacity building and information exchange for which the orchestrator's technical expertise is an enormous asset" (ibid.).

At the same time, the O-I-T-model assumes the orchestrator lacks the resources to address the policy problem alone.

> "We should therefore expect orchestration to be facilitated by mid-levels of resources: enough to convene and facilitate, but not enough to implement solutions of sufficient scale on their own" (Hale and Roger 2014, 68).

Although "analytically distinct, material and ideational support are often joined in practice" (Abbott et al. 2012, 9). Important forms of support in orchestration consist of convening, agenda setting, assistance, endorsement, and coordination (Abbott et al. 2012, 9–11). Orchestrators strategically convene and connect selected intermediaries based on their privileged access to other governance actors. Their influence can even lead to the creation of new intermediaries. Orchestrators also facilitate conventions at later stages of the political process, e.g. as hosts. Orchestrators also steer intermediaries through agenda setting. Their agendas provide guidance on priorities and plausible measures. Intermediaries build on such agendas to define their own strategies, and benefit from the legitimacy of the orchestrator when their activities coincide with the agenda. This helps to obtain funding by third parties and cooperation by targets. Sometimes, orchestrators can also provide direct assistance in the form of material support for intermediaries. Often their financial and administrative resources are limited, but their assistance enhances the visibility, authority, and network of intermediaries. Ideational support can have a similar impact through endorsement. Orchestrators formally recognize intermediaries as competent, legitimate actors in a given policy field and thereby increase the pressure on target actors to comply with interme-

diary policy. Last but not least, orchestrators provide coordination so as to synchronize the activities of intermediaries, thereby enhancing their effectiveness.
Support has a twofold impact:

> "Support strengthens the intermediary, while providing the orchestrator modest leverage over its actions. Material support strengthens the intermediary's operational capacities; ideational support – such as guidance, formal approval or political endorsement–enhances the intermediary's effectiveness and legitimacy vis-à-vis targets. By conditioning its support, and by assuming ideational leadership over the intermediary's problem definitions and policy priorities, the orchestrator can nudge the intermediary toward governance goals that are compatible with its own goals. Orchestration thus simultaneously empowers intermediaries and provides the orchestrator soft influence over them" (Abbott et al. 2016, 4).

Agency, understood here as "the genuine choice made by actors among possible and plausible options" (Bukowski, Piattoni, and Smyrl 2003, 7), cannot be ascribed to either the orchestrator or the intermediary alone:

> "Like hierarchical governance solutions, orchestration involves focal, public institutions directing sub- and non-state actors toward a common goal. But like horizontal solutions, orchestration is negotiated between the various parties, and must ultimately respond to the needs of all. It is 'top-down' in that the agency at the state or IO [international organization] is a notable and distinguishing feature and their public authority plays a key role, but it is also 'bottom-up' because it seeks to unlock the agency of transnational actors to help provide public goods transnationally. It therefore falls, interestingly, between existing ideal types of transnational governance" (Hale and Roger 2014, 64).

A look at the initiation of orchestration arrangements further highlights their non-hierarchical nature. Abbott et al. (2014, 3) compare orchestration of states by international governmental organizations and, to their surprise, find

> "that orchestration is not only a top-down technique initiated by an IGO or other orchestrator; it often operates bottom-up, with intermediaries and even targets actively seeking orchestration."

This is because intermediaries benefit from orchestration in terms of their policy goals and their organizational interest as they obtain additional legitimacy and empowerment (Abbott et al. 2014, 10–11).

The willingness of a governor to rely on orchestration is only one side of the coin; the other consists in its ability to do so (Abbott 2017, 755). In order to succeed, an orchestrator needs authority and legitimacy in the eyes of the actors it seeks to influence based for example on its institutional position, past successes, expertise, or moral qualities. It further needs strong relationships with the actors it aims to involve, and sufficient available material and ideational resources to provide them with support. Finally, the governing organization must be willing, and able, to act collectively as an orchestrator. In many respects, these prerequisites depend on an orchestrator's "focality", i.e. the centrality of its role in a given policy field. An institution

> "…may be seen as the most legitimate forum for an issue to be addressed due to, for example, the representativeness of its governance, the breadth of its constituency, its history of engaging with the issue, or other reasons. It may also achieve focality due to its unparalleled expertise or unique operational capacity. Whatever the reason, focal IOs [international governmental organizations] can identify concerns, catalyze action and bring coherence to governance activities. They can therefore encourage actors to pursue certain paths of governance rather than others" (Hale and Roger 2014, 67).

Focal institutions are accepted as unavoidable partners in tackling a given collective action problem. Public recognition enhances the ability of an institution to act as an orchestrator:

> "Explicit public support from a policymaker or group of policymakers can also increase a transnational scheme's public exposure in the eyes of the actors it seeks to coordinate" (Hale and Roger 2014, 67).

In particular, this impacts the ability of an institution to orchestrate by convening, along with public endorsement.

The primary limitation of orchestration derives from the capacities of the intermediary:

> "…the orchestrator must select intermediaries based on their governance goals, and therefore may have to compromise on their capabilities. Given its lack of hard control, the orchestrator depends on intermediaries that are intrinsically motivated to work in concert with the orchestrator. [...] Yet compatible motivations do not imply adequate capabilities. Often the orchestrator must work with intermediaries that are partly or completely incapable of performing the required tasks. The orchestrator will try to improve intermediary performance through appropriate forms of support, but these measures can fail" (Abbott et al. 2016, 5).

Consequently, the effectiveness of orchestration depends on the capacities and preferences of intermediaries. Are they able to implement their commitment, or will the governance arrangement remain an empty shell? Will they maintain interest over time, or withdraw openly or reduce their engagement to symbolic gestures? However, as long as viable alternatives such as delegation are not available, even deficient orchestration might be the better than no governance at all (Abbott et al. 2016, 9).

Only recently have scholars begun to discuss the democratic legitimacy of orchestration more vividly. On the one hand, orchestration lacks electoral accountability and "potentially allows governors [...] to pursue goals outside their mandate" (Abbott et al. 2016, 9–10). Orchestration not only serves to attain policy goals, but also organizational interest of the participants. Governors employ orchestration to expand into new areas and to gain prominence and expertise without major investments (Abbott et al. 2014, 8), as shown for the European Commission with regard to environmental policy in a case study building on the O-I-T-model by Mattli and Seddon (2015). On the other hand, the interdependence of governance actors in orchestration makes it impossible for a single one of them to exercise unconstrained authority, and favors collective deliberation and mutual learning (Abbott et al. 2016, 9–10). For more systematic assessments of the democratic legitimacy of individual orchestration arrangements, Bäckstrand and Kuyper (2017) suggest to check for participation, deliberation, accountability, and transparency.

The O-I-T-model as introduced here does not exclude the study of multi-level systems, but has been developed as part of international relations theorizing. So far, it focused on international institutions as orchestrators, and national and transnational actors such as NGOs, business organizations, public-private partnerships, transgovernmental networks and other international governmental organizations as intermediaries. Targets are either states or private entities. In this understanding, orchestration either serves to *manage* states by creating and enforcing rules of conduct, or to *bypass* states, i.e. to substitute or complement state action by the provision of regulation or public goods (Abbott et al. 2012, 3–4). This conceptualization of the O-I-T-model has been criticized for focusing on – and promoting – private actors as intermediaries (Schleifer 2013, 534). In contrast, the O-I-T-model is now extended from the international level to the study of territorial governance, that is, it is applied to a new type of target: subnational governments.

Subnational governments have been included as intermediaries to an earlier application of the orchestration model in a policy paper. The preliminary assessment of non-state climate action launched at the 2014 UN Climate Summit in New York pointed out that, in contrast with conventional wisdom on transnational climate governance, most climate actions are undertaken by states and international organizations. In the sample under study, subnational governments such as cities and regions turned out to be the third most prominent partner ahead of private and civil society actors (Chan et al. 2015a). Despite this reference to the importance of subnational governments in transnational climate governance, the policy paper lacks a discussion of the conceptual implications and specificities of this kind of intermediary for orchestration. This gap is now addressed by applying the O-I-T-model to local climate policy coordination where orchestration serves to manage local climate policy by creating and promoting goals and proceedings for municipal action.

3.2 Orchestration through the Covenant of Mayors

With the Covenant, the European Commission aims to incentivize and provide selected instruments for local sustainable climate and energy policy. The Covenant operates on a voluntary basis and addresses municipal targets indirectly, contradicting the assessment by Hale and Roger (2014, 80) according to which the EU, albeit a 'natural' orchestrator, has not yet employed orchestration effectively in climate policy. But why did the European Commission decide to rely on orchestration? Theoretically, the same goal could be pursued by (hierarchical) regulation; provided the member states of the EU agreed on mandatory local climate planning legislation. In the absence of such consensus, the European Commission could have relied on cooperation, but the number of target actors to be addressed, i.e. local governments, overshoots the capabilities of the European Commission. Thus, orchestration represents a second best solution. Also, orchestration corresponds to the level of resources available. The European Commission is able to convene and facilitate local climate policy through the Covenant, but its means would not suffice to comprehensively fund and support local climate policy. In order to better address municipalities, subnational governments have been involved as intermediaries as they are more suited than the European level to provide tailored assistance.

How can the O-I-T model be applied to the Covenant in view of further developing orchestration theorizing on the one hand and improving our understanding of the Covenant on the other? An answer to this question first requires a detailed understanding of how the Covenant is set up and why it challenges common concepts of climate policy analysis (3.2.1). The subsequent discussion then depicts the roles and perspectives of the Covenant Office as a governor, of inter-municipal associations that act as voluntary intermediaries on its behalf, and of municipalities that are the actual targets of this governance arrangement (3.2.2).

3.2.1 Introducing the Covenant of Mayors

The Covenant aims to stimulate sustainable energy policy at the municipal level in the EU. The EU initiative launched in 2008 provides a platform for local and regional authorities willing to engage in local climate action and sustainable energy policy. Based on the voluntary involvement of various governmental and other actors at multiple levels, it is designed primarily to promote municipal climate action. Signed to date by nearly 7,200 municipalities that have submitted more than 5,600 action plans, it covers approx. 225 million inhabitants (Covenant of Mayors 2016d). On the occasion of a relaunch in 2015, the Covenant has integrated Mayors Adapt, the affiliated initiative on urban adaptation to climate change set up in 2014, and has been renamed Covenant of Mayors for Climate & Energy. Starting January 2017, the Covenant launched the Global Covenant of Mayors for Climate & Energy together with the Compact of Mayors, a similar initiative from the UN-level.[21]

This section explains in more detail how the Covenant is designed (3.2.1.1) and how its creation and development relate to EU climate and energy policy and to discussions over territorial governance more broadly (3.2.1.2). It concludes by discussing to what extent current conceptualizations give consideration to the specificities of the Covenant (3.2.1.3).

3.2.1.1 The Covenant's Mode of Operation

The Covenant combines unilateral commitments by participating municipalities, specific targets and measures to achieve them, and scrutiny of reports. The EU initiative builds on the involvement of non-municipal governments at the EU, national and subnational levels, and other stakeholders (Ballesteros Torres and Doubrava 2010). This setup entails five key features: EU sponsorship, voluntary commitments, compliance control, hybrid membership, and explicit reference to EU multi-level governance.

---

21   The Compact of Mayors is a joint initiative by the C40 Cities Climate Leadership Group, ICLEI – Local Governments for Sustainability (ICLEI), United Cities and Local Governments (UCLG), and UN Habitat, the UN program for human settlements.

Firstly, the Covenant has been founded and is funded by the European Commission. Thus, it is no bottom-up initiative. But the European Commission is not involved directly in the day-to-day management of the initiative. Instead, it established the Covenant' Office (CoMO) in Brussels, run by a consortium of networks of local and regional authorities. This consortium is led by Energy Cities (a network of European local authorities on sustainable local energy) and includes Eurocities (an advocacy group of big European cities), the Climate Alliance (Europe's biggest municipal climate network), the Council of European Municipalities and Regions CEMR (Europe's biggest association of local and regional authorities) and the European Federation of Agencies and Regions for Energy and the Environment FEDARENE (regional and local organizations which implement, coordinate and facilitate energy and environment policies). As a result, adherence is free of charge, but the running of the consortium is subject to contractual terms with the European Commission.

Secondly, participation requires commitment to predefined targets and proceedings. All Covenant signatories commit to the EU's CO_2 reduction objective, and engage in a three-step process: 1. establish a Baseline Emission Inventory (BEI); 2. adopt a Sustainable Energy Action Plan (SEAP);[22] and 3. report regularly on implementation based on distinct Covenant methodology and tools. Initially, the Covenant pursued the EU's goal to reduce emissions 20% by 2020. Since the Covenant's relaunch in 2015, new signatories now commit to the EU's target of 40% greenhouse gas reduction by 2030 and additionally pledge to adopt a joint approach to climate mitigation and adaptation.[23]

Thirdly, compliance with these commitments is not only required for sustained membership, but also systematically controlled based on regularly scheduled reporting. BEIs, SEAPs, and progress reports on implementation and SEAP updates, are checked for their plausibility and their ability to reach the above mentioned objectives by the European Commission's Joint Research Centre (JRC), the European Commission's in-house science service. Delays can be granted upon request by the CoMO, but failure to submit, or to provide inquired rejection of submitted documents, may result in suspension of signatory status.

---

22   Since the relaunch of the Covenant of Mayors in 2015, it refers to Sustainable Energy and Climate Action Plans (SECAPs).
23   Out of the total number of Covenant signatories, 540 have committed to the new Covenant of Mayors for Climate and Energy Covenant of Mayors (2016d).

Fourthly, the Covenant involves a broad range of actors from the public and private sectors based on various forms of participation. The Covenant's hybrid membership (Andonova, Betsill, and Bulkeley 2009, 58–62) includes governments from above the municipal level as well as non-governmental actors. Initially subsumed in the category of 'supporting structures', they are now included as Covenant Coordinators and Covenant Supporters. National and territorial authorities such as ministries, provinces, and inter-municipal associations act as Covenant Coordinators providing

"strategic guidance, financial and technical support to signatories. Network [sic] of local authorities, known as Covenant Supporters, commit to maximize the impact of the initiative through promotional activities, liaison with their members and experience-sharing platforms" (Covenant of Mayors 2015b) (see figure 3.4).

Level of governance	Type of actor	
	Governmental	Other
European	Sponsor (European Commission)	Associated Partners (e.g. associations)
National	Covenant Coordinators (e.g. territorial authorities)	Covenant Supporters (e.g. municipal networks)
Territorial		Local and Regional Energy Agencies
Municipal	Signatories	

Figure 3.4: Typology of Covenant membership
Source: Author, based on Covenant of Mayors 2015b.

In addition, associations and federations from related sectors including renewable energies or energy-efficient construction are invited to share their expertise and broker partnerships between Covenant participants. Covenant participants cooperate within national Covenant Clubs.[24]

---

24   National clubs to the Covenant were initiated with support of Networking the Covenant of Mayors (NET-COM), a project for Covenant promotion and capacity-building on the national level funded by the Commission running from June 2011 to November 2013. Covenant Clubs exist in Austria, Austria, Belgium, Bulgaria, Catalonia, Croatia, France, Germany, Italy, the Netherlands, Poland, Romania, Slovakia, Serbia, Sweden, and Ukraine; in addition, there are the Covenant Clubs of the Union of the Baltic Cities (Denmark, Estonia, Finland, Germany, Latvia, Lithuania, Norway, Poland, Russia, Sweden), and of the Capital Cities of South East Europe NETCOM (2017).

Last but not least, the Covenant is not conceived as just some additional EU-project or as another TMCN alongside bottom-up initiatives, but as an 'official movement' that provides a platform for existing networks, municipalities, authorities from other levels of government, local and regional energy agencies, related international organizations and other actors to connect in view of the EU's climate and energy targets. The Covenant's design is a response to the lobbying efforts of local and regional governments, acknowledges their contribution to implementing EU climate and energy policy, and reflects their perception of European multi-level climate and energy governance as an integrative and cooperative trans-local action space with both a vertical and horizontal orientation (Benz et al. 2015).

3.2.1.2 Situating the Covenant of Mayors within European Climate Governance

The Covenant has to be understood as a program that is part of EU climate policy more broadly; EU sponsorship of the initiative has certain implications as compared to bottom-up or state initiatives. Hsueh and Prakash (2012, 461) find that the design of voluntary programs necessarily depends on the governance level that launches them: "Because actors establish voluntary programs with specific objectives, program design reflects sponsoring actors' resources and institutional context." In particular, they argue, the benefits a voluntary program can offer depend on its sponsor. The authors further suggest that the link between sponsorship and design should be taken into account in studies of voluntary programs, in particular in the EU, its member states, and respective subnational authorities (Hsueh and Prakash 2012, 446). Correspondingly, the perspectives of different participants in the Covenant are carefully distinguished in the following, namely those of the European Commission as its principal, the CoMO as its dedicated orchestrator, territorial governments as intermediaries, and municipalities as target actors. In order to provide the necessary background for this undertaking, this section explains how the Covenant relates to the European climate and energy targets, and to the roles of the European Commission and local governments in European climate governance respectively.

With the adoption of the climate and energy targets, the EU had laid the groundwork for its leadership ambitions in global climate governance in the run-up to the COP 15 in Copenhagen, but it still lacked the legislation to implement these policy goals (see section 2.1). In parallel to the preparation of the EU's climate and energy package adopted in December 2008, the European Commission's Directorate-General Transport and Energy (DG TREN)[25] pursued other options to implement the 20-20-20 targets. In light of its limited competences with regard to local-level implementation (see section 2.3), an attractive backdoor for influence consisted in incentivizing subnational actors to participate on a voluntary basis. Such an approach to climate and energy governance was in line with the European Commission's White Paper on European governance (2001). This white paper responded to increased criticism at the time about the long-standing democratic deficit of EU policy making, and to calls for alternatives to regulatory modes of governance (Knill 2008a). It made several suggestions for strengthening subsidiarity, including improved citizen participation through regional and local democracy. For this purpose, the white paper suggested to enhance participation by local-government associations at earlier stages of policy making so as to ensure that European legislation can be adapted more flexibly to local conditions in policy implementation.

Lobbying efforts of local authorities and TMCNs had long pointed out the important role of the local level in many climate- and energy-related sectors and decisions, and obtained support at DG TREN. The renewables faction at DG TREN included local authorities in its funding scheme IEE. In view of reaching the EU 2020 targets, it supported sustainable energy projects from areas such as renewable energy, energy-efficient buildings, industry, consumer products and transport. More than that, principal administrator Pedro Ballesteros Torres became a supporter of the idea to establish a climate- and energy-related network of mayors.[26]

The Covenant's creation has to be understood with regard to internal EU politics and external relations. The European Commission aimed at the implementation of the 20-20-20 targets in the absence of competences for hierarchical regulation or other forms of direct intervention at the local level, at an improved involvement of the local level in European governance, and at obtaining reliable data on local climate mitigation. It also aimed to legitimize its leadership claims

---

25 Since February 17, 2010, DG TREN of the European Commission has been split up to form the Directorate-General for Mobility and Transport and the Directorate-General for Energy; the Covenant is handled by the latter.

26 Local-level advocates included political leaders from the inter-municipal associations under study, namely Gérard Collomb, president of Greater Lyon and of Eurocities, and Eckart Würzner, Lord Mayor of Heidelberg, a city from Rhine-Neckar Region, and president of Energy Cities. Their role will be discussed in more detail in chapter 5.

in international climate negotiations, and to showcase what it promotes as an exemplary instance of multi-level governance. The Covenant is open to participants from outside the EU. Regional offices have been established to attend to Covenant participants from Eastern Europe and the South Caucasus (2011), the South Mediterranean (2012), and Sub-Saharan Africa (2016). Within the Global Covenant initiative, five additional regional Covenant offices in North America, Latin America and the Caribbean, China and South-East Asia, India and Japan were established in 2017 (Covenant of Mayors 2017b).

Internal benefits European politics	External benefits International politics
• Implement climate policy locally • Strengthen subsidiarity • Measure local climate mitigation	• Increase credibility of leadership claims in UNFCCC negotiations • Promote European mode of governance

Figure 3.5: Benefits of the Covenant of Mayors for the European Commission
Source: Author

The Covenant first figured in the European Commission's energy efficiency action plan of 2006 as a priority measure to be implemented in 2007. Originally, the Covenant was intended to bring together

> "the mayors of 20-30 of Europe's largest and most pioneering cities [...] to exchange and apply best practices thereby improving energy efficiency significantly in the urban environment, where local policy decisions and initiatives are important" (European Commission 2006a, 18; 24)

with a focus on transport (European Commission 2006b, 28–29).

For local and regional authorities as well as for transnational municipal climate networks (TMCNs), the creation of the Covenant constituted a success in terms of institutional recognition as the European Commission acknowledged their significance and contribution in sustainable energy policy. In the course of Covenant creation, they lobbied for a more inclusive design addressing various levels of government and other types of actors (interview with G. Magnin, 17 June 2013). In fact, the initial approach with its narrow focus on pioneers, exchange of experiences, and energy-efficient transport was quickly abandoned in favor of a broader design as depicted above.

The Committee of the Regions credited the Covenant's "long term commitment, citizens' involvement, and ambitious and integrated climate action" for its ability to be "a model also for the other continents" and for it to "be a start of a global movement" (Committee of the Regions 2014) and refers to the Covenant as an exemplary multi-level governance arrangement. In a similar vein, the Covenant prides itself as "an emblematic example of multi-level governance and bottom-up action in Europe" (Covenant of Mayors 2013a), "a groundbreaking movement […] also because of the substantial boost it has given to multi-level governance in Europe and beyond" (Covenant of Mayors 2014b, 14). The notion of multi-level governance used here does not refer to conceptualizations from political science theorizing, but to the white paper on multi-level governance by the Committee of the Regions. It builds on the principles of good governance as determined in the White Paper on European Governance (European Commission 2001), and defines multi-level governance as

> "coordinated action by the European Union, the Member States and local and regional authorities, based on partnership and aimed at drawing up and implementing EU policies" (Committee of the Regions 2009, 12).

In this understanding, multi-level governance is a community method that aims to increase participation and efficiency in European policy making.

Within the norm triangle of European energy policy – sustainability, competitiveness, and energy security (Knodt and Piefer 2014, 225–27) – the Covenant emphasizes sustainability with a focus on emission reductions. Signatories commit to meeting or exceeding the EU's CO_2 reduction objectives. Since its relaunch in October 2015, the focus of the Covenant has been broadened so as to also pursue the norms of energy security and competitiveness. The initiative – renamed Covenant of Mayors for Climate & Energy – now rests upon three pillars: climate mitigation and adaptation and secure, sustainable and affordable energy. The implementation of SEAPs has been referred to, by the European Commission in its energy security strategy, as a means to enhance energy efficiency and thereby reduce external energy dependency (European Commission 2014b, 7–8).

Summing up, the Covenant's creation traces back to a meeting of minds between representatives of the European Commission and local authorities lobbyists at a moment in time when climate mitigation was a priority in international and European politics. Opening a backdoor for the European Commission to engage local-level authorities in implementing the EU's climate and energy targets, local authorities conversely perceive the Covenant as empowering them to act on climate change and more generally to obtain a place within the European multi-level system. With regard to the analysis of the types of excludable benefits offered to Covenant participants, it is important to keep in mind that the European Commission lacks the competences and resources to intervene directly at the local level (see 2.3). Thus, it largely depends on intermediaries for providing municipalities with tangible benefits.

3.2.1.3 Uncertain Drivers, Agency, and Impact

Some attempts have been made to make sense of the Covenant arrangement in theoretical terms and to understand its mode of governance. But conceptualizations of the Covenant as a TMCN, a voluntary club, and as an instance of transnational environmental regulation fail to account for its multi-level architecture and neglect participants' actorness, i.e. the high degree of autonomy exposed by Covenant participants (Bendlin 2016). So far, references to the Covenant as an instance of orchestration only served as illustrative examples without actually applying the O-I-T-model to its analysis (Abbott 2017; 2018; Gordon and Johnson 2017). As a result, the actual driving factors behind the Covenant's numerical success, the distribution of agency between actors within this governance arrangement, and its impacts in terms of governance and goal achievement remain unclear.

The Covenant has been referred to in the literature as a transnational municipal climate network (TMCN) (Bansard, Pattberg, and Widerberg 2016; Busch 2015; Fünfgeld 2015; Hoffmann 2011; Hoppe, van der Vegt, and Stegmaier 2016; Lee 2013; van der Heijden 2018; Widerberg, Pattberg, and Kristensen 2016).[27] This assessment holds true to the extent that the Covenant aims to promote local climate action, primarily addresses the municipal level, and aims to network participants across levels of governance and national borders. Indeed, the Covenant performs all three typical functions of TMCNs as described in the international relations literature: information sharing, capacity building and implementation, and rule setting (Andonova, Betsill, and Bulkeley 2009, 63–66). Information sharing mainly takes the form of newsletters and so-called benchmarks of excellence shared through the Covenant website. Capacity building and

---

27 The Covenant's design strongly resembles that of TMCNs (Kern and Bulkeley 2009); its consideration in this perspective can therefore be justified in particular cases depending on the focus of study (Busch, Bendlin, and Fenton 2018; Hakelberg 2014).

implementation occur through workshops, webinars, and complementary EU-projects. Rule setting builds on reporting obligations and benchmarking.

Another interesting insight from TMCN literature consists in a heads-up against overly optimistic accounts of network impacts. Research on TMCNs has shown that network membership does not necessarily make a difference in terms of GHG emission reductions (Bulkeley and Newell 2010, 66). In fact, the impacts of network membership depend on

> "the level of engagement with the network among its members and their ability to access the resources provided by the network. The experience of several local authorities suggests that the process of translating a rhetorical commitment to climate protection into effective policies and programs for controlling GHG emissions is far from straightforward" (Betsill and Bulkeley 2006, 144).

These findings are in line with scholarly criticism of multi-level governance theorizing to mistake mobilization for influence (Jeffery 2000).

While the above mentioned literature mainly assesses TMCN functions based on secretariat activities, Busch, Bendlin, and Fenton (2018) developed an alternative typology based on municipal use of TMCNs. Municipalities rely on TMCNs for 1. enabling internal mobilization, 2. formulating emission reduction goals, 3. institutionalising climate trajectories, 4. enabling direct exchange, and 5. offering project support. Interestingly, the authors find that TMCN membership impacts are stronger in laggard rather than pioneering municipalities and relate to internal political processes within member municipalities rather than external interactions.

The TMCN perspective fails to correctly account for the Covenant's mode of operation. Admittedly, some of the Covenant's features occur in other TMCNs as well (Hakelberg 2014). The Climate Alliance also has a particular target; member municipalities commit to reducing their CO_2 emissions by 10% every five years in view of reaching a 50% reduction of per-capita emissions by 2030 and a per-capita GHG emission level of 2.5 t per year on the long run.[28] ICLEI's Cities for Climate Protection Program also includes a step-wise process, and Energy Cities also goes back to an EU initiative. Yet, only the Covenant combines these features, and the Covenant's mode of operation differs from TMCNs at least in four respects. It is EU-sponsored; it is run by a consortium that includes

---

28 Climate action in view of this target is supposed to follow the principles to be fair, local, resource-saving, regenerative, and diverse in order to tackle climate change in with a "holistic and truly sustainable approach" (Climate Alliance 2016).

TMCNs; it entails a voluntary commitment system under the supervision of an external agency, the JRC; and it aims to provide a platform for existing networks rather than establish an additional or competing one (cf. 3.2.1.1). What might at first look like any other TMCN turns out to be a unique supranational initiative aiming at incentivizing and providing instruments for municipal climate action on a comparable, verifiable basis.

The TMCN perspective also tends to overestimate network impact as compared to municipal agency, i.e. municipal actors' abilities to choose between different climate policies and forms of institutionalization. Categorizations of TMCNs based on the functions they perform, such as secretariat activities or publications, are mostly rooted in an international relations perspective. They are interested in the interplay of networks with other international actors, or they take a climate policy perspective, looking for example at network activities and how they contribute to diffusing climate policy innovations (Andonova, Betsill, and Bulkeley 2009; Busch 2015; Hakelberg 2014). The use made of TMCNs by individual municipalities is not at the center of attention. In contrast, excludable benefits in the form of support for participants are at the core of the O-I-T-model. Conceptualizing the Covenant as in instance of orchestration highlights the motivations of Covenant intermediaries such as inter-municipal associations, and provides a lens to further explore empirically to what extent the latter are able to translate the support they receive into tangible benefits for municipalities, the target actors within this governance arrangement.

In a similar vein, attempts to conceptualize the Covenant as an instance of transnational environmental regulation fail to examine in due detail whether and to what extent the incentives offered to participants actually match local needs and impact behavior. Heyvaert (2013) builds her argument upon the assumption that the Covenant provides participants with "a significant amount of support". To illustrate this point, she refers to the functions and services performed and advertised by the CoMO without any consideration of the actual use made by signatories of offers such as guidance documents and networking opportunities, or of the availability and significance of excludable funding opportunities for Covenant signatories from different countries and backgrounds. An assessment built on the offer by the Covenant as opposed to its use by municipalities entails the risk to fall for self-praise by interested parties such as the CoMO or the European Commission. Avoiding this pitfall requires studying in more detail what kinds of support are actually offered and used within the Covenant, and with what results. A creditable conceptualization needs to provide categories for empirical research at the local level, or the exercise remains largely an intellectual pastime.

Another weakness of a TMCN perspective consists in the fact that it most often entails a conceptualization of network membership as an independent variable and examines membership's impact on climate policy performance and diffusion (Andonova, Betsill, and Bulkeley 2009; Hakelberg 2014). In this view, membership is a causal explanatory factor for local authorities to act on climate change. In contrast, research on cities' motivation to join TMCNs has shown that membership depends on local rather than national circumstances (Lee 2013) and results in different uses of membership (Busch 2015). Hence, a macro-level perspective neglects reasons for local authorities to join networks in the first place as well as potential attempts to use the network for pursuing their own policy goals. This entails a tendency to overestimate network impact when assessing causal relations between network membership and local climate policy performance, and more generally, a risk to obscure instances of agency at the local level. Local authorities do not obediently carry out a supranationally-defined mission in total harmony with the European Commission, and assuming this would hardly advance our understanding of global climate governance. Rather, it would be yet another example of top-down views of local governments in multi-level governance theory (Jeffery 2000).

In contrast, this book adopts a critical approach to macrodeterminism building on Bukowski et al. (2003). In their analysis of the determinants of territorial governance, they refute "a general devaluation of politics in the sense of choice and decision making in favor of structural, or at least exogenous, variables" (Bukowski, Piattoni, and Smyrl 2003, 1–19), namely globalization and Europeanization as variables 'from above', and institutions (in their broader sense including political culture) as variables 'from below'. Building on a series of case studies from the regional level of government, they conclude that territorial governance – despite intervening, external variables – is actually determined by agency. Without neglecting constraints limiting actors' choices "in scope, in time, and in effect" (Bukowski, Piattoni, and Smyrl 2003, 7), they argue that individual territorial institutions can emerge as a result of agency in territorial politics despite the seemingly convergent trends of globalization and Europeanization, and territorial identity can be constructed in ways not predestined by political culture (Bukowski, Piattoni, and Smyrl 2003, 1–19). Thus, "[t]he challenge for the researcher is to find the choices that matter" (Bukowski, Piattoni, and Smyrl 2003, 7). This approach entails an actor-centered perspective that focuses on the motivations of local governments to decide for Covenant signature and to sustain participation over time. It accounts for the possibility that local climate policy performance might be linked to other factors, and that Covenant adhesion might be an expression rather than a driver of local climate action.

Vice versa, a conceptualization of the Covenant as an orchestration arrangement also highlights the fact that despite the proliferation of voluntary non-state action in climate governance, many of these initiatives rely on orchestration and that most of the orchestrators consist of states or international governmental organizations. 'Traditional' actors remain key drivers of transnational climate governance (Hale and Roger 2014, 79). This raises the question what role subnational governments actually assume in governance arrangements such as the Covenant, and whether the rise of local climate policy also reflects a rise of local governments in global climate governance.

Another approach to understanding the Covenant has built upon the literature on so-called green clubs, voluntary environmental programs aiming to incite companies to exceed regulatory requirements (Prakash and Potoski 2012). For voluntary programs to be effective, they need to respond to two main challenges, recruitment and shirking: Why should target actors join a voluntary program, and why should they actually implement their commitment (Potoski and Prakash 2009)? Adherence to voluntary programs has been shown to be motivated by excludable benefits, i.e. benefits that are available to program participants only, while the policy goals pursued typically contribute to the common good, e.g. climate mitigation. In this line of argument, the Covenant been conceptualized as a "voluntary club" (Dolšak and Prakash 2016). From this point of view, the guiding questions for studying the Covenant consist of what it offers to, and demands from, local governments. The club perspective sheds light on the motivations and benefits of participants, but neglects the Covenant's inherent multilevel architecture. Covenant participants at all levels – sponsor, orchestrator, intermediaries and targets – operate in a multi-level system and depend on each other's resources in multiple ways. Understanding why and how resources are used jointly in local climate governance is crucial.

Summing up, the Covenant has been conceptualized in different ways that all have certain weaknesses. Although TMCN research provides an interesting point of departure for assessing the Covenant's functioning and membership, this perspective distorts the initiative to a considerable extent, namely because of its EU sponsorship. Also, it entails contradicting indications with regard to the Covenant's impact on local climate policy as it tends to overestimate network influence while at the same time warning against mistaking mobilization for effective GHG emission reductions. A conceptualization as transnational environmental regulation adopts an inadequate top-down view on the Covenant and fails to account for the interactions of participating actors in the European multi-level system. In a similar vein, a club perspective has the virtue of highlighting the importance of excludable benefits for participants, but neglects that actors depend on each other in multiple ways. All these approaches suffer from a lack of empirical groundwork, namely with regard to the motivations and strategies of Covenant participants from the local level.

In order to pave the way for an empirically-based discussion, the next section explicates how the Covenant is to be understood as an orchestration arrangement, and what this conceptualization entails for an understanding of the behavior of participating governance actors. This approach is intended highlight the kinds of resources exchanged between interdependent participants in the multi-level governance arrangement and to account for the agency of participating local governments.

3.2.2 Conceptualizing the Covenant of Mayors

The conceptualization of the Covenant as an instance of orchestration provides the basis for an empirically-based discussion in terms of policy goals, resources exchanged, and impacts on local climate governance.

EU sponsorship of the Covenant is accounted for by differentiating between the governance arrangement as a whole and the role of the orchestrator. In this perspective, the CoMO acts as an orchestrator on behalf of the European Commission. Notably, the latter did not mandate an initiative among existing voluntary commitment systems or TMCNs to be supported, but established its own scheme with pretensions to represent an overarching platform which it continues to fund, and to shape. As put forward by Hale and Roger (2014), at earlier stages of governance, an orchestrator may initiate a transnational governance scheme and continue or cease to play an active role at later stages. At later stages, an orchestrator may rather intervene in a given transnational governance arrangement in view of shaping or supporting the initiative, often by strengthening one initiative or certain actors therein rather than others.[29]

Obviously, this genesis has implications for the agency of the orchestrator. The European Commission as a governor (Abbott et al. 2015) relies on delegation to the CoMO which is run by a consortium of networks of local and regional authorities. Based on a contract of limited duration, a principal-agent relation is established, i.e. the CoMO is granted conditional authority for managing the Covenant on behalf of the European Commission (Hawkins et al. 2006). A tendency for orchestration and delegation to overlap in practice depending on the degree and form of interdependence has been acknowledged by the originators of orchestration theorizing (Abbott et al. 2016, 4). They concede that orchestration is likely not to occur in a pure and exclusive form, but "mixed and blended into hybrid forms" (Abbott et al. 2012, 6). The distinctions between modes of governance as introduced in Figure 3.2 are to be understood as ideal types whose actual appearance is a matter of degrees (Abbott 2015) – degrees of '(in)directness' and 'hardness' (Abbott et al. 2012, 6). Also,

> "governance arrangements frequently combine multiple modes: for example, both agents and intermediaries may take on the role of orchestrator within chains of governance" (Abbott 2015, 488).

With these precisions of the O-I-T-model in mind, the CoMO can be conceptualized as an agent and an orchestrator at the same time (see figure 3.6).

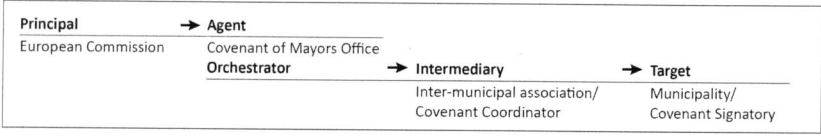

Figure 3.6: Covenant of Mayors as an orchestration arrangement
Source: Author; based on Abbott et al. 2012.

Orchestration has been described to serve to either manage *or* bypass states (Abbott et al. 2012, 3–4). This plays out more complexly in the case of the Covenant. It manages local governments by creating and enforcing rules of conduct in view of the European climate and energy targets that have been negotiated among and

---

29   Examples of shaping by orchestrators include the support by the World Bank for C40 and ICLEI (Hale and Roger 2014, 64).

adopted by EU member states. At the same time, the Covenant bypasses national governments, i.e. substitutes and complements state action by the provision of regulation and public goods. To this, intermediaries have been crucial for the promotion of the Covenant among European municipalities.[30] The response of the latter has exceeded initial expectations about the numerical development of the Covenant, but is far from universal.

The following sections subsequently depict the implications of an orchestration perspective for understanding the behavior of the CoMO as an orchestrator (3.2.2.1), Covenant Coordinators as intermediaries (3.2.2.2), and Covenant signatories as target actors (3.2.2.3).

3.2.2.1 The Covenant of Mayors Office as an Orchestrator

The CoMO represents the Covenant's secretariat. Based in Brussels, it has about 16 staff that either act as helpdesk to Covenant participants or are in charge of communication, events, and mobilizing finance (Covenant of Mayors 2016b). Each of them is not employed by the European Commission, the Covenant's sponsor, but by one of the networks of local and regional authorities that have been mandated with the running of the initiative: Energy Cities, the Climate Alliance, Eurocities, CEMR, and FEDARENE (see 3.2.1.1). This peculiar design raises the question why it was chosen, and what it implies for the ability of the CoMO to orchestrate local climate policy in the EU.

With the CoMO, the European Commission preferred to initiate an orchestrator rather than to shape the present landscape of transnational sustainable energy governance by choosing an existing initiative to support (Hale and Roger 2014). An explanatory factor for this choice consists in the absence of a clearly focal initiative. Several national and transnational municipal climate networks operate within EU member states. Obviously, a national network would not have been a good fit to make an offer to all local governments across Europe. But also TMCNs have regional core areas depending on their history and struggle to expand beyond the language area of their origin. For example, Energy Cities is particularly spread in France, the Climate Alliance in Germany. Despite the theoretically transnational setup, each single network has thus not achieved the coverage required to become an attractive orchestrator on behalf of the European

---

30   Membership numbers show that the Covenant of Mayors is more successful where it can rely
      on the intermediary of Covenant Coordinators (see 4.2.1).

Commission. In contrast, more general-purpose networks such as CEMR have broader coverage, but lack the technical expertise for local climate policy. By mandating a consortium of associations with the overall coordination of the Covenant instead of upgrading one particular scheme, the European Commission can rely on the resources of all participating associations, including their expertise, credibility, and networks.

The Covenant plays a relatively focal, i.e. central role in European climate and energy governance.[31] Its focality strengthens the orchestration ability of the CoMO. To date, the Covenant disposes of the largest and broadest constituency among TMCNs. It is open to local governments from all EU member states and beyond and is not historically anchored in one or even a few member states. Run by a consortium of local and regional authorities associations, it represents a meta-network endorsed by the EU and enjoys particular legitimacy and power to convene European local governments. By help of its reporting methodology, the Covenant intends to catalyze and also streamline local climate governance. Uneven municipal responses in EU member states in terms of becoming Covenant signatories and complying with reporting obligations indicate that the CoMO's orchestration ability depends on the respective national context. This includes varying degrees of focality depending on the existence of competing schemes and the anchorage of associations from the CoMO consortium.

The CoMO's ability to orchestrate also depends on its intellectual and material resources, i.e. on the benefits it can provide to participants. Hsueh and Prakash (2012) have shown that these benefits depend largely on the orchestrator's governance level. They demonstrate that in the US context, federal voluntary programs emphasize reputational benefits because federal programs can rely on well-known federal institutions such as the Environmental Protection Agency (EPA), greater coercive power, and potential economies of scale in program promotion across states. Operating at the European level, the CoMO can rely on the reputational benefits of endorsement by the European Commission that officially acknowledges every Covenant signatory's commitment. The CoMO also benefits from the EU's agenda-setting power in that its targets directly derive from EU legislation. The high number of targeted municipalities, and by now of Covenant participants, allows for economies of scale in program promotion.

---

31   For an introduction to the concept of focality as a key determinant of a governor's ability to orchestrate, see 3.1.3.

When it comes to material resources, the budget and personnel of the CoMO in itself are very restricted. But it can build on the resources of associations from the CoMO consortium, and on support by the JRC. The Covenant disposes of unparalleled operational capacities for monitoring member activities as compared to TMCNs thanks to the support by the JRC that is in charge of SEAP analysis and feedback, monitoring, and the technical helpdesk for Covenant signatories.

An additional challenge derives from the CoMO's stakeholder relations. The work of the CoMO is subject to contractual terms with the European Commission that defines its priorities and expectations per contract duration in a target agreement. The networks running the CoMO and the European Commission share many policy preferences with regard to sustainable energy policy and to active participation of local governments in European governance. Hence, the CoMO is likely to enact its mission without major instances of bureaucratic drift (Majone 2001). CoMO staff is at the same time confronted with the expectations of Covenant participants and particular members of their respective delegating network. These expectations need not coincide with those of the European Commission or the network they are affiliated to (Hooghe and Marks 2001a, 78–79). It is up to the consortium to bridge potential gaps between both worlds by serving as an interface and reconciling its main stakeholders' expectations.[32]

Summing up, the CoMO has strengths and weaknesses alike with regard to its orchestration ability. It can build on its affiliation to the European level and relative focality, its close ties to knowledgeable and well-networked associations of local and regional authorities, and support by the JRC and accompanying projects, but is challenged when stakeholder expectations diverge, and suffers from limited material resources and uneven access to targets in different national contexts. This situation – resources that suffice to convene and facilitate, but do not allow for addressing all European municipalities at a time – is a natural seedbed for orchestration, i.e. the involvement of intermediate actors to enhance one's governance capability (Hale and Roger 2014, 68).

---

32 The agency of the CoMO and possible agency slack in the form of shirking or slippage (Hawkins et al. 2006, 8) to the detriment of the European Commission are not at the core of this inquiry but would constitute an interesting issue to research. For a discussion of the scopes of action available to agents of the European Commission on the example of the EEA, see Saerbeck (2014).

3.2.2.2   Covenant Coordinators: Inter-municipal Associations as
          Intermediaries

The Covenant relies on the voluntary collaboration of Covenant Coordinators, public authorities that act as intermediaries on its behalf. Their primary role consists in providing operative assistance to Covenant signatories from their territory. Initially, they figured as so-called supporting structures, a category of membership that subsumed nearly all non-municipal Covenant participants. Over time, differentiations between different kinds of supporting structures were introduced in order to account for their respective sectors, levels, and scopes of action. Covenant Coordinators now consist of national authorities such as ministries and national energy agencies, and territorial governments such as regions, provinces, and inter-municipal associations. To date, 191 public authorities from the national and subnational level act as Covenant Coordinators (Covenant of Mayors 2016c). The Covenant refers to "numerous agreements signed between respondent Coordinators and other sub-national public authorities regarding the implementation of the Covenant in their territory", namely the support of municipalities in developing their SEAPs, as an expression of the Covenant's "multilevel governance dimension" (Covenant of Mayors 2013c, 6).

The role of Covenant Coordinators within the orchestration arrangement consists in providing "strategic guidance, financial and technical support to municipalities signing up to the Covenant of Mayors" (Covenant of Mayors 2014a, 1). Their more operational intellectual and material resources complement the CoMO's capacities. Within the typology of typical benefits offered by voluntary environmental programs as developed by Hsueh and Prakash (2012), their offer corresponds to state-level programs. In contrast to– programs at the federal-level – here, the EU-level, these emphasize more tangible benefits such as information, technical assistance, and funding because they have more detailed information on targets' needs, tend to pursue co-benefits such as economic development or energy security, and seek to influence federal legislation by experimentation (Hsueh and Prakash 2012, 461–62).

Earlier job descriptions still referred to Covenant Coordinators as supporters of a particular target group, namely "municipalities signing up to the Covenant of Mayors but lacking necessary skills and/or resources to fulfil [sic] their commitments" (Covenant of Mayors 2013c, 1). Experience showed that support mattered not only for small, rural, or poor municipalities. The reference to a somewhat flawed target group has been deleted in more recent job descriptions (Covenant of Mayors 2014a). The decisive impact of Covenant Coordinators on incentives for Covenant signature is highlighted by the fact that a large part of Covenant signatories stem from parts of Italy and Spain where many provinces provide extensive guidance on developing SEAPs, including financial support (Covenant of Mayors 2013c, 2). Also in countries with fewer Covenant Coordinators, it is on their territory that Covenant membership is particularly common among municipalities. For example, out of 81 Covenant signatories from France, 34 are supported by Rennes Metropolis, one of the inter-municipal associations studied.

Covenant Coordinators accompany Covenant signatories in implementing their commitments. The CoMO expects them to carry out a number of functions, namely the promotion of the initiative, the facilitation of Covenant-signatory interactions, the provision of tailored technical assistance to Covenant signatories by means of workshops or the provision of and information on various tools and methodologies, the mobilization of other local actors' expertise, support in obtaining funding, and the transmission of Covenant signatories' priorities with regard to funding and legislation to the EU-level (Covenant of Mayors 2014a, 14). In practice, Covenant Coordinators have developed a series of activities to implement this mission, as documented regularly in reports by the CoMO (Covenant of Mayors 2014a). Many Covenant Coordinators use their focal position to actively promote the Covenant among member municipalities via events and publications. Some make use of their ability to hire specialized staff for acting as an administrative interface between Covenant signatories and the Covenant. By assisting Covenant signatories in submitting reporting documents to the CoMO and managing their Covenant website profiles, they significantly enhance municipalities' cost-benefit calculation of Covenant membership. By conveying information between Covenant signatories and the JRC, they contribute to easing the JRC's remarkable overload with verifying Covenant reporting. About two thirds of Covenant Coordinators rely on internal expertise or their ability to commission external agencies for adapting Covenant methodology to the local context and provide Covenant signatories with specific tools, methods, or guidelines for BEIs or SEAPs. Beyond strictly Covenant-related activities, more than half of all Covenant Coordinators build on their coordinating role in territorial governance more broadly by convening working groups for municipal delegates or staff to develop local capacities and foster occasions for networking and learning among member municipalities, and sometimes other relevant territorial actors.

Although "financial support for the implementation of SEAPs is paramount" (Covenant of Mayors 2014a, 10), only a minority of Covenant Coordinators directly funds SEAP-related projects of Covenant signatories via subsidies and grants or by creating a dedicated fund. Mostly, Covenant Coordinators rather support signatories in identifying, applying to or executing other sources and forms of funding such as EU funds, joint projects, energy performance contracting and others. This is hardly surprising in the light of Covenant Coordinators' budgets. According to a 2013 survey among Covenant Coordinators, most respondents had no or only a small (<50.000€/year) budget for this activity other than their staff; the working time of internal staff therefore represents the primordial material resource deployed (Covenant of Mayors 2014a, 2–4).

To put it in a nutshell, Covenant Coordinators serve as facilitators for local climate policy and as an interface between Covenant signatories and the CoMO, filling in where the resources of Covenant signatories and of the CoMO do not suffice for fulfilling their mission. In light of this ambitious set of tasks, smooth communication with the Covenant Office is all the more important. There are signs of a certain reserve between Covenant Coordinators and the CoMO. A survey conducted by the CoMO among Covenant Coordinators showed that interactions with the CoMO remain limited (Covenant of Mayors 2014a). For instance, the majority of Covenant Coordinators does not attend Covenant events such as the annual ceremony, thematic workshops and webinars; many are not at ease in dealing with the Covenant's website; and more than a third does not communicate its activities to the CoMO in view of best practice exchange or self-promotion. At several instances, Covenant Coordinators have failed to report on their commitment's implementation and have been suspended, i.e. their profile has been temporarily removed from the Covenant's website. Other Covenant Coordinators have failed to supervise even a single Covenant signatory. Some still figure on the Covenant's website, others have disappeared from the website – such as the Hérault département and the Territorial Community of Corsica, former French Covenant Coordinators considered for a case study at an earlier stage of this research project – or, just like Greater Lyon, are now listed as Covenant signatories instead. The apparent contradiction between these indications for implementation difficulties and the obvious impact of Covenant Coordinators on the Covenant's numerical success calls for independent empirical research in order to understand Covenant Coordinators' motivations, the modes of governance they rely on for implementation, and to assess whether they are able to live up to the challenging role ascribed to them within this orchestration arrangement.[33]

The resources of Covenant Coordinators in terms of information about municipal needs, expertise on local climate and energy policy, dedicated person hours, and funding schemes are decisive for the implementation of their commitment. Obviously, even within the group of territorial (as opposed to national) Covenant Coordinators, their endowment with these resources varies importantly, namely across levels of government – inter-municipal associations, provinces, and regions – and EU member states. In order to allow for comparability, all cases stem from the former category.[34] They benefit from maximum proximity to municipalities and hold relevant planning responsibilities, but have limited capacities with regard to regulation and funding. The analysis of how they implement their commitment to the Covenant is at the heart of this book.

3.2.2.3 Covenant Signatories: Municipalities as Target Actors

During the Covenant's creation, the initial idea to create a relatively self-contained network of mayors from pioneering big cities was quickly abandoned (see 3.2.1.2), the name stuck. Signatories embody the Covenant's raison d'être. To be precise, not all Covenant signatories are represented by mayors. From its beginning, the Covenant was designed to also comprise public governments other than municipalities. While signatory status as such, at first, was a privilege of municipalities, it has since been opened for local governments such as districts. They can now hand in, in the name of their member municipalities, a joint SEAP that also comprises individual municipalities' actions. This development – although the Covenant's website continues to refer to cities or urban areas on most occasions – reflects the shift from a metropolitan focus towards an inclusive approach that also accounts for contributions of small villages.

In quantitative terms, the Covenant has largely exceeded the expectations of its founding fathers, and met the objectives of the first target agreement with the CoMO consortium at its very first signature ceremony (interview with P. Ballesteros Torres, 30 May 2012). Covenant signatories figure prominently on the Covenant's website; profiles include administrative information such as con-

---

33   Reports published by the CoMO itself might fail to triangulate information obtained from Covenant Coordinators or to ask fundamental questions about the Covenant's design and performance.
34   For a detailed discussion of case selection, see section 4.2.

tact details, submitted documents, i.e. BEIs, SEAPs, and implementation reports, and contributions to the best practice database, so-called Benchmarks of Excellence. Covenant signature requires a decision by the respective local assembly, and entails a commitment to the EU's climate and energy targets and to a three-step reporting scheme. A BEI, a SEAP, and regular implementation reports are to be submitted in a timely fashion to the CoMO for verification by the JRC. Failure to report satisfactorily may result in suspension or exclusion from the Covenant.

In light of the supranationally-defined obligations entailed by Covenant signature, the question arises why cities would want to join at all. As discussed above (2.3; 3.2.1.3), excludable participation benefits are decisive because the European Commission has no competence to directly intervene within member-states' administrations. With the Covenant, the European Commission offers them a network for sustainable energy policy (problem solver), a platform and audit for profiling and identity building (stage), a chance to anticipate European or national legislation (threat) or to overcome domestic hindrances set by national policies or the political system (alternative), or a convenient framework to implement unavoidable European and national policy targets (duty) (Wolffhardt et al. 2005, 94–95). Whether local governments learn about this offer, decide to participate, and are able to use it successfully, depends importantly on the resources at their disposal, either their own or those of a Covenant Coordinator. Case studies will explore in more detail how municipalities benefit from Covenant participation, and to what extent Covenant Coordinators intervene in and facilitate this process.

3.3 Expected Theoretical Contributions

An improved understanding of local climate policy coordination requires special attention to strategies for coping with interdependencies within the multi-level governance system (see chapter 2). To date, multi-level governance constitutes a general theoretical framework of high abstraction, but with little explanatory or predictive power when studying a particular case. Instead, middle-range theories are required that "are deliberately limited in their scope", attempting "to explain different subclasses of general phenomena" and "to formulate well-specified conditional generalizations of more limited scope" (George and Bennett 2005, 266).

The concept of orchestration is a recently developed model of indirect governance (Abbott et al. 2012; Abbott 2015; Hale and Roger 2014). Based on the O-I-T-model, a theoretical framework was developed for studying the Covenant, an EU-initiative that aims to stimulate and coordinate local climate action. It thereby probes Hale and Roger (2014, 80) claim that the EU has not yet made effective use of its orchestration ability for climate governance. In contrast, one of the founding fathers of the O-I-T-model casually refers to the Covenant as one of several voluntary climate commitment organizations that "could help local governments [...] to develop experimental programs focused on urban policies" (Abbott 2017, 757). As orchestrating experimentation is particularly challenging (see 3.1.2), this suggestion raises the question whether the CoMO actually fulfills the prerequisites for being a strong orchestrator. The following case studies of intermediary and target behavior serve to address that question on a sound empirical basis beyond mere conceptual considerations.

Orchestration theorizing is relatively recent and remains under construction. As is often the case in the social sciences, it does not yet possess the predictive and explanatory power and the clarity required to allow for rigorous testing, which is why a refinement and further theory building are preferable (George and Bennett 2005, 182). This book contributes to theory building by refining and extending orchestration theorizing in two ways.

Firstly, it follows up on attempts to transfer the O-I-T-model from the study of global governance to the study of European multi-level governance (Schleifer 2013). In doing so, it accounts for the impacts of sponsorship on the design of voluntary climate programs (Hsueh and Prakash 2012). In particular, it clarifies what kinds of incentives governance actors at the European and at the subnational level can offer when requesting municipalities to act on climate change.

Secondly, it extends the O-I-T-model from the study of international organizations and nation states to the study of subnational governments. Adopting a macrodeterminism-critical approach (Bukowski, Piattoni, and Smyrl 2003), special attention is paid to the actual experience of intermediaries, and the intermediary-target relationship. The Covenant is studied as a governance arrangement not only from the perspective of the governor, i.e. the European Commission and its agent, the CoMO, but also from a bottom-up perspective in order to account for the agency of all governance actors involved. As Pasquier and Weisbein (2013) point out, studies of the local level in territorial governance help to substantiate governance concepts and cool the overheated theoretical debate

(Pasquier and Weisbein 2013, 277). This book thereby contributes to refining theoretical propositions regarding governance actors other than the governor that participate in orchestration arrangements, in particular intermediaries.

4 Methodological Approach

As put forward in the purpose statement (see section 1.1), the value of this research project is instrumental – as opposed to intrinsic – in that it is not "conducted out of pure interest and curiosity in the particular case [...] [but] done with a purpose in mind" (Thomas 2011, 98). Its aspiration is to be relevant to both real-world problems and political science, namely the advancement of orchestration theorizing. Its research questions (see section 1.3) are answered within the framework of a holistic multiple-case design by means of country comparison and cross-case analysis of five inter-municipal associations that act as Covenant Coordinators: Stuttgart Region and Rhine-Neckar Region from Germany, and Greater Lyon, Rennes Metropolis, and Val d'Ille from France.

This chapter explains why and how research has been conducted within a multiple-case study research design, a research design that has been praised for

> "its particular utility involving intergovernmental issues. The design provides a researcher an opportunity to focus on the unique qualities of specific state or local jurisdictions and, at the same time, allows for the consideration of aggregate patterns" (Agranoff and Radin 1991, 223).

The chapter begins with a presentation of the case study approach, its development and analytical stance (4.1). It then explains why and how the above-mentioned cases from Germany and France were chosen (4.2). Subsequently, it introduces expert interviews and triangulation as central means of data collection and evaluation (4.3). It concludes with a discussion of methodological challenges and their impacts on the scope of this research (4.4).

4.1 Case Study Research Design

Why has a case study research design been chosen and how exactly has it been composed? A case study approach was chosen for its strengths in dealing with phenomena like those at the focus of this research project (4.1.1). The case study design evolved in an iterative process, including specifications of the initial re-

© Springer Fachmedien Wiesbaden GmbH, part of Springer Nature 2020
L. Bendlin, *Orchestrating Local Climate Policy in the European Union*,
Energiepolitik und Klimaschutz. Energy Policy and Climate Protection,
https://doi.org/10.1007/978-3-658-26506-9_4

search question (4.1.2). This design allows for combining exploration and explanation based on analytic generalization (4.1.3).

4.1.1 A Case Study Approach to Research

With its case study approach, this research project accounts for its explorative nature and its search for in-depth understanding rather than a strictly comparative analysis aiming for probability predictions (Thomas 2011, 37). It mainly builds on approaches to case study research design as suggested by Yin (2014) and Thomas (2011) who provide detailed guidance for case study research.[35] Other approaches to case study research prove less appropriate as they rely primarily on systematic comparison, statistical generalization, and variable-driven hypotheses.[36] But the existing literature, including orchestration theorizing, did not provide sufficiently detailed theoretical predictions and reliable variables for systematic case selection and theory testing. Given the limited amount of empirical research into the Covenant, such an approach would not have been sufficiently open-ended to account, as intended, for unexpected findings in the field. More than that, the transfer of principles from quantitative research based on regression analysis has been rejected as "often simplistic, misleading, and inappropriate as a guide for designing social inquiry" by a growing body of new methodological literature from different strands (Mahoney 2010, 122; see also Steinberg 2015, 155).

In contrast, and in line with the purpose of this book, Yin (2014) and Thomas (2011) highlight the discrete strengths of case study research as a tool for understanding social phenomena, and provide a set of research strategies that allow for valid analytical generalization (see section 4.1.3). A case study research strategy accounts for "the complexity and developmental nature of many public administrative events and practices" (Agranoff and Radin 1991, 204), and is best suited to study contemporary phenomena whose boundaries to context are difficult to define clearly prior to the inquiry because of the researcher's limited understanding of the situation, and over which the researcher has little or no control – particularly when the research questions take the form of 'why'- and 'how'-

---

35   In contrast, other handbooks of reference rather provide an overview of qualitative methods (see for example Babbie 2013, Creswell 2013a; 2013b).

36   For instance, representativeness remains the gold standard of social research to King, Keohane and Verba (1994), often referred to as KKV.

questions (Yin 2014, 14). Another advantage over variable-centered research designs is linked to the fact that

> "[e]mpirically derived, theory-oriented case studies are particularly suited for discovering equifinality and developing typological theory for the phenomenon in question" (George and Bennett 2005, 241).

This allowed to explain why and how different circumstances in Germany and France and in the five individual cases caused the same outcome in terms of Covenant participation.

The case study approach also helps to 'stay real' (Thomas 2011, 7), i.e. to avoid overly abstract conceptualizations of local climate governance and overly optimistic accounts of local governments' roles in climate mitigation. These represent recurrent weaknesses in the climate governance literature when it prematurely refers to transnational regulation without establishing a sound empirical basis first (cf. for example Heyvaert 2013).[37]

4.1.2   Results of Iterative Revision

Case study research has been characterized as a "linear but iterative process" (Yin 2014, 1). While authors differ on how to differentiate phases in case study research (George and Bennett 2005; Thomas 2011; Yin 2014), they agree that "some iteration is often necessary to ensure that each phase is consistent and integrated with the other phases" (George and Bennett 2005, 73). Iteration enables case study research to "generate analytical and descriptive dimensions of the research topic simultaneously" thanks to an "interplay between inductive and deductive reasoning"; initial hypotheses and research design might be modified "dramatically [...] as the research progresses" (Agranoff and Radin 1991, 216–17).

That is also true for the making of this book. The research project started from the proposition to study the Covenant of Mayors as a site of European climate governance (Bulkeley and Newell 2010, 106), i.e. as an "arena[.] of governance in which actors interact and make decisions" (O'Neill 2009, 6). Hence, the initially considered case study design was an embedded (or 'nested') single case design with eight Covenant Coordinators as embedded units of analysis (Yin 2014, 62–63). These represented the totality of territorial Covenant Coordinators

---

37   See also section 4.4.1 for a discussion of risks for normative distortions in climate governance research.

in Germany and France at the time: Stuttgart Region, Rhine-Neckar Region (from Germany), Greater Lyon, the Hérault Département, Rennes Metropolis, the Community of Communes of Val d'Ille, the Territorial Community of Corsica, and the Nord-Pas de Calais Region (from France).

Before finalizing the research design, tentative research in the Stuttgart Region, Greater Lyon, and the Hérault Département served to gain familiarity with the subject (George and Bennett 2005, 88). Inquiry into Stuttgart Region quickly showed that considering the Covenant as a site of inter-municipal climate policy coordination meant to overestimate its impact on local climate governance. Subsequently, inquiry into Greater Lyon confirmed results from Stuttgart that in practice, the Covenant commitment only plays a subordinate role for inter-municipal associations' participation in multi-level climate governance. The third and last preliminary inquiry into the Hérault Département revealed the large extent to which subnational scopes of action depend on the level of government. I revised the case selection accordingly (see section 4.2) and excluded the regional governments Hérault Département, the Territorial Community of Corsica, and the Nord-Pas de Calais Region from the research project in order to concentrate on the local level.

Furthermore, the pilot studies also made me realize that the Covenant's limited significance for local climate policy was not a disappointment, but an interesting finding in itself, all the more in light of its contrast to enthusiastic references to the Covenant in existing literature. Appreciatory feedback from scientific conferences about my preliminary results encouraged me to stick to the Covenant as the object of study, but with a different approach. With the O-I-T-model, I found a theoretical framework that enabled me to account for Covenant participants' self-will in using the initiative as a resource in policy making. In order to capture their individual approaches to local climate policy coordination, I finally adopted a holistic multiple case design (see figure 4.1). (Yin 2014, 62–63).

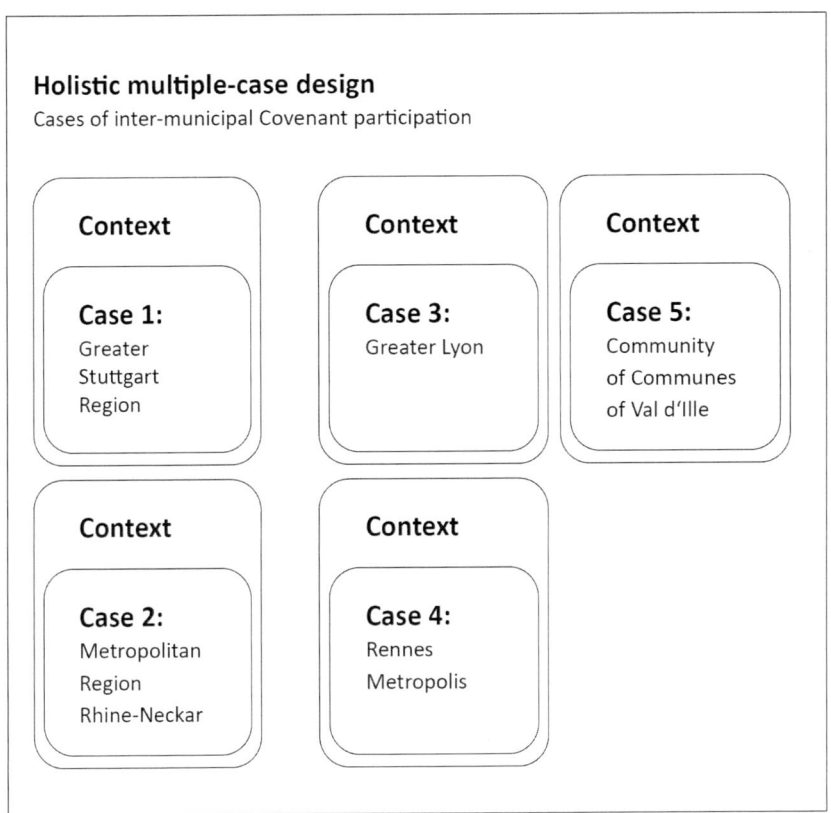

Figure 4.1: Holistic multiple-case design
Source: Author; based on Yin 2014, 50, figure 2.4.

4.1.3 Analytic Generalization

This book is both exploratory and explanatory in nature (Thomas 2011; Yin 2014). It is partly exploratory because little is known to date about the role of inter-municipal associations in local climate policy, or even with respect to multi-level governance structures more generally. But it can also build on findings from adjacent fields and on preliminary evaluation as provided by the Covenant itself. Preliminary findings from initial fieldwork helped to identify the O-I-T-model as a promising explanatory framework for further analysis. Orchestration theorizing has not reached the stage of a full, logically complete typological theory, but enables theory-driven induction and the use of deductive theoretical ideas within an empirically derived, theory-oriented case study research design (George and Bennett 2005, 240–41). Cases are not treated as illustrations – as in illustrative-demonstrative case studies (Thomas 2011, 111–35) – but as actual sources for insight and analysis. Such a combination of theory building and theory testing is a common approach to case study research (Thomas 2011, 112–18). More exploratory separate case reports are provided in chapter 5, followed by a more explanatory multi-case analysis in chapter 6 that attempts to wrap up the individual case stories in an overall story to explain the phenomena observed (Agranoff and Radin 1991, 206).

This approach includes a claim for providing generalization, understood here as "a logical argument for extending one's claims beyond the data, positing a connection between events that were studied and those that were not" (Steinberg 2015, 153). Analytic generalization is used for achieving conclusions at a superordinate conceptual level (Yin 2014, 41). The actual processes within each case were carefully traced, analyzed, and compared, and cross-case analysis enables analytic generalization.

In this way, results qualify for being transferable to resonance groups of cases beyond the cases studied here. Steinberg (2015, 168) defines resonance groups

> "as categories of social problems and processes that share many characteristics in common across borders. In this sense, they are connected; a finding that applies to one of them often sheds light on many of them."

A first type of a potential resonance group exhibits similarities that are linked to domestic factors rather than deliberate exchange between governance actors (Steinberg 2015, 170). Building upon this course of argument, chapter 7 will demonstrate that the proliferation of Covenant participation and local climate policy initiatives more broadly give reason to assume that conclusions are also relevant to other countries with national-level low-carbon policies and a significant mobilization of local governments for climate mitigation, and to local governments with a similar position within the political system. A second type of a potential resonance group consists of "policies or programs surrounded by actor networks whose members routinely exchange information and ideas across borders" (ibid.). These include other orchestration arrangements and voluntary commitment systems, and to a limited extent also TMCNs (see section 7.2).

4.2 Case Selection

The holistic multiple case study design was chosen in view of conducting cross-case analysis. This section first justifies the comparison of cases from Germany and France (4.2.1). Moreover, it explains how the cases were selected from the totality of German and French Covenant Coordinators (4.2.2), and how they were bounded so as to enable focused research (4.2.3).

4.2.1 Comparing Germany and France

The selection of cases from two countries, Germany and France, follows a replication logic (Small 2009; Yin 2014). Two very different national frameworks for local climate governance, including territorial organization and national-level policies, facilitate theoretical conceptualization (Yin 2014, 41).[38] This choice is based on the assumption that despite Europeanization, persistent differences in domestic frameworks entail corresponding differences in local (climate) governance (Börzel 2001; Keating 2008; Pasquier, Simoulin, and Weisbein 2013). Accordingly, similar results across cases are expected within each country.

Local scopes of action depend to a large extent on territorial organization, i.e. the repartition of responsibilities and resources between levels of government within the political system (Hooghe et al. 2016; Hooghe, Marks, and Schakel 2010). Keating (2011, 52–55) points out that subnational governments' capacity for policy-making depends on constitutional competences, available expertise, and leadership, while their capacity for policy implementation depends on administrative organization and intergovernmental as well as public-private networks. This general remark is in line with findings from local climate policy research about enabling and hindering institutional factors such as local governments' competences, capacities, and integration into municipal climate networks for learning and innovation (Alber and Kern, 2; Schreurs 2008, 352–53). Hence, studying cases from countries with different forms of territorial organization will

---

38   Strategies for systematic case selection from comparative politics could not be applied to the present inquiry because they require more advanced levels of theory development and a broader universe of cases to choose from. Consequently, strategies for causal inference as developed within that literature, such as statistical generalization, cannot succeed here. This situation supersedes an extended discussion of the prerequisites and pitfalls of causal inference in a comparative research design. For an overview of recurrent criticisms and problems with regard to causal inference in comparative studies, see Rohlfing (2009, 140–50). He also provides potential responses and solutions.

be helpful in detecting impacts on local climate policy coordination. Germany and France provide a particularly contrasting background for local-level case studies.[39]

In both countries, national governments attempt to transform the national energy system via extensive political projects known as the 'Energiewende' (energy turn or transition) in Germany and the 'transition énergétique' (energy transition) in France. This increases the likelihood of encountering actual needs and initiatives for local climate policy coordination in the case studies. The two countries have very different starting positions for local climate policy with regard to the energy system and national climate and energy policies. Germany has engaged in a phase out of nuclear energy, but relies extensively on coal at least transitionally (Schröter 2017; Setton and Helgenberger 2016); countries' reliance on carbon has been shown to significantly impact their climate policies (Harrison 2015; Houle, Lachapelle, and Purdon 2015). Germany's frontrunner role in renewable energy development has been linked to national legislation as well as decentralized initiatives. In contrast, France relies heavily on nuclear power for electricity, and has a very centralized energy supply system with fewer opportunities for decentralized initiatives. Although both countries engaged in an energy transition, their differences further broaden the resonance group of country cases for conclusions.

In addition to the above comparativist considerations, Germany and France also are of particular relevance in practical terms for European climate policy and EU politics. Constituting two of Europe's largest national economies,[40] they account for nearly one third of the EU-28's GHG emissions (Eurostat 2016a; 2016b). A low-carbon energy transition in Europe can hardly be achieved without them (Schreurs 2016b). Furthermore, the so-called Franco-German 'couple', 'tandem' or 'motor' has played a central, if not at times dominant role in European integration since its beginnings (Hendriks and Morgan 2001; Janning 2005; Krotz and Schild 2015; Woyke 2004). Franco-German consensus has often given direction to European decision-making (Demesmay, Koopmann, and Thorel 2013; Webber 2005). Hence, successful orchestration of local climate governance is hardly conceivable without these two countries.

---

39   See chapter 5 for a detailed discussion of national frameworks.
40   Germany has the largest GDP in the EU. France ranks third after the United Kingdom. As soon
      as the latter will have implemented its decision to leave the EU, France will rise in rank accordingly.

Up to now, divergent visions for sustainable energy often prohibit mutual understanding between Germany and France despite their special bilateral relations and their shared engagement in an energy transition (Glastra and Rüdinger 2016). But an ever growing body of translation efforts, bilateral conferences and workshops mirrors the political and economic interest in understanding these neighbors and potentially learning from each other's experiences.[41] Summing up, the comparison of cases from Germany and France is promising in terms of theoretical conceptualization and for transferability of conclusions to large resonance groups of other cases, and in terms of problem relevance for climate mitigation and European politics.

In order to realize this potential, this research project required detailed knowledge about and access to the field, namely

"the question of special language skills and special historical and sociological knowledge of cases, for lack of which comparativists are often justly criticized" (Eckstein 1975, 122).

In this regard, I was able to build upon my education in a Franco-German study course of political science with a focus on European studies in terms of background knowledge on the respective political systems, language skills, and access to key informants. German is my mother tongue, and I speak French fluently after studying in France for two years. Hence, I was able to consult German and French documents and literature, and to conduct interviews in French. This is particularly important as many of my interviewees do not speak another language fluently and would not be at ease when interviewed in English. Interviewing experts in their mother tongue enabled me to gather their information as exact and differentiated as possible (Kruse 2015, 312–17; Lauth, Pickel, and Pickel 2009, 183).

Admittedly, the inclusion of a third country or even more countries might have allowed for additional insights or enhanced transferability of results. The requirements for soundly covering another country would have been extensive (Steinberg 2015, 165). Studying a third EU member state would have remained

---

41   Examples include Tandem, a project run by Energy Cities and the Climate Alliance on behalf of Germany's and France's environmental agencies that aimed at fostering climate policy partnerships between municipalities from the two countries, and the Franco-German Office for the Energy Transition, an association founded in 2006 by ministries, companies, and research institutes that offers information and business contacts brokering in the renewable energies sector.

inadequately superficial, or it would have increased considerably the sheer work-load of fieldwork and analysis beyond the scope of a dissertation project carried out by a single researcher.

A final justification is appropriate for having chosen to cover two countries with moderate numbers of Covenant participants. In contrast, studying another pair of EU member states would have allowed for covering the vast majority of Covenant participants. The engagement of Italian and Spanish provinces as Cov-enant Coordinators has resulted in large numbers of Covenant signatures by mu-nicipalities from their territories. The two countries account for about two thirds of all Covenant Coordinators and three quarters of all Covenant signatories.[42]

The preponderance of Covenant participants from Italy and Spain make these two countries extreme cases,[43] suited to provide insights about conditions that facilitate Covenant participation (Yin 2014, 52).[44] But this would not have helped to understand why Covenant participation is less pronounced in the ma-jority of EU member states. Also, studying Italy and Spain would hardly have allowed for focusing on local climate policy coordination in a narrower sense because the vast majority of Italian and Spanish Covenant Coordinators consist of subnational governments from the regional level such as provinces and regions – as opposed to inter-municipal associations from the local level.

4.2.2 Selecting the Cases

While cross-country comparison sheds light on domestic frameworks for local climate policy coordination and prevents the researcher from overlooking their influence, Germany and France do not constitute the cases studied here. Instead, the case studies consist of five local governments from the two countries: Stuttgart Region, Rhine-Neckar Region (from Germany), Greater Lyon, Rennes Metropolis, and the Community of Communes of Val d'Ille (from France).[45] This research design allows for differentiating between local and national ("country-

---

42   Calculation based on profiles listed on the website of the Covenant of Mayors (2016c; 2016i).
43   Extreme cases are also referred to in the literature as 'unique', 'unusual' or 'special' cases.
44   Readers interested in these extreme cases are advised to consult the quantitative analysis of factors motivating Spanish Covenant participants by Pablo-Romero, Sánchez-Braza, and Ma-nuel González-Limón (2015), and the reports by Di Martino (2013; 2015) that build on case studies of Covenant participation in the Italian Emilia-Romagna Region.
45   It should be noted that this choice was made based on Covenant membership as of October 2012, for participants can be (and have been) suspended or excluded in case of non-compli-ance.

level") specificities as it accounts for local governments' agency in climate governance, i.e. their genuine choice between different policy options:

> "Examining subnational territories provides an additional measure of leverage and control in the construction of focused comparisons. [...] It effectively neutralizes country-level institutional variables that may explain divergent response to similar structural constraints. If we do indeed see differences in governance responses to these constraints (embodied in institutional construction or change, resource allocation, economic planning designs, etc.) we can be more certain that these differences are attributable to political agency" (Bukowski, Piattoni, and Smyrl 2003, 2–12).

Local climate policy coordination happens among tiers of government that belong to the "forth level" in the EU multi-level system (Panara and Varney 2013b). The specificities of territorial organization in Germany and France make it difficult to determine a comparative design in which levels of government perfectly correspond (Wollmann and Bouckaert 2006, 12–14). Case selection from the totality of German and French Covenant Coordinators required identifying those that are superordinate to municipal governments, but still belong to the local level.[46]

This raised the question of how to compare levels of subnational government despite the difficulties in establishing an unambiguous typology of local government. Existing categorizations depend very much on the criteria applied such as constitutional provisions or the position within the multi-level system (e.g. Bertrana, Xavier, and Heinelt 2013; Frenzel 2013; Hooghe, Marks, and Schakel 2010). Due to the multi-dimensional variation of local government across Europe, no simple universal indicator is available:

> "...some of the more noticeable differences relate to the scale (the number of inhabitants), the position of the local authority as a second, third, or fourth tier in the administrative system, its character and scope (the functions it performs in the whole of the state administration, including the amount of local government expenditure in relation to total public expenditure), and the degree of autonomy and discretion it enjoys in exercising its functions" (Hulst and van Montfort 2007b, 1).

Also, formal provisions have to be assessed in view of their actual implementation. For instance, the organization of local government might vary within one country because it is not a national, but regional responsibility as in Germany; or constitutional provisions might be of little practical significance as in France

---

46   Notably, there are no Covenant National Coordinators (CNCs) in both Germany and France such as national-level ministries or agencies.

where the general competence of municipalities was long limited by administrative supervision by superordinate governments, and continues to be restricted by a lack of administrative and financial capacity (Hulst and van Montfort 2007b, 1–2).

In the absence of an unambiguous typology of local government in the EU and the limited comparative literature on inter-municipal associations, I established a tentative correspondence between German and French levels of government based on the EU's Nomenclature of territorial units for statistics (NUTS), including its definition of Local Administrative Units (LAU) (European Commission 2008; 2015; European Union 2015).[47] This scheme, a hierarchically structured division of member state territories into subunits for statistical processing, informs EU cohesion policy and more generally serves statistical purposes. NUTS favors institutional divisions, i.e. subunits build on subnational governments (normative regions) or administrative units (analytical or functional regions) where applicable. NUTS 1 represents major socio-economic regions with a population of 3 to 7 million inhabitants, NUTS 2 stands for basic regions for the application of regional policies with 800,000 to 3 million inhabitants, and NUTS 3 consists in small regions for specific diagnoses with 150,000 to 800,000 inhabitants. Additionally, Eurostat collects data on two levels of LAU. LAU 1 refers to local associations of administrations; LAU 2 corresponds to the municipal level. At least one of the two has to build on actual administrations.

The comparison of territorial organization of Germany and France by help of the NUTS classification revealed that Covenant Coordinators stemmed from different levels of government (see figure 4.2). The local level consists of municipalities, districts and regional associations of municipalities in Germany, and of municipalities and intercommunalities in France.[48] Regional associations (*Regionalverbände*) from Germany such as the Stuttgart Region and Rhine-Neckar Region are located between rural and city districts (*Landkreise* and *Kreisfreie Städte*) on the one hand (NUTS 3) and governmental districts (*Regierungsbezirke*) on the other (NUTS 2).

---

47   Based on the Regulation (EC) No 1059/2003 of the European Parliament and of the Council of 26 May 2003 on the establishment of a common classification of territorial units for statistics (NUTS), NUTS is regularly updated in order to account for changes and reforms within member states.

48   Regional associations are not the only forms of institutionalized inter-municipal cooperation in Germany (see section 5.1.1.3).

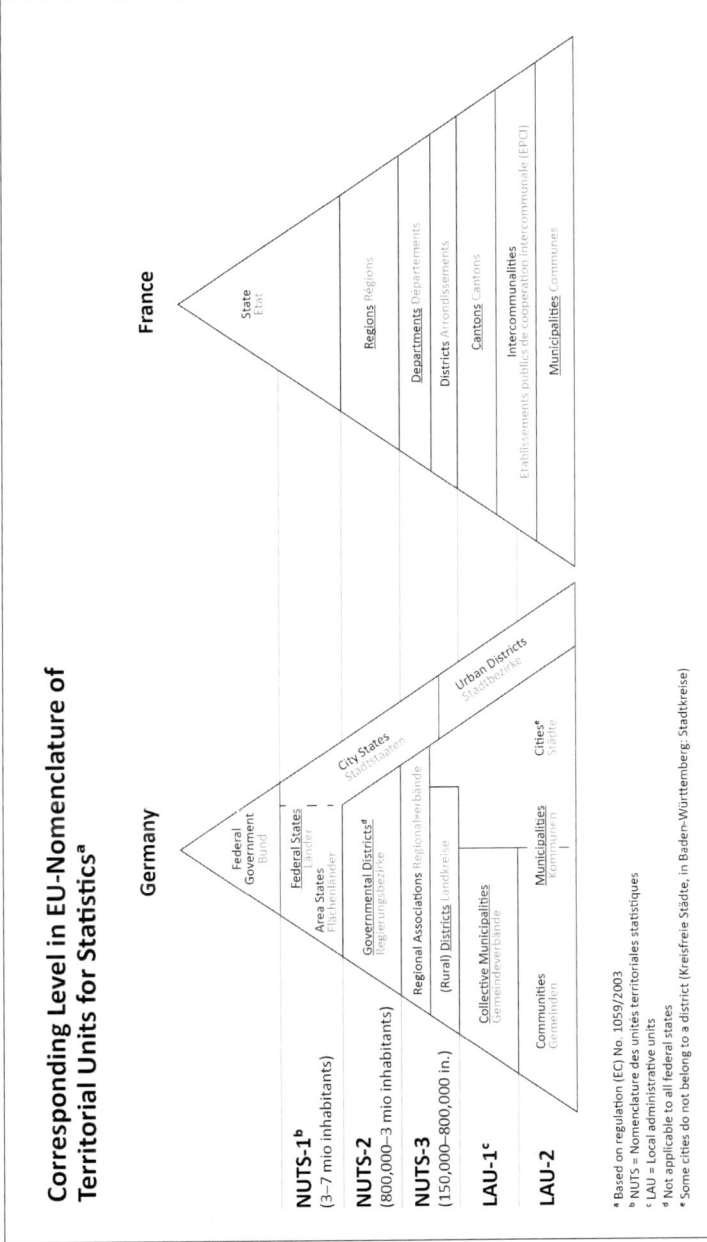

Figure 4.2: Territorial organization of Germany and France in comparison Source: Author; based on NUTS classification (European Union 2015).

In France, the Territorial Community of Corsica, and the Nord-Pas de Calais Region belonged to NUTS 2, the Hérault Département to NUTS 3. The intercommunalities (*établissements publics de coopération intercommunale*, EPCI) Greater Lyon, Rennes Metropolis, the Community of Communes of Val d'Ille are located between municipalities (LAU 2) and cantons (LAU 1). This led to an exclusion of Corsica and Nord-Pas de Calais.

The pilot studies (see section 4.1.2) helped to refine and finalize the case selection process. The approach of the Hérault Département to local climate policy coordination differed importantly from those of Stuttgart Region and Greater Lyon in that it was not characterized by proximity and inter-municipal cooperation, but superordinate supervision and consultancy. In France, the departmental level represents a longstanding unit of unitarian state administration and is not perceived as part of the 'local bloc' at all, but as the lower tier of regional government (Martínez Soria 2007, 1024; see also Hooghe, Marks, and Schakel 2010). Hence, I objected comparisons of rural and city districts to the departmental level (cf. Bertrana, Xavier, and Heinelt 2013), and decided not to include the Hérault Département.[49]

In contrast, the pilot studies further demonstrated that the German regional association and the French intercommunality shared many features and functions although the former tend to exceed the latter in terms of population and territory. Established in response to functional pressures on municipal self-rule, both types of territorial government associate existing municipalities rather than forming a distinct level of government. Also, they are governed by member municipalities, i.e. their assemblies consist of municipal delegates (Hörster 2007, 903). As a result, many of their missions are carried out in close cooperation with member municipalities, and they are still in the process of developing a shared identity. They can be conceptualized as inter-municipal associations, a corresponding type of subnational government. Inter-municipal associations play an increasingly important role within the Covenant; the CoMO observes an increased interest for participation and points out that they accounted for nearly 20 % of territorial Covenant Coordinators by 2013 (Covenant of Mayors 2014a, 1–2). Correspond-

---

49   Interestingly, Corsica, Nord-Pas de Calais and Hérault never succeeded in associating with any
     Covenant signatory and disappeared from the Covenant's website soon after the case selection.
     This might indicate their lack of proximity to the municipal level and insofar validated my
     selection.

ingly, I decided to study all Covenant Coordinators in Germany and France that can be defined as cases of inter-municipal associations.[50]

How can this case selection process be characterized in terms of research methodology? Through studying several cases having spatial, but not temporal variation, this research design builds on the comparative method (Gerring 2007, 28). A targeted selection of cases as required for controlled comparison, the case study design most similar to experimentation – i.e. "two cases that resemble each other in every aspect but one" (George and Bennett 2005, 152)– was not chosen for lack of both a sufficiently broad universe of cases in the field (Gerring 2004, 351–52), and reliable variables in the literatures on climate policy coordination and orchestration arrangements. Since the research project is explorative, selecting most-likely and least-likely cases for theory testing was also not an option for it would have required a more advanced state of middle-range theory building, namely predictions in orchestration theorizing of sufficient precision to be testable.

The limited number of available cases also would not have allowed for identifying one or several typical case(s) (Gerring 2007, 91) or for "the familiar alternative of using statistical analysis to achieve the functional equivalent of an experiment" (George and Bennett 2005, 152). In such a situation "where a small number of experimental or exemplary projects are being investigated there will be no choice but to investigate the universe" (Agranoff and Radin 1991, 210). This entailed the rare opportunity to study an exhaustive sample (Gerring 2007, 22).[51]

This case selection neither pretends to follow Mill's method of agreement or of difference, nor a most-similar or most-likely design logic (Przeworski and Teune 1970). In other words, the research design and case selection approach are not variable-driven. King, Keohane, and Verba (1994) might have criticized them as biased for being based on the dependent variable, i.e. the Covenant Coordinator. But this represents an appropriate focus in view of the research problem (Collier and Mahoney 1996, 68–69). Most likely, *some* multi-level interaction and coordination attempts – as focused on by the theoretical framework –

---

50   An even closer match in terms of size, and the position in the multi-level system, might have been achieved by comparing intercommunalities with rural districts (cf. Hulst and van Montfort (2007a)). But no such cases were available, and they would not necessarily have allowed for studying the importance of rescaling processes to the same extent.

51   Without intending to refer, by means of the term 'sample', to representativeness for some population of cases (Gerring 2007, 88).

will be observable. In this sense, Covenant Coordinators could be considered 'crucial' cases for a multi-level governance approach to European climate and energy policy (Eckstein 1975, 123).

Within the resulting 'sample', each case has strong particularities. To name but a few, Stuttgart Region and Rhine-Neckar Region are both polycentric, but Stuttgart Region is located within the *Land* of Baden-Württemberg while Rhine-Neckar Region spans across Baden-Württemberg, Hessen, and Rheinland-Pfalz. Rennes Metropolis and Val d'Ille are neighboring *intercommunalités*, but the latter is of a significantly more rural character. In contrast, Greater Lyon constitutes a particularly integrated intercommunality. Cases also differ importantly with regard to the number of Covenant signatories involved and the duration of Covenant participation. This diversity allows for revealing an enlightening range of circumstances, processes, and relationships.

4.2.3 Bounding the Cases

The exploration inter-municipal climate policy coordination within the Covenant of Mayors in the case studies is subject to the methodological specificity (and challenge) of case study research that "phenomenon and context are not always sharply distinguishable in real-world situations. [...] A case study inquiry copes with the technically distinctive situation in which there will be many more variables of interest than data points" (Yin 2014, 17). As depicted above, this is why the case selection approach is not variable-driven. This does not supersede a clear definition of a case. In contrast, it necessitates careful bounding of cases so as to define the scope of each case study in territorial and temporal terms, in form and content in order to establish a distinction between phenomenon and context and to obtain the necessary precision in defining its cases (Yin 2014, 31).

The territorial boundaries of each case study are obviously defined by inter-municipal associations' territory – without intending to exclude the consideration of intergovernmental relations with superordinate levels of government whose territory exceeds that of the respective case. The core study period time frame of the parallel study spans from Covenant signature in 2009 to the end of municipal legislative periods in early 2014.[52] This time frame is extended by desk research on developments before and after the core study period, including the run-up to Covenant signature, notably prior experience in climate policy and European af-

---

52   March 23/30 in France, May 25 in Baden-Württemberg/Germany.

fairs, and developments after the field research period. With regard to the latter, the main indicator for instances where further research is needed to obtain a full picture is change within Covenant participants' website profiles: updates by Covenant participants as well as their suspension or reclassification by the CoMO.

Covenant Coordinators are not studied as a 'whole', but in terms of inter-municipal governments' local climate policy coordination in connection with their commitment to the Covenant (see section 1.1). This includes their elected legislative (political) bodies; those services or departments within their executive (administrative) institutions that have been assigned the task of climate policy coordination (as opposed to departments that coincidentally deal with climate-related issues); and – if applicable –their agents, e.g., Local Energy Agencies.

This approach conceptualizes inter-municipal associations as collective, albeit non-monolithic actors. The underlying assumption is that territorial units can perform actions beyond the mere combination of individual actors' preferences and decisions, i.e. collective actions driven by institutional structures of that territorial unit.[53] This conceptualization builds upon the criteria for subnational governments' collective actorness in European governance as developed by Le Galès (2011): 1. a system of collective decision-making, 2. common interests and the perception of such, 3. integration mechanisms [of interests], 4. internal and external representation, and 5. innovation capacity. He further argues that members of collective actors under construction mobilize to obtain recognition for their collective actorness and, for a lack of clear concepts of actorness, rely to a large extent on legitimation by the European Commission and its agencies, for instance by applying for EU funds (Le Galès 2011, 69–73; 408-413).

This conceptualization informs methodological choices and the ensuing analysis. It points our attention to the processes that make up inter-municipal associations' collective actorness, and to the risk that would be associated with treating them as monolithic. In particular, the equation of inter-municipal preferences and actions with those of its leaders and deputies would risk overlooking the mechanisms of individuals' integration into collective action. To remedy this risk, special attention will be given to the perspectives of member municipalities of inter-municipal associations, and of administrative staff employed at the intercommunal and communal level.

---

53 This refutes certain strands of sociology in the Weberian tradition of methodological individualism which refuse to consider groups of individual actors and organizations as collective actors in studies of urban governance.

The five case studies rely on the concepts from the O-I-T Model as introduced in chapter 3. Firstly, they examine in what respect the policy goals pursued by the respective inter-municipal association coincide with those of the Covenant. This overlap of interests, the O-I-T-model assumes, is a prerequisite for the cooperation of the orchestrator and its intermediary (Abbott et al. 2016, 1). Secondly, the case studies explore the kinds of material and ideational support by the Covenant that have attracted, and been delivered to, the respective inter-municipal associations. The O-I-T-model considers the exchange of resources to be at the core of the orchestrator-intermediary relationship, and expects support from the orchestrator to consist of convening, agenda setting, assistance, endorsement, and coordination (Abbott et al. 2012, 9–11). In addition to examining the orchestrator-intermediary relationship accordingly, case studies look into the intermediary-target relationship and show how the respective inter-municipal association has actually assumed its role as a Covenant Coordinator. Which targets has it addressed, which instruments has it used, and what reference to the Covenant has it made in this process? Cross-case analysis then assesses the role and strength of the CoMO, the orchestrator in this governance arrangement.

4.3 Qualitative Interview Research

The questions why and how inter-municipal associations engage in local climate policy coordination, and what modes of governance result (see section 1.3), can only be answered when taking the understanding of the people involved into account. Where information cannot be gathered by standardized, macro-level procedures alone, expert interviews represent the method of choice for data collection. This is often true in political science where "elite actors will thus often be critical sources of information about the political processes of interest" (Tansey 2009, 483), in particular when processes cannot be assessed only on the basis of their output and outcome (Lauth, Pickel, and Pickel 2009, 166–67). While context knowledge on multi-level climate governance in general could also be derived from other sources, expert interviews were the main or only means to access operational knowledge on inter-municipal climate policy coordination, i.e. to gain an understanding of people's proceedings and their scope of action in a particular case (Meuser and Nagel 2009, 470–72). Also, expert interviews conform to the case study research design that aims at generalizations about causal mechanisms, not about distinct populations of people, actors or cases. Expert interviews intend to understand rather than explain, use open rather than closed questions, are evaluated in an interpretative rather than statistical way (Lauth, Pickel, and Pickel 2009, 167–68), and allow staying true to the case(s):

> "The nature of interviewing also allows interviewers to probe their subjects, and thus move beyond written accounts that may often represent an official version of events, gather information ab out the underlying context and build up to the actions that took place" (Tansey 2009, 485).

Therefore, half-structured expert interviews served as a central means of data collection. This section explains how experts disposing of relevant operational knowledge have been identified and selected (4.3.1), and how expert interviews were conducted by help of interview guidelines (4.3.2). It concludes with an account of how these and other data were analyzed, and how their validity was enhanced by means of triangulation (4.3.3).

4.3.1 Selection of Expert Interviewees

Expert interviewees sampling in case study research should not rely on random techniques but use targeted non-random sampling approaches in order to ensure that "that the most important and influential actors are included in the sample, and that testimony concerning the key process is collected from the central players involved" (Tansey 2009, 490). In general, experts are identified not based on their societal status or profession, but on their exclusive insider-knowledge obtained through the function they perform as active participants in problem-solving within the field under study (Gorden 1987).[54] At the same time, experts' knowledge and their problem definition are necessarily framed by their professional position. Therefore, expert selection must cover a diversity of perspectives (Meuser and Nagel 2009, 468).

Because expert selection requires background knowledge about relevant functions and actors within the field, experts have been chosen in a two-steps process following Tansey's (2009) plea for combining purposive and snowball sampling based on both positional and reputational criteria as the optimal method of interviewee selection for process tracing.

At an early stage of the research project, expert interviews with collaborators of the European Commission and the Covenant refined my understanding of

---

54   In this sense, they become experts by the researcher's definition, which is why the term 'expert interviewing' is more accurate here than 'elite interviewing' (cf. Tansey (2009)).

the EU-initiative under study and helped to specify research questions. The pilot studies enabled me to obtain more profound background knowledge as to which kinds of experts are relevant. For example, French local energy agencies turned out to implement important parts of inter-municipal climate mitigation programs, their representatives thus being indispensable interviewees. Also, the openness of middle-level interviewees during the pilot studies confirmed experience reported in the methodological literature that more information might be gathered through interviews at a middle level in the organization rather than with its top leaders due to constraints coming along with their political function (Lauth, Pickel, and Pickel 2009, 170). The pilot studies also pointed me to highly politicized aspects of the ongoing National Energy Transition Debate in France, thereby highlighting areas where it has been of particular importance to account for ideological diversity in interviewees' selection (Lauth, Pickel, and Pickel 2009, 170). Furthermore, pilot studies informed the revision of research questions as well as interview guidelines as I "[mined] the views of case participants or others who experienced the case for hypotheses" (van Evera 1997, 70).

Subsequently, I established a list of relevant experts from different subgroups of actors for each case as well as the French and German contexts (Small 2009, 13). Based on desk research, this list followed – with all objections to statistical vocabulary in mind (Small 2009, 28) – a sampling for range logic. Relevant groups of experts included technical (administrative) staff in charge of the Covenant within municipal and inter-municipal administrations, political leaders who had actively supported the Covenant at the municipal and the inter-municipal levels, as well as representatives of Covenant Supporters from Germany and France (see section 3.2.1.1). The Covenant contact persons as listed on the Covenant website were the obvious technical staff to inquire. In some cases, though, this information was misleading.[55] Where necessary, more than one officer was interviewed. Relevant political leaders included delegate mayors or councilors in charge of the Covenant within municipal and inter-municipal assemblies, and presidents of inter-municipal assemblies. The relevance and completeness of this list has been tested by snowball sampling, i.e. asking interviewees whom they

---

55   The alleged Covenant contact person of Rhine-Neckar Region, an officer in charge of European affairs, had only been involved in the preparation of Covenant signature, but not in implementation; the one of Greater Lyon was actually in charge of coordinating stakeholders' contributions to implementing the inter-municipal climate plan, but was not aware of any function or task with regard to the Covenant.

recommend for further interviews. In this way, the final list of interviewees resulted from intentional in-network selection (Small 2009, 14). Case studies thus cover networks of actors *as they actually interact*, not as the researcher expected them to be related to each other prior to the investigation.

Obtaining an appointment was not always easy, a common challenge in interview-based research (Agranoff and Radin 1991, 225).[56] In the beginning, the first contact was via e-mail in order not to appear intrusive. A personalized letter and an attached letter of recommendation served to valorize my inquiry. Since I found that follow-up phone calls where more successful, I called first, and then sent an e-mail with further explanations on my inquiry once the targeted expert had already agreed in principle to meet. This worked out well for technical staff, but political leaders were only available via their secretaries who generally asked for e-mail inquiries. The last resort was to approach people at conferences or to be introduced to them in person by other interviewees, e.g. colleagues. In total, I conducted 47 interviews with 53 individuals between May 2012 and February 2015 during several field research trips.

4.3.2 Conduct of Expert Interviews

In preparation of the expert interviews, I established a set of interview guidelines, but no questionnaire with standardized closed-response questions. The kind of information needed and the openness of expert interviewees were more likely to be obtained in a more conversational fashion (Agranoff and Radin 1991, 222). Half-structured interviews are sufficiently open and flexible (Tansey 2009, 486 87), especially when aiming at operational knowledge which is less explicit and thus needs to be reconstructed indirectly from the expert's narration. Therefore, without being a narrative interview, expert interviews can contain narrative passages which allow for the reconstruction of implicit operational knowledge. Interview guidelines were applied in a flexible manner, i.e. not as a standardized course of action, but rather an inventory of issues to be covered. This also allowed for being open to unexpected issues raised by interviewees that provided valuable insights and topics for later interviews (Meuser and Nagel 2009, 472–76).

---

56   For a discussion of the challenge of scheduling the desired expert interviews, and its implications for validity, see section 4.4.3.

Questions aimed at how decisions are taken and actions carried out, with a focus on institutional rather than personal knowledge (Lauth, Pickel, and Pickel 2009, 169). The line between private and expert knowledge is hard to draw, though, since private experience can shape professional assessments. In order to uncover potential biographical influences without shaping the conversation into a biographical interview (Meuser and Nagel 2009, 469), I included a biographical question in the beginning of the conversation. Being neither difficult nor politically sensitive but emphasizing my interest, it allowed interviewees to warm up to the situation.

Although also interested in context knowledge, interview guidelines mostly focused on operational knowledge (Meuser and Nagel 2009, 472).[57] Interview guidelines provided an opening question for each area of interest as well as a list of potential further questions as to facilitate the conversation.[58] It also helped to keep track of issues covered during the interview. In most cases, few follow-up questions were required to obtain the necessary information. Interviews largely consisted of experts' accounts with few interventions from my side. In view of also formulating policy recommendations, it has been natural to also ask interviewees about their advice for future policy and program development "near the end of the conversation, leaving key informants with a feeling that their expertise is sought and valued" (Agranoff and Radin 1991, 222).

As Meuser and Nagel (2009, 472–76) point out, status and gender play an important role in this regard because experts, more than other interviewees, tend to value the researcher's formal status (such as her academic degree). They further report that female researchers can find themselves in a situation where they are not accepted as equally competent – which can sometimes be turned into an advantage by inciting particularly detailed or unreserved explanations by interviewees. I made use of this effect by opening the conversation with rather general questions and following the flow of experts' accounts whenever possible without neglecting the interview guidelines. Only near the end of the conversation did I ask detailed follow-up or more critical questions.

---

57   Author's translation; original quote: *"Die Experten geben Auskunft über die Bedingungen ihres eigenen Handelns. Dessen Maximen, Regeln und Logiken sind der Gegenstand des Forschungsinteresses."*

58   For each interview, I brought a printout of interview guidelines to check of issue areas and questions in French, German, or English depending on the language of the conversation.

Interviews were usually conducted at interviewees' workplace, either in their personal office or a conference room.[59] In some cases, I was confronted with unexpected participants, i.e. colleagues who the interviewees thought could provide additional information, or interns whom they thought could learn something from the conversation without contributing themselves. When it was impossible to schedule an interview in person, I resorted to the phone or, preferably, to Skype.[60] I normally asked for a 45 minutes appointment, but we sometimes overran this by up to another 45 minutes.

Questions referring to previously evaluated documents or prior interviews reassured interviewees about the seriousness of our conversation and motivated detailed accounts. As Agranoff and Radin (1991, 228) point out, the informed discussion mode "serves to bridge the gap between the academic and practitioner. It demonstrates to administrators that academics are interested in studying *their* problems". At some instances, my inquiries pointed new staff to their organization's Covenant commitment and to their role as a contact person in the first place, a task they had not been instructed about by their predecessor or colleagues (e.g. interview with L. Ponsar, 13 June 2013), or our conversation informed interviewees about previously unknown Covenant proceedings such as joint SEAP submission (e.g. interview with S. Dallinger, 29 January 2015). In order to limit my interference with the object of study, I did not bring up such issues myself, but kept potential remarks or suggestions for after the completion of this book. All in all, interviewees seemed to enjoy a conversation with a competent interlocutor interested in their perceptions.

Interview research entails special ethical obligations towards the participants, in particular their informed consent. "The nature and purpose of the study, including its methods [and] expected benefits" (Thomas 2011, 69–70) were presented both in written form in the cover letter and orally at the beginning of each interview. At the end of each interview, when informants were aware of what information exactly they had shared, they were given the opportunity to determine the degree of confidentiality their information was to be treated with. Obviously, anonymity cannot be provided in expert interview research for it is based

---

59   Some interviewees suggested meeting at a restaurant instead. Although this complicated recording, I generally accepted the suggestion in order not to compromise the appointment, and interviewees' comfort.

60   A video call approximated the personal interview with regard to a personal relation being established and visual information being collected during the conversation.

on personal communication (Babbie 2013, 36). Also, it was in my interest as a researcher to enhance the validity of my argumentation by relating information provided to the informant's position. Thus, each participant was presented with a consent form detailing how I may use her information.[61] In order to allow for follow-up questions (Thomas 2011, 69–70), participants additionally received my business card. With the exception of one municipal staff from Rhine-Neckar Region, all interviewees agreed to being mentioned by name or affiliation. Where applicable, interviewees were given the opportunity to object to verbatim citations. For this purpose, they received an e-mail with a detailed account of the planned citations. I received sporadic specifications, but no objections.

4.3.3 Data Evaluation and Triangulation

Half-structured expert interviews served as a central means of data collection. This section depicts how they were evaluated, and explains why and how interview data were triangulated with document analysis and participant observation.

I recorded all interviews with a Dictaphone, including phone interviews, with only four exceptions.[62] During the interviews, I took notes in the form of bullet points along with indicating at which time during the conversation a topic was discussed. Afterwards, I documented surroundings and main statements of each interview in a one-page or so memo before writing the actual transcript, preferably directly after the conversation. When my initial impressions deviated from the transcript, e.g. when I over- or underestimated the importance given to some aspects during the conversation, this contortion pointed me to personal biases, for instance when I was overly impressed by interviewees' accounts or appearance.

Differently from other forms of social research interviews, expert interviews need not be analyzed sequentially because their meaning derives from the institutional context rather than their position within the narration. Instead, the evaluation consists in six subsequent steps: transcribing, paraphrasing, coding, topical comparison, conceptualization, and theoretical generalization (Meuser and

---

61   The consent form was presented in French, German, or English depending on the language of the conversation.

62   Exceptions consist of 1. one confidential interview where the expert objected a recording, 2. one interview where the recording failed, 3. one interview where my Dictaphone was blocked and we used the interviewee's iPhone instead, and 4. one interview conducted via Skype where I used a special Skype recording software.

Nagel 2009, 476–77). In contrast to other forms of qualitative interviews, expert interviews need not necessarily be transcribed verbatim. As long as text sequences can still be unambiguously located within the transcript, the latter may also be selective under certain circumstances: "when surrounding conditions and the interviewee's behavior are not that important for one's own topic, it is reasonable to limit the transcript to the level of contents. Sometimes the taking down of bullet points can be sufficient" (Lauth, Pickel, and Pickel 2009, 173–74).[63]

My pilot studies confirmed these accounts. Accordingly, I decided to transcribe the remaining interviews mainly with regard to contents, i.e. to merge the steps of transcription and paraphrasing. I kept verbatim transcripts of particularly telling passages in view of including direct citations when composing the case study reports. As Thomas (2011, 7) put it: "You should, in a case study, be able to smell human breath and hear the sound of voices", and interview material is a primordial means to achieve this. In contrast, I only noted bullet points for off-topic interview passages so as to ensure that I could come back to analyzing them in more detail should they turn out to be relevant at a later stage. The paraphrased transcripts served as a basis for coding. Coding followed Dey's (1993, 65) iconic advice that qualitative data evaluation requires an open mind, but not an empty head; it was informed by ex ante knowledge as established during the development of interview guidelines, but did not rely on pre-existing coding schemes. Initial categories were thematic and indicated that a certain issue appeared in a given passage (Kuckartz 2010, 86). Refined codes emerged from the analysis in an iterative bottom-up process. Subsequently, the coded interview material was compared first within, and then across cases in order to establish the case study reports presented in chapter 5 and the cross-case analysis developed in chapter 6. The conceptualization of findings showed that a perceived match of policy goals, and support from the Covenant decisively shaped Covenant participation, which led to the adoption of the O-I-T-model as a theoretical framework for further generalization.

Although expert interviews are key sources for the following case studies, they are triangulated with each other, with document analysis, and participant observation. Triangulation consists of combining different approaches to the

---

63   Author's translation; original quote: *"Gelegentlich ist es aber, wenn z. B. die Rahmenbedingungen und Verhalten des Interviewten nicht von so großer Bedeutung für die eigene Thematik sind, ebenfalls sinnvoll, das Transkript nur auf die inhaltliche Ebene des Interviews zu reduzieren. So kann gelegentlich das Notieren von Stichpunkten ausreichend sein."*

study object in order to counterbalance individual approaches' weaknesses, and to obtain additional insights from the combination of different perspectives so as to support later generalization (Lauth, Pickel, and Pickel 2009, 205). It aims to verify the validity of inferences from data about "a particular definition, theme, assertion, hypothesis, claim, etc." by examining the phenomenon by help of "multiple data sources, multiple investigators, multiple theoretical perspectives, multiple methods, or all of these" (Schwandt 1997, 163).

While triangulation is essential to increase the reliability of results in any kind of social research, four aspects make it particularly important here. Firstly, it is constituent of the case study research method that starts from "[thinking] small but drill deep, using different methods and drilling from different directions. [...] This collation of methods provides the triangulation" (Thomas 2011, 68). It is both a characteristic and a strength of case study research that it "relies on multiple sources of evidence, with data needing to converge in a triangular fashion"(Yin 2014, 17; see also Thomas 2011, 216). Secondly, the explorative character of the research project makes triangulation a particularly well-suited method for it is sufficiently flexible and open-ended (Lauth, Pickel, and Pickel 2009, 207). The state of the art simply does not allow for choosing a single entry point for analysis; what aspects are relevant and influential is only discovered during the research process.

Thirdly, triangulation is all the more important for expert interview-based research (Lauth, Pickel, and Pickel 2009, 207). Interviewed politicians might have an interest to over- or underestimate their role in a particular process, civil servants might have a habit to under-represent their role in general, and participants might simply have forgotten certain details when interviewed years after the events (Tansey 2009, 486–87). Lauth, Pickel, and Pickel (2009, 182–83) identify three main weaknesses: Expert selection can be difficult, especially when the researcher lacks detailed background knowledge, and interviewees' responses might be affected by a social desirability bias or colored by subjectivity.

Conversely, expert interviews can also be used to "corroborate what has been established from other sources" (Tansey 2009, 484) – a fourth reason to employ triangulation, here for the verification of information issued by inter-municipal associations or other levels of government. Archival documents, if available at all, are often biased, namely presenting official accounts rather than informal processes or actual disagreement. If documentation is abundant, interviewees can help to identify the most useful sources (Tansey 2009, 486). They can also point to instances where decisions taken or programs announced might not actually be implemented.

Two complementary strategies for triangulation are applied. Interviewees from different levels of government as well as from non-governmental organizations are included in expert selection in order to account for diverse perspectives. Information is sought to be confirmed by at least two sources independently, e.g. more than one interviewee or another source that have no identifiable common interest, or a mixture of both.

An obvious alternative source consists of secondary literature on climate policy, territorial governance and adjacent fields in the inter-municipal associations under study. Additional primary sources consist of document analysis and participant observation. Documents analyzed with regard to each case include the relevant Covenant website profiles and related material such as BEIs, SEAPs, and benchmarks for download, news posts and press releases, official documents such as deliberations and climate or energy plans, and promotional material linked to the Covenant. Publications of Covenant Supporters and national Covenant clubs from Germany and France helped to understand cases' national contexts. Participant observation represented a third goal of field research trips alongside expert interviews and library visits. I attended 4 municipal conferences dealing with local climate policy – including two embedded sessions of the national Covenant clubs – and, on these occasions had several informal background conversations with municipal staff and elected officers.[64]

Participant observation also helped to counterbalance a certain imbalance between expert interviews in Germany and France. Far more French than German organizations acts as Covenant Supporters (cf. section 3.2.1.1). Interviews with their staff provided me with valuable background information on local climate governance in France. In Germany, conferences helped to get a comparable overview of actors and issues in local climate governance.

4.4 Methodological Challenges

Both the problem addressed and the way it was addressed methodologically entailed specific challenges for the validity of research. This section depicts chal-

---

64   Transatlantic Urban Climate Dialogue 2012 (Stuttgart, Germany), Assises nationales de l'énergie 2014 (Dunkerque, France), Municipal climate protection conference 2014 (Lübeck, Germany), 100% Renewable Energy Regions 2014 (Kassel, Germany).

lenges during the research process, how they were dealt with, and what they imply for the scope of this research. Although normative distortions constitute a recurrent temptation in (militant) climate policy research, an alert approach to the study object allowed for preventing biases (4.4.1). Limitations that derive from the limited time frame of researching contemporary phenomena were tempered by extending the study period wherever possible (4.4.2). Further challenges arose with regard to the study of and to data collection by means of expert interviews, especially in rural settings (4.4.3).

4.4.1   Risk of Normative Distortions

There is a tendency in research on climate policy, sustainable development, transformation, and related fields to implicitly adopt a normative perspective. "Because of the urgency of climate change, most climate issues are investigated not simply to understand what is occurring but to identify opportunities for intervention and remediation" (Purdon 2015, 2). Often, the researcher assumes that 'more' climate mitigation or a 'more sustainable' policy would be preferable. This research project has not been entirely free of such tendencies in that it departed from the assumption that orchestrating local climate policy in view of effective climate mitigation was a desirable political goal. Also, research in the field mentioned above exhibits a preference for political innovation. Departing from the assessment that change is necessary, innovation is both welcomed and at the focus of many inquiries. Researchers may want to be part of the change they urge for but encounter difficulties to do innovative research because developing new methodological approaches is a challenging undertaking that might also receive skepticism from peers in the scientific community. In such a situation when it is hard to conduct research in an innovative way, a temptation can arise to study innovative cases instead. Such an ex ante classification of a case as being innovative can limit the researcher's openness or, if you like, objectivity in analyzing it. This book does not succumb to the temptation of celebrating its subject but clearly points out limitations of the Covenant. A critical appraisal was facilitated by adopting a local perspective. Examples include in-depth case studies that revealed the individual, functional logics of inter-municipal associations acting as Covenant Coordinators, and the large share of interviewees who had a local rather than a Covenant background. This approach uncovered that innovative climate governance and policies at the local level can only be traced back to the Covenant to a very limited extent.

4.4.2 Limitations of Studying Contemporary Phenomena

Evaluating recent and contemporary phenomena can be difficult. In studying cases of Covenant participation, a limitation arises from the fact that the core study period only covers five years (2009–2014). Certain theoretical approaches and analytical or evaluative perspectives were thus impossible to apply at this stage. Examples include theoretical frameworks designed to examine policymaking in complex, e.g. multi-level systems over longer periods of time such as Sabatier's advocacy coalition framework (Sabatier 1988; Sabatier and Jenkins-Smith 1993; Sabatier and Weible 2014a), and the evaluation of realized emission reductions. To date, less than one out of six Covenant signatories has submitted a monitoring report (Covenant of Mayors 2016f), i.e. data on implementation is still only fragmentarily available. But the evaluation of emission reductions realized within inter-municipal associations and the assessment of potential links to Covenant membership will always be hampered by methodological challenges in emissions accounting (Bader and Bleischwitz 2009; Ibrahim et al. 2012). Also, restrictions deriving from the short duration of inter-municipal associations' active involvement in the Covenant were tempered by additional desk research on developments before and after the core study period. This included an examination of the run-up to their Covenant commitment and their prior record in climate governance, and of recent developments such as electoral change or suspension from the Covenant that occurred after the field research period up to October 2016.

4.4.3 Pitfalls of Expert Interview-based Research

As argued in section 4.3, experts' operational knowledge on political processes and their way to make sense of local governance is crucial for the study of local governments since written documents often do not exist, are inaccessible, or fail to provide a holistic picture. But an research project that relies on expert interviews to a significant extent necessarily depends on experts' disposition to grant a research interview. Consequently, unsuccessful interview inquiries can represent a considerable setback for research.

The dependency on individual experts' cooperation was particularly strong because the number of people actively involved in one particular case was limited, so no other person could have provided the same kind of information. Although the vast majority of interview inquiries were actually successful, a certain

distortion of interview data derives from the fact that politicians were more reluctant than staff to meet for an interview. This applied in particular to those
politicians who were still in office at the time of my inquiry. If they or their
secretariat replied at all, a lack of time was the most common reason given for
declining my inquiry. In contrast, deselected politicians obviously had more time
and also seemed interested in sharing their view of the former legislative period
so as to preserve their political heritage. Potential distortions include a neglect of
party politics which are likely to seem less important to staff than politicians.
Some political leaders were unwilling to be interviewed. Although they played
an important role in local climate governance and for the Covenant in particular,
their personal motivation and assessment of the Covenant experience was not
available to the researcher even though it would have been an interesting plus for
the respective case study. Examples include Gérard Collomb, president of
Greater Lyon, and Heidelberg's Lord Mayor Eckart Würzner. In order to deal
with this methodological challenge, no assumptions about the personal viewpoint
of local leaders were made, and staff were inquired all the more carefully about
potential tensions within the local assembly or with the respective delegates in
charge, and about resulting challenges facing the administration. Fortunately, the
Covenant is not a highly politicized issue in local politics as compared to other
topics in climate policy. For instance, Covenant signature is often voted for unanimously. Also, staff are more directly in touch with the CoMO and, if applicable,
the respective Covenant Coordinator than politicians. Hence, they are more apt
to report on the actual Covenant experience. In sum, the difficulty in conducting
expert interviews with high-ranking local leaders was regrettable, but not prohibitive.

A particular challenge arose from unsuccessful interview inquiries in studying the case of Val d'Ille. This experience illustrates the difficulties for case
studies of small and rural local governments more generally. The limited number
of political and administrative staff restricted the number of possible interviewees. At the same time, staff – as is often the case at smaller administrations –
were obviously very busy and therefore reluctant to grant an interview. In addition, marketing considerations seemed to play a minor role in comparison to
larger local governments. Thus, self-display in academic publications is less of
an argument for dedicating working hours to meeting researchers. President Daniel Cueff, Vice-president Christian Roger, and the officer Soazig Rouillard did
not respond to interview inquiries. Among the expert interviews informing this

case, the interview with Michel Janssens was the only one from the inter-municipal association itself; other interviewees stemmed from neighboring or superordinate governments or networks. The total number of relevant official documents is limited, too, all the more as Val d'Ille favors palpable actions over detailed planning and documentation. In addition, few academic publications on local climate policy and inter-municipal governance in Val d'Ille were available while whole volumes deal for example with metropolitan governance and climate policy in Greater Lyon (Galimberti et al. 2014; 2014; Jouve 1998; Pastille Consortium 2002; Rocher 2016). This reflects the larger universe of cases of rural local governments as compared to (large) cities and metropolises, but also a preference for urban and metropolitan governments in research on local climate policy and local governance. At least in part, this preference is possibly due to difficulties in researching small local governments. Since fewer data could be collected, triangulation for the case of Val d'Ille was not fully possible which, to a certain extent, restricts the reliability of findings for this case. In particular, information about the importance of personal leadership by Cueff could not be verified based on various independent sources. Thus, the analysis largely depends on interview data from Michel Janssens, a technical staff at the inter-municipal administration. But where triangulation was possible, none of his statements turned out to be inaccurate, which gives reason to believe that his testimony is viable.

5 Inter-municipal Coordination in European Local Climate Governance

To date, we know little about the role and potential of inter-municipal associations in climate governance – but they can and do engage in local climate policy coordination. The local level is conceptualized here as consisting of municipalities and their associations (see section 4.2.2). Their scopes of action for climate policy are significantly determined by domestic institutional and political frameworks. Local governments carry out European, national and regional regulation, and depend on domestic provisions for their own initiatives. How then could an enabling policy framework for local climate action be achieved? Alber (2009, 11–12) argues that it would have to combine different modes of governance. She shows that self-governing, i.e. the horizontal collaboration within a region, state, or network, mainly reaches pioneers. For broad participation, other governance modes are required that rely on vertical relations between local and superordinate governments. National governments can govern through enabling, i.e. provide information and guidance for local climate policy. In contrast, governance by authority could rely on mandatory climate planning. Potential caveats of this strategy consist in unfunded mandates, insufficient competences, and a lack of local participation when binding targets are determined at superordinate levels. As long as climate policy is mostly a voluntary task for local governments, governing through provision remains central. Rather than case-by-case project funding that requires advance provision, Alber concludes, comprehensive funding schemes can accompany local governments in climate policy making and bring about significant numbers of emission inventories and climate action plans.

This chapter examines inter-municipal climate policy coordination within the contrasting national frameworks of Germany and France. Successively, it presents the findings from field research into five cases of inter-municipal associations acting as Covenant Coordinators from Germany (5.1), and from France (5.2).[65] Each country section first examines the domestic framework for inter-

---

65   For a discussion of the benefits of Franco-German comparison, see section 4.2.1.

© Springer Fachmedien Wiesbaden GmbH, part of Springer Nature 2020
L. Bendlin, *Orchestrating Local Climate Policy in the European Union*,
Energiepolitik und Klimaschutz. Energy Policy and Climate Protection,
https://doi.org/10.1007/978-3-658-26506-9_5

municipal climate policy coordination in terms of territorial organization, super-ordinate climate and energy policies, and local governments' experience in climate governance (5.1.1 and 5.2.1). It then provides self-contained reports on the respective cases: Rhine-Neckar-Region (5.1.2) and Stuttgart Region (5.1.3) from Germany, and Greater Lyon (5.2.2), Rennes Metropolis (5.2.3), and Val d'Ille (5.2.4) from France. The main findings from the five cases will be summarized (5.3) so as to inform cross-case analysis in the following chapter.

5.1 Covenant Coordinators in Germany

Covenant participation in Germany is restricted to a limited number of mostly larger municipalities. As early as 2008, the cities of Kiel, Heidelberg, and Bielefeld decided to sign the Covenant; shortly after, the eleven largest cities of Germany followed, including Berlin, Hamburg, and Munich (Deutsche Gesellschaft für Internationale Zusammenarbeit 2013, 13). Since 2012, the number of signatures is more or less stagnating with now 63 Covenant signatories. Larger cities are still prevailing; 32 German Covenant signatories have more than 100,000 inhabitants, only seven have less than 20,000 (Covenant of Mayors 2016h). Twelve are currently put on hold for a delay in reporting (Covenant of Mayors 2017a). Rhine-Neckar Region and Stuttgart Region remained the only Covenant Coordinators from Germany; both are metropolitan regions that encompass considerable agglomerations.

The vast majority of German Covenant signatories (55) are listed on the Covenant website as being supported by Climate Alliance, a member of the CoMO consortium (see section 3.2.1.1), followed by ICLEI, the Forum European Energy Award, and two regional organizations (Covenant of Mayors 2013b; 2016e). German Covenant signatories have been particularly active in submitting best practices for publication on the Covenant website (Deutsche Gesellschaft für Internationale Zusammenarbeit 2013, 14). With the creation of a national Covenant Club in October 2011 in Heidelberg, they aimed to enhance the dialogue with inter-municipal umbrella organizations, federal ministries, and the EU (Klima-Bündnis, Energy Cities, and Klimaschutz- und Energieagentur Baden-Württemberg 2014, 8).

The share of German local governments among Covenant participants is relatively low when considering their considerable mobilization for voluntary local climate policy in general. Many have engaged in local climate policy and

participate in European and international networks, partnerships, and commitment systems since the early 1990s, and supporting federal schemes have given
local climate policy an additional push since 2008 (see e.g. Beermann 2014;
Busch 2015; Hakelberg 2014). The following exploration of inter-municipal climate policy coordination within the Covenant in Germany pays special attention
to the question why Covenant participation is not more common in Germany,
especially among rural local governments. The section starts with an overview
of the German framework for local climate governance in terms of polity and
policies (5.1.1). Case study reports from two metropolitan regions, Rhine-Neckar
Region (5.1.2) and Stuttgart Region (5.1.3), show that orchestration of local climate policy in Germany is hampered by a mismatch of local needs and the Covenant's offer for support that traces back to prior records in local climate policy
making and competing national schemes.

5.1.1   The German Framework for Inter-municipal Climate Governance

Germany has a longstanding record in environmental protection. After a period
of considerable industrial growth, increasing environmental problems gave rise
to environmental policy as a new policy field around 1970 (Jänicke and Mez
2000, 596). Although a federal environmental ministry was only established in
1986, Germany soon became a leader in this field (Jänicke 2006; Sommerer
2011).[66] Repeatedly, EU policies helped to overcome domestic hindrances in
multi-level games; in contrast, the German federal system itself enables horizontal rather than vertical competition (Jacob et al. 2016). It was also among the first
states to develop a national climate policy, and accommodates a broad and vivid
landscape of (not only governmental) local climate action. Local climate governance in Germany benefits from municipal self-government, ambitious national
and regional climate policies, decentralized energy production, and municipalities' longstanding record in sustainable development and climate policy. This
section explains why the German framework is both an asset and a challenge for
inter-municipal associations that still have to assert their place in territorial governance and local climate policy. After an introduction to superordinate climate
policies in Germany (5.1.1.1), it discusses the position of local governments
within Germany's political system in general (5.1.1.2) and the role of inter-mu-

---

66   For a detailed policy analysis of German environmental policy, see Böcher and Töller (2012).

nicipal associations in particular (5.1.1.3). Finally, it examines the broad and vivid landscape of local climate governance (5.1.1.4).

5.1.1.1   National and Regional Climate and Energy Policy

Germany's climate policy can be characterized as technology-driven, combining innovative climate policies with measures that promote its climate protection industry, an approach that has been described as ecological modernization and ecological industrial policy (Jänicke 2017a). In European climate governance, Germany plays an ambiguous role. As Europe's largest national economy and by far the most populated EU member state, it stands for the largest share of GHG emissions within the EU (see section 4.2.1). Per-capita emissions, and to a lesser extent the GHG emissions intensity of the German economy, have long been above the European average (European Environmental Agency 2007, 22–23). Early GHG emission reductions were partly due to so-called wall-fall profits in the course of unification since 1990 and the ensuing modernization, and in parts decline, of industry in the 'new' *Länder* (Schleich et al. 2001). Driven by influential veto players from the car and coal-based power industry, Germany has lagged behind on implementing market mechanisms such as the EU-ETS (Jänicke 2017a, 119), and successfully objected against more ambitious EU legislation on GHG emission from cars (Hey 2010b). The nuclear phase-out entails tensions with the desire to reduce the share of coal-fired power in the energy mix (Renn and Marshall 2016; Schröter 2017; Setton and Helgenberger 2016). The coherence of German climate and energy policy is further challenged by its significant institutional fragmentation and ensuing interdepartmental tensions, in particular between the Ministry for the Environment, Nature Protection and Nuclear Safety (*Bundesministerium für Umwelt, Naturschutz und Reaktorsicherheit*, BMU)[67] and the Ministry of Economics and Technology (*Bundesministerium für Wirtschaft und Technologie*, BMWi) (Jacob and Kannen 2015).[68]

---

67   Since December 2013, the ministry is also in charge of construction (now called *Bundesministerium für Umwelt, Naturschutz, Bau und Reaktorsicherheit*, BMUB).

68   Further ministries involved in climate policy making include the Ministry for Transport, Construction and Urban Development (*Bundesministerium für Verkehr, Bau und Stadtentwicklung*, BMVBS) for transportation, the Ministry for Education and Research (*Bundesministerium für Bildung und Forschung*, BMBF) for technology, and the Ministry for Consumer Protection, Food and Agriculture (*Bundesministerium für Verbraucherschutz, Ernährung und Landwirtschaft*, BMVEL) for agriculture and forestry.

But Germany also has a history of progressive environmental politics (see e.g. Jänicke 1997; Weidner and Mez 2008). It achieved its GHG emission reduction target under the Kyoto protocol of 21 % by 2012 as early as 2007, accounts for 75 % of EU-15 GHG emission reductions under the burden sharing agreement, and has been referred to as – and claimed to be – a climate frontrunner leading by example (see e.g. Huber 2013; Jänicke 2017a; Michaelowa 2008; Schreurs and Thiberghien 2007; Steinbacher and Pahle 2016). Although parliamentary representation of the Green Party since 1983 played a role in this process, climate mitigation is supported across party lines, albeit for different reasons (Fischer 2017; Jänicke 2005; 2017a; Schreurs 2016a; Wurzel 2010),[69] and by public opinion (Arnold et al. 2016). As early as 1987, concerns over energy security in view of elevated energy imports and future economic costs of climate change motivated the establishment of a Parliamentary Enquiry Commission; its first report informed Germany's first climate strategy in 1990 that was updated in 2000 and 2005 (Carlarne 2010, 199–202; Jacob and Kannen 2015, 8; Jänicke 2017a, 117; Michaelowa 2008, 144). Ever since, Germany repeatedly pushed for ambitious climate policies at the European and international level. Examples include its active advocacy for ambitious provisions of the Kyoto protocol, including the 1995 Berlin mandate (Andresen and Agrawala 2002, 47), its double EU and G8 presidencies in 2007 (Jänicke 2017a; Wurzel 2010), and its support for the creation of the International Renewable Energies Agency in 2009 (Carlarne 2010). Domestically, Germany successfully promoted the development of renewable energy based on a set of policies (Jacobsson and Lauber 2006), including a mandatory feed-in tariff since 1991 that supported technology development and led to the proliferation of decentralized and locally owned energy production (Lauber and Mez 2004), and was subject to extensive diffusion worldwide (Mendonça, Jacobs, and Sovacool 2009, 80), Germany also introduced an ecotax (Wurzel et al. 2003), and engaged in a low-carbon energy transition (*Energiewende*) (Quitzow et al. 2016; Schreurs 2016b).

Germany aims for a GHG emission reduction of 40 % by 2020, 55 % by 2030, 70 % by 2040, and 80–95 % by 2050. The Integrated Energy and Climate Program (*Integriertes Energie- und Klimaprogramm*, IEKP), an action program with 29 measures adopted in 2007 illustrates Germany's desire at the time to play

---

69   As an exception, the recently ascended rightwing populist *Alternative für Deutschland* (AfD) has voiced fundamental objections to climate mitigation (Götze and Kirchner 2016).

a forerunner role in global climate governance (Jacob and Kannen 2015, 8). In this context, the BMU launched a key federal funding scheme for local climate action: the National Climate Protection Initiative (*Nationale Klimaschutz-Initiative*, NKI). The NKI aims to support municipalities, consumers, companies, and social and cultural institutions in climate mitigation projects; a line of funding under the NKI that specifically targets municipalities will be discussed in the next section.

After the change of government in 2009 and a decline in support from chancellor Angela Merkel, though, IEKP implementation lost momentum to a large extent (Jacob and Kannen 2015). The Energy Concept (*Energiekonzept*) adopted in 2010 put special emphasis on the promotion of renewable energy and energy efficiency. A legislative package for the energy transition that took effect in 2011 included provisions for facilitating the promotion of energy efficiency and renewable energy at the local level, namely with regard to planning and building law.[70] Additionally, a Climate Action Program (*Aktionsprogramm Klimaschutz*) consisting of more than 100 individual measures was launched in 2014 so as to prevent Germany from falling short of its 2020 target. Following the adoption of the Paris agreement, the federal government engaged in a consultation process with local governments and other stakeholders and adopted, in November 2016, a Climate Action Plan (*Klimaschutzplan 2050*) in view of reaching GHG neutrality by 2050.

The *Länder* participate in German climate policy making in various ways that will be depicted in the next section. Their preferences in climate policy making are often linked to regional interests and capacities as determined by geography, dominant car or coal-based industries, or shares of renewable energy production (Monstadt and Scheiner 2016; Ohlhorst 2015).[71] Several *Länder* engaged in renewable energy promotion as early as in the 1980s, ahead of the federal level (Schönberger and Reiche 2016, 30). As of the second half of the 1990s, the majority of the *Länder* had voluntary climate and energy plans in place that built on targets established at superordinate levels (Jörgensen 2012, 108). Since the early 2000s, all 16 *Länder* have a climate plan, most with targets for long-term climate

---

70　　Gesetz zur Stärkung der klimagerechten Entwicklung in den Städten und Gemeinden; Gesetz zur Förderung des Klimaschutzes bei der Entwicklung in den Städten und Gemeinden. For a more detailed overview of national energy policies that impact local scopes of action, see Barbey (2012).

71　　For a discussion of coordination requirements of the energy transition between the Länder, see Schreurs and Steuwer (2015).

planning, often accompanied by financial incentivizing measures (Deutscher Bundestag 2005; Jörgensen 2002, 15–16; Schönberger and Reiche 2016, 32), and concrete climate policies have to some extent replaced *Länder*-level strategies for sustainable development (Jörgensen 2012, 107). Some *Länder* established ministries for climate policy or energy transition. Policies range from building renovation, awareness campaigns, technology development, public procurement, and low-carbon transportation, to the promotion of renewable energies, including renewable heat (Finck 2014, 452–56; Jörgensen 2012, 115). Competences with particular relevance for local scopes of action include provisions for wind energy siting and regulation for municipal energy companies (Schönberger and Reiche 2016, 36).

In view of assessing *Länder*-level climate policy, Weidner and Mez (2008, 371) argue that in implementation responsibilities for national-level regulation are limited as compared to other fields of environmental policy, which limited the occurrence of implementation deficits linked to particular interests (Weidner and Mez 2008, 371). With regard to the energy transition, though, inward perspectives in climate and energy policy making risk boosting macroeconomic costs and wasting opportunities for multi-level reinforcement (Ohlhorst 2015). Although the increased number of *Länder* governments involving the Green Party since 2011 has given a push to renewed climate policy leadership also at the European level (Jänicke 2017a, 126), the development of climate plans at *Länder* level is still in an experimental phase (Wick 2015). Jörgensen (2012) concludes that the *Länder* do not fully seize their possibilities for climate policy experimentation as most policies are driven by socioeconomic co-benefits, and many remain programmatic, if not symbolic.

Baden-Württemberg, home to both German inter-municipal associations studied below, has a longstanding record in climate policy. It is Germany's third-biggest Land in terms of population and territory.[72] Situated in Germany's South-West and seated in Stuttgart city, Baden-Württemberg is a particularly affluent region with important machine building, vehicle and metal construction industries and correspondingly high energy consumption. It adopted a first climate strategy already in 1994, and has an acclaimed Renewable Heat Act that prescribes the use of renewable heat on existing and newly constructed buildings

---

72 Quite extraordinarily, Rhine-Neckar Region exceeds *Länder* borders and also comprises some municipalities from Rhineland-Palatinat and Hesse.

(Jacob and Kannen 2015, 12). Climate policies were further strengthened in 2011 when Baden-Württemberg became the first *Land* to be governed by a coalition led by the Greens, along with the Social Democrats.[73] The climate protection act adopted in 2013 and the Integrated Energy and Climate Concept (*Integriertes Energie- und Klimakonzept*, IEEK) adopted in 2014 aim to reduce Baden-Württemberg's energy consumption by 50% and to raise the share of renewable energy up to 80% so as to reduce GHG emissions by 90%, all by 2050, with an interim goal of 25% GHG emission reductions by 2020. The translation of the IEEK into binding laws, a participatory approach and the pooling of responsibilities for environment, climate, and energy within one ministry have promise for an integrated and coherent regional climate policy (Jacob and Kannen 2015). Building upon prior policies, the Green-Red government increased investment in renewable energy research and succeeded to approximate the regional share of renewable energy to the federal average, but as yet failed to counteract an increased share of coal-based power in the regional energy mix, including the regional company EnBW (Wurster 2016).

5.1.1.2 Local Governments within the Federal System

Territorial organization in Germany is characterized by a federal system as codified in the Basic Law (art. 20 GG). German federalism has been described as unitarian because it aims at providing for uniform living conditions across territorial units (Große Hüttmann 2010). Due to close interlinkages between the federal level and the 16 *Länder*,[74] it is also referred to as cooperative federalism (for a classic conceptualization, see Scharpf, Reissert, and Schnabel 1976). German federalism is often criticized for a lack of problem-solving capacity, transparency, and ability for reform (see e.g. Rowe and Jacoby 2013). It also inspired the joint decision trap model of EU decision making (Scharpf 1985; 1988; 2006; see section 2.3). Although the term suggests otherwise, critiques of cooperative federalism typically do not refer to voluntary forms of cooperation, but to institutionalized interdependencies in decision making (Kropp 2010). The latter re-

---

73 After 57 years of Christian-Democrat participation in government, one factor for this exceptional electoral outcome consisted of the nuclear accident in Fukushima about two weeks before that also led to Germany's nuclear phase-out insert Fischer (2012); Roth (2013).

74 This account focuses on interlinkages between levels of government, but forms of joint decision-making in Germany are not limited to vertical cooperation; they include horizontal fora of coordination between the *Länder* (see e.g. Hegele and Behnke (2013).

sisted repeated attempts during the last decade to disentangle German federalism; on the positive side, joint decision making increases input-legitimacy and provides opportunity structures to sustain political attention for societal problems (Kropp and Behnke 2016).

The *Länder* are responsible for the implementation of most federal laws, exercise residual competences, and have legislative power for selected fields of public policy. Hooghe, Marks, and Schakel (2010) find that they represent the most autonomous level of subnational government in Germany, with a relatively elevated degree of self-rule and the highest possible ranking in terms of shared rule, i.e. participation in national policy making. The *Länder* have veto power in national environmental legislation within the second chamber of parliament, the *Bundesrat*;[75] climate-related (concurrent) *Länder* competences include economic affairs, transportation, waste, and nature conservation (Finck 2014, 452).

Further layers of government are determined by the *Länder* and thus vary slightly across Germany, especially in the city states Berlin, Bremen, and Hamburg.[76] As an additional regional tier of government Baden-Württemberg, Bavaria, Hesse, and North Rhine-Westphalia maintain governmental districts (*Regierungsbezirke*). The diverse landscape of municipal government in Germany generally comprises rural and urban districts (*Landkreise* and *Kreisfreie Städte*), collective municipalities (*Gemeindeverbände*), and municipalities (*Gemeinden*) (see figure 5.1). Several waves of territorial reform since the 1960s aimed to counteract municipal fragmentation and to ensure functional administrative units by municipal mergers and by creating collective municipalities (Bogner 2007). The resulting diversity of today's landscape of municipal governments is linked to the strong position of major cities, many of which maintained their independence (Seele 2007, 1061–62). In addition, various kinds of inter-municipal associations have been established in order to facilitate cooperation.

---

75 For a discussion of the veto threat under current constitutional provisions, see Stecker (2016).
76 The specificities of city states have been depicted in Figure 4.4; for a detailed introduction see Mann and Püttner (2007)).

Districts (total)	Thereof		Municipalities (total)	Thereof	Level of collective municipalities		
	Urban	Rural		Cities[a]	total[b]	Collective municipalities	Nonaffiliated	Affiliated
	Districts						Municipalities
402	107	295	11,092	2,060	4,540	1,269	3,195	7,895

Figure 5.1: Number of local governments in Germany

[a] Including urban districts.
[b] Including collective municipalities, municipalities, and areas without municipal administration.
Source: adapted from Statistische Ämter des Bundes und der Länder 2016.

German municipalities have a longstanding right to self-government that traces back to legislative autonomy of medieval towns (Martínez Soria 2007; Saldern 1999). Art. 28 II GG guarantees the right to self-government for municipalities and their associations on condition they respect superordinate regulation. This includes an institutional guarantee for a municipal level of government within Germany's polity, but not for the individual municipality. Local democracy varies importantly because it is subject to state legislation; most cases covered here feature different forms of mayoral dominance in municipal government (Delcamp 1995, 132). Municipal and district assemblies are elected every four or five years; mayors' direct election has contributed to personalizing local leadership (Keating 2011, 54). Municipalities are responsible for services of general interest (*Kommunale Daseinsvorsorge*) and obliged by construction law (*Baugesetzbuch*) to take climate and air into consideration in development planning (*Bauleitplanung*). Depending on *Länder* provisions, they may take action on all public affairs that are local in nature and are not assigned to another level of government.[77] Voluntary climate policies include energy management in public buildings and building renovation, awareness campaigns, public procurement, low-carbon mobility, energy efficient street lightning, the promotion of renewable energies, and climate-friendly waste disposal (Finck 2014, 452–54).

---

77   In contrast to some *Länder* constitutions that include lists of municipal responsibilities, the constitution of Baden-Württemberg, the *Land* most relevant to the cases studied in this research, does not further specify local affairs and public services (see art. 71).

At the same time, municipalities and their associations are integral parts of the administrative system and execute *Länder* regulation. Since 2006, the federal level may no longer entrust them with any mandatory task (Article 84 I 7 GG); voluntary participation of local governments in federal programs remains possible. Municipalities partake in tax revenues, but can also raise local taxes as included in the constitutional right to self-government, fees for municipal services, and receive compensatory funds aiming at balancing irregularities (Frenzel 2013, 118–19; for a detailed discussion see Zimmermann 2016). Additional mandates are to be funded accordingly by the state; the exact financial requirements and the connectivity between legislation and cost absorption are recurrent objects of intergovernmental dispute (Nierhaus 1999, 27). Moreover, many municipalities struggle with financial straits, especially urban districts and municipalities within economically weak areas (Brand 2015; Karrenberg and Münstermann 1999; Naßmacher 2015).

For purposes of interest representation at the regional, national, and – especially since the early 1990s – European level, municipalities are organized in three umbrella organizations (*Kommunale Spitzenverbände*) with their respective regional sections, and European offices in Brussels: the Association of German Cities (*Deutscher Städtetag*), the German County Association (*Deutscher Landkreistag*), and the German Association of Towns and Municipalities (*Deutscher Städte- und Gemeindebund*).[78] They have some statutory possibilities to participate in policy making within national and regional parliaments and ministries, and to a limited extent within the EU's CoR (Henneke 2007b; Seele 2007). Examples of lobbying efforts in the field of climate and energy policy include position papers calling for more support for local climate policy from the European, national and regional level (Deutscher Städtetag 2014), and for accelerating grid expansion in the context of the energy transition (Deutscher Städte- und Gemeindebund 2014). Other than interest representation, municipal umbrella organizations distribute information on relevant legislation, programs, and funding schemes, and provide a platform for exchange, including topical working groups and conferences on environmental protection, climate mitigation and adaptation, and energy policy.

The diversity of local government in Germany is further spurred by various forms of inter-municipal cooperation that will be introduced in the next section.

---

78   For separate introductions, see Articus (2007); Henneke (2007a); Landsberg (2007).

5.1.1.3   Inter-municipal Cooperation within Regional Associations

Inter-municipal associations take on different forms and functions in the *Länder* (Heinz 2007, 101). In general, municipalities and districts associate to carry out certain tasks, be it planning, e.g. for land use (*Bauleitplanung*), or public services such as waste management. Mandatory tasks defined by public law may be supplemented by voluntary transfers of municipal responsibilities. Common forms include new public authorities and various kinds of territorial cooperation ranging from regional associations as introduced in more detail below, to special purpose associations (*Zweckverbände*) and associations of administrations (*Verwaltungsgemeinschaften*).

Other than the search for cost-efficient service provision in view of tight municipal budgets, an important driver for inter-municipal cooperation consists in the perceived need for improved cooperation especially in metropolitan areas (Desage 2011, 285). As Heinz (2007, 97-99) argues, fragmented territorial structures do not match the necessities of economic development and transport planning or larger-scale projects with respect to technical or cultural infrastructure. The built environment of metropolitan areas has grown continually but uncoordinated as core-cities and their environs increasingly compete to offer similar functions. At the same time, inner-metropolitan migration reduces the resources of core-cities while a peri-urban "fat belt" (*Speckgürtel*) emerges. But inter-municipal cooperation is also driven by newer factors. Financial support schemes by superordinate levels increasingly require regional cooperation. Metropolitan regions strive for eye height in negotiations with third parties. Simplified structures seem helpful to succeed in growing inter-municipal competition entailed by globalization and Europeanization. Also, the deficits and inadequacies of existing forms of inter-municipal cooperation call for innovation.

In response to these challenges, a diverse landscape of metropolitan cooperation has emerged, as Heinz (2007, 102-103) depicts, including neighborhood associations in charge of land-use planning and regional planning associations.[79] Genuine multi-sectoral approaches for metropolitan cooperation beyond mere

---

79   Examples of the latter include Kommunalverbund Niedersachsen/Bremen, Region Hannover, Zweckverband Großraum Braunschweig, Zweckverband Raum Kassel, Regionalverband Ruhr, Regionaler Planungsverband Leipzig-Westsachsen, Regionaler Planungsverband Oberes Elbtal/Osterzgebirge, Planungsverband Region Chemnitz, StädteRegion Aachen, Regionalverband FrankfurtRheinMain, Planungsverband Region Nürnberg, Verband Region Rhein-Neckar, Regionalverband Saarbrücken, Verband Region Stuttgart, Regionalverband Südlicher Oberrhein, and Planungsverband Äußerer Wirtschaftsraum München.

planning have been intensely debated, but rarely established, and even fewer have been retained. This is because their creation requires exceptional administrative organization and they do not fit into the German political-administrative system. The emerging ad hoc solutions suffer from dependencies which result from an overlap of responsibilities with municipalities, e.g. with regard to public transport or waste, and a lack of their own tax revenues. The situational set-up of inter-municipal cooperation typically results in unstable approaches; they tend to be dismantled or strengthened whenever the political or personal constellations change.

Baden-Württemberg's regional associations of municipalities (*Regionalverbände*) – two of which are subject to case studies below– belong to the second category. They are endowed with regional planning by state law and mediate between federal state planning and municipal land use planning. In this way, regional associations of municipalities contribute to the implementation of the counterflow principle (*Gegenstromprinzip*). This principle of German building law stipulates that lower levels of land use planning have to be heard in superordinate planning on the one hand, and have to comply with superordinate requirements on the other. Municipalities are represented within the inter-municipal assembly by their delegates; the latter elect an honorary president, and a full-time director who heads the inter-municipal administration for a period of eight years. Climate-related competences include economic development, transportation, the allocation of wind turbines, and regional energy planning.

Both regional associations under study are recognized by the federal Conference of Planning Ministers (*Ministerkonferenz für Raumordnung*, MKRO) as European metropolitan regions (EMRs). As such, they participate in the German Metropolitan Regions Initiative (*Initiativkreis Europäische Metropolregionen in Deutschland*) and the network of European Metropolitan Regions and Areas (METREX). Their specific features will be discussed in more detail in the respective case study reports. Rhine-Neckar Region spans across *Länder* boundaries and comprises local governments from Rhineland-Palatinate and Hesse. Stuttgart Region is not only in charge of planning, but also of implementation for certain tasks and has a directly elected assembly. In contrast to other forms of inter-municipal cooperation, metropolitan regions are particularly affected by the above-mentioned pressures for inter-municipal cooperation beyond the scope of their state-level planning mandate (Fürst 2004; Ludwig 2009). In its 2016 update of the overall concept of regional policy, the MKRO calls for further strengthening EMRs' role in national and European governance, and defines climate adaptation and renewable energy promotion as central tasks for regional planning (Ministerkonferenz für Raumordnung 2016).

5.1.1.4 Local Climate Governance

German municipalities must implement and respect superordinate regulation such as EU water regulation or national construction legislation on energy efficiency that have certain implications for climate mitigation. But genuine local climate policy is a voluntary task that is subject to local political will and capacities, namely financial resources. Since the early 1990s, pioneering municipalities mobilized to take action in the fields of energy, transportation, urbanism, waste management, waste water, and public procurement, often with a focus on cost reductions by energy savings that can in turn fund additional measures (Kern et al. 2005). Early climate plans often developed within the framework of Local Agenda 21 activities (Weidner and Mez 2008, 372). Prevalent municipal climate networks include the Climate Alliance (493 out of 1720 members from Germany) and ICLEI (21 out of 158 European members in Germany) that both have their main office in Germany (ICLEI 2017; Klima-Bündnis 2017). Since 2003, 320 municipalities have engaged a European Energy Award (EEA) procedure (European Energy Award 2016b).[80] German municipalities regularly participate in competitions that decorate outstanding local climate policy. Among nine laureates of the EU's European Green Capital Award to date, two are from Germany (Hamburg 2011, Essen 2017). National competitions include the Climate Protection Capital (*Bundeshauptstadt im Klimaschutz*) by German Environmental Aid (*Deutsche Umwelthilfe*), an environmental association.

An asset of local climate action in Germany derives from decentralized energy production (Klima-Bündnis, Energy Cities, and Klimaschutz- und Energieagentur Baden-Württemberg 2014, 35–36). Energy cooperatives play an important role in developing local low-carbon governance strategies (Deutsche Gesellschaft für Internationale Zusammenarbeit 2013; Schröder and Walk 2013). Many municipalities hold energy companies (*Stadtwerke*).[81] This enables them to "directly influence investments and pricing and thus have an impact on emis-

---

80   In Germany, the EEA program is managed by Länder ministries that are supported by regional EEA offices which are coordinated by a national office (*Bundesgeschäftsstelle*).
81   After a phase of privatization, the energy sector has undergone extensive re-municipalization since the turn of the millennium (Hall, Lobina, and Terhorst 2013).

sions, provided the right priorities are being set" (Collier and Löfstedt 2013, 12). In a similar vein, municipalities' scopes of action with regard to low-carbon transportation benefit from the ownership of municipal mobility companies (Schönberger and Reiche 2016, 38).

Assessments of local climate action in Germany diverge significantly. Some scholars refer to municipalities as the "real policy leaders" (Huber 2013) on climate change in Germany. But there long was no national policy in place to ensure comprehensive local climate action. Hence, only forerunner municipalities engaged in climate mitigation, and those that did still failed to fully tap their potential – shortcomings ascribed to financial restrictions, competing priorities for voluntary municipal action, failure to integrate climate mitigation across municipal departments, and a neglect of measures in fields other than energy policy (Kern et al. 2005; Sippel and Jenssen 2009). Notably, there was no national scheme at the time to provide funding and standards for local climate action. This changed in the context of the IEKP.

Since 2008, the BMU supports municipal climate action within the framework of the NKI based on the Local Authority Guideline (*Kommunalrichtlinie*).[82] To date, the Local Authority Guideline has supported more than 10,000 projects from more than 3,000 municipalities (Bundesministerium für Umwelt 2016). These include energy saving projects, investment measures, the employment of local climate managers, and the development of municipal climate protection concepts (*Kommunale Klimaschutzkonzepte*, in the following referred to as municipal climate plans). The environmental ministry conceptualizes a municipal climate plan as a strategic document aiming to structure and inform local climate action by the municipal government and other actors by establishing an overview of GHG emission reduction potentials and available options for action, determining targets and indicating possible measures (Bundesministerium für Umwelt 2014). Municipalities can choose between a partial or integrated (i.e. complete) climate plan with regard to land use management, public buildings, street lighting, private households, companies, mobility, energy, water, waste, and – optionally – climate adaptation. Municipal climate plans are typically developed in a participatory process by an external agency within a timeframe of one year (Klima-Bündnis, Energy Cities, and Klimaschutz- und Energieagentur Baden-

---

82   Guideline on promoting climate action projects in social, cultural and public organisations
      (Richtlinie zur Förderung von Klimaschutzprojekten in sozialen, kulturellen und öffentlichen
      Einrichtungen).

Württemberg 2014, 19). Only municipalities with a municipal climate plan are eligible for funding for a municipal climate manager for three, extendable by two years. The Local Authority Guideline significantly increased the diffusion of municipal climate plans in Germany (Klima-Bündnis, Energy Cities, and Klimaschutz- und Energieagentur Baden-Württemberg 2014, 26).

A project funded by the BMU from 2007 to 2014 was instrumental in promoting renewable energy at the local level in rural areas. The initiative 100% Renewable Energy Regions (100%-Erneuerbare-Energien-Regionen, 100ee) established a network of more than 150 rural local governments (districts, municipalities, associations of municipalities) aiming for energy neutrality by promoting renewable energy and built on a number of citizen-driven initiatives known as plus-energy villages (Weidner and Mez 2008, 373). The EU-project 100% Renewable Energy Sources (RES) Communities running from 2012 to 2015 under the IEE funding scheme aimed to connect 100ee with adjacent networks in Europe and to promote the Covenant among its participants.

Although a minimum of financial leeway is required for any voluntary policy, Schönberger and Reiche (2016) find that subnational governments in Germany do not engage in renewable energy promotion as a luxury activity out of (financial) surplus – as suggested by earlier accounts from Kern et al. (2005) – but as a means to seize economic opportunities, overcome the budgetary crisis, and create employment. They further highlight that neighboring municipalities often cooperate on climate policy such as "establishing energy-related advisory agencies or defining priority areas for wind power" (Schönberger and Reiche 2016, 33).

This also applies to Baden-Württemberg. A comprehensive network of more than 30 district-level local energy agencies is maintained jointly by municipalities with local companies and energy suppliers. The regional-level Climate and Energy Agency (*Klimaschutz- und Energieagentur Baden-Württemberg*, KEA) based in Karlsruhe coordinates and networks local energy agencies on a voluntary basis. Regional funding schemes for local climate policy include support for EEA procedures, and start-up funding for local energy agencies. Some of Germany's municipal climate frontrunners such as Freiburg, Heidelberg and Stuttgart stem from Baden-Württemberg.

5.1.2 Rhine-Neckar Region

The case of the Rhine-Neckar Regional Association (*Verband Region Rhein-Neckar*, in the following referred to as Rhine-Neckar Region) shows how an orchestrator-intermediary relation can fail despite a promising start. Rhine-Neckar Region politically embodies the densely populated and highly industrialized metropolitan region Rhine-Neckar with more than 2,350,000 inhabitants (Verband Region Rhein-Neckar 2016b). Reaching from Bavaria to France, the metropolitan region encompasses urban hubs, smaller towns and villages, ample wine-growing districts, farmland and unspoiled nature (Barbey 2012, 309). Situated at the border triangle of Baden-Württemberg, Hesse, and Rhineland-Palatinate, the polycentric urban region (Meijers 2007) comprises eight urban districts, including regional metropolises such as Heidelberg and Mannheim, and 280 district-affiliated municipalities within seven rural districts (see figure 5.2).

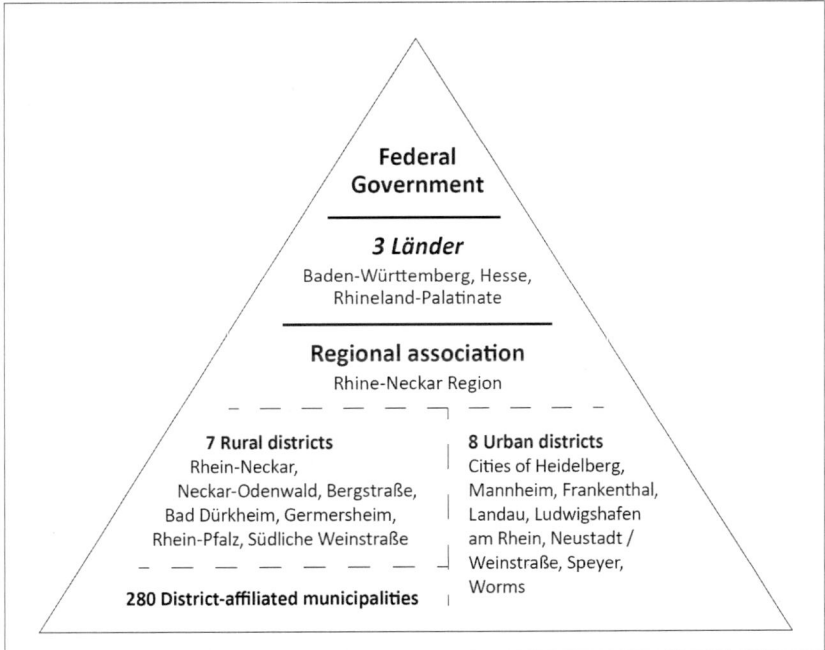

Figure 5.2: Territorial organization of Rhine-Neckar Region
Source: Author.

The cross-border inter-municipal association corresponds to regional associations (Regionalverbände) in Baden-Württemberg and to regional planning communities (Regionale Planungsgemeinschaften) in Rhineland-Palatinate. The origins of transboundary regional planning in the metropolitan region trace back to an inter-municipal working group created in 1951.[83] In 1969, a first interstate treaty between Baden-Württemberg, Hesse, and Rhineland-Palatinate established a regional umbrella organization for spatial planning.[84] Rhine-Neckar Region in its current form, i.e. territory, institutions and responsibilities, was set up by a second interstate treaty concluded in 2005.[85] As of January 1st, 2006, Rhine-Neckar Region succeeded three so far separate authorities: the previous regional umbrella organization for spatial planning, the regional planning community Rheinpfalz from Rhineland-Palatinate, and the regional association Rhein-Neckar-Odenwald.

The regional assembly (Verbandsversammlung) of Rhine-Neckar Region comprises the lord mayors of the urban districts, the district chief executives of the rural districts, the mayors of affiliated municipalities with more than 25,000 inhabitants, and additional delegates. They elect the honorary chairman (Verbandsvorsitzende*r), who also represents the regional association and heads its administration. In addition, the regional assembly determines a full-time regional director (Verbandsdirektor*in). According to art. 3 § 5 of the 2005 interstate treaty, Rhine-Neckar Region is in charge of economic development, landscape planning, transport and energy planning, events of regional significance, and tourism. The Metropolitan Region Rhine-Neckar Corporation (Metropolregion Rhein-Neckar GmbH, in the following referred to as metropolitan corporation) intervenes operationally in several fields including economic development, energy, and the environment.[86]

Although a joint Covenant initiative of Rhine-Neckar Region and the regional metropolis Heidelberg initially showed great promise, the number of signatories later stagnated and even decreased. The Covenant initiative fit well with the agenda of climate and energy policy at Rhine-Neckar Region and was ac-

---

83   Kommunale Arbeitsgemeinschaft Rhein-Neckar GmbH.
84   Raumordnungsverband Rhein-Neckar.
85   Staatsvertrag zwischen den Ländern Baden-Württemberg, Hessen und Rheinland-Pfalz über die Zusammenarbeit bei der Raumordnung und Weiterentwicklung im Rhein-Neckar-Gebiet, signed on July 26, 2005.
86   Stakeholders of the metropolitan corporation are Rhine-Neckar Region, a registered society for regional cooperation (*Zukunft Metropolregion Rhein-Neckar e.V.*) and several chambers.

tively supported by Heidelberg, a local climate policy pioneer and an early Covenant signatory (5.1.2.1). But local actors were soon disappointed with the workload of reporting and the lack of dedicated funding. Consequently, local climate policy hardly refers to the Covenant anymore (5.1.2.2). This development indicates that a momentary coincidence of policy goals between several levels of government might not translate into effective orchestration. Under certain circumstances, joint policy goals do not benefit from joint efforts. The case thus raises the question whether effective or unsuccessful orchestration should be assessed in terms of the orchestrator-intermediary relationship or in terms of the shared policy goals (5.1.2.3).

5.1.2.1 Teaming up for Local Climate Policy

In order to understand the approach of Rhine-Neckar Region to local climate policy coordination on behalf of the Covenant, it is important to understand the starting position of both the regional association and Heidelberg as its ally in Covenant promotion, an *"altogether lucky situation in society as a whole"* (interview with S. Dallinger, 29 January 2015).[87] At the time, Rhine-Neckar Region was in the process of preparing its regional energy plan (Regionales Energiekonzept). The Covenant seemed to represent a welcome tool for networking member municipalities and structuring a regional initiative for local climate and energy policy. Heidelberg had already signed the Covenant and took on a leading role in promoting its signature within the metropolitan region. Regional energy planning and the quest to conduct implementation at the municipal level coincided with the high-profile commitment of a local Lord Mayor for the Covenant. This situation brought about a meeting of the minds that resulted in a considerable number of Covenant signatories from Rhine-Neckar Region.

The origins of climate policy at Rhine-Neckar Region date back to a climate protection concept initiated in 2000. In order to develop this inter-municipal document, Rhine-Neckar Region hired a half-time officer in charge of climate policy for the first time; the seed of today's team working on climate and energy issues at Rhine-Neckar Region and at the metropolitan corporation. Albeit adopted by the regional assembly, the climate protection concept was barely implemented. Taking stock of emissions rather than prescribing detailed actions, but also for lack of political priority, the climate mitigation concept fell into oblivion (inter-

---

87   Author's translation; original quote: *"glückliche gesamtgesellschaftliche Situation"*.

view with J. Eustachi and A. Finger, 3 February 2015). Today, climate and energy policy in Rhine-Neckar Region builds on the constituent interstate treaty of 2005. Art. 3 § 5 n° 3 assigns the regional association with a coordinating role in transport planning, transportation management, and energy supply based on regional development plans. In addition, Rhine-Neckar Region is in charge of regional planning for wind energy development. When assigned with these new responsibilities, Rhine-Neckar Region adopted a mandatory wind energy concept, introduced voluntary measures such as a renewable energies concept and a biomass management scheme, and established a regional atlas of public and private climate protection projects (MVV Energie AG 2007). The new responsibilities are now carried out jointly by the regional association itself, the metropolitan corporation, and the registered society StoREgio.

The main climate and energy policy instrument of the regional association consists of its regional energy plan. The metropolitan corporation has a special department for energy and environment that deals in particular with energy efficiency and renewable energy.[88] It supports the Cluster Network Energy and Environment, which connects public, economic and academic actors from the local to the state level with networking activities.[89] The registered society StoREgio deals with the development of smart grids, a subarea of the regional energy plan.[90]

The strong roles of the metropolitan corporation and of business-oriented programs within inter-municipal policies indicate the importance of economic development for inter-municipal climate and energy governance. Rhine-Neckar Region perceives the energy transition as a challenge for economic development and aims to support local businesses and municipalities in seizing economic opportunities (Verband Region Rhein-Neckar 2016e).

Although climate and energy considerations play a role in regional planning more generally, Rhine-Neckar Region did not engage actively in local climate policy coordination before the development of its regional energy plan. Only then was the regional association given a coordinating role by the regional assembly

---

88   Fachbereich Energie und Umwelt.
89   Metropolitan support for the *Clusternetzwerk „Energie und Umwelt"* includes the yearly Regional Conference Energy and Environment (*Regionalkonferenz Energie und Umwelt*) and the business meeting Energy Forum Rhine-Neckar (*Energieforum Rhein-Neckar*).
90   *Cluster-Verein StoREgio Energiespeichersysteme e.V.* For a detailed overview of their climate and energy activities, see Verband Region Rhein-Neckar (2016a).

and engaged in the accompaniment of municipal measures, (model) projects,[91] and monitoring.[92] Commissioned in 2009[93] and adopted by the regional assembly in 2012, the regional energy plan aims to pursue energy security, affordable energy and environmental protection at the same time – which is fully in line with the objectives of the Covenant. The regional energy plan lists recommendations for 75 measures towards the 2020 objectives with regard to energy efficiency, renewable energies and their integration to the grid, and transport (Verband Region Rhein-Neckar 2012b). It referred to Covenant promotion and assistance to member municipalities as priority measures for making the Rhine-Neckar Region a pioneer in sustainable energy policy (Verband Region Rhein-Neckar 2012b, 177).[94] In 2013, Rhine Neckar Region hired a regional climate manager with the help of funding under the Local Authority Guideline. She is in charge of managing the implementation of measures from the regional energy plan, networking regional actors, and public communication (Verband Region Rhein-Neckar 2016c).

A challenge for Rhine-Neckar Region results from the heterogeneity of the metropolitan region in terms of municipal size, prior experience in local climate and energy policy, and the mayor's political preferences. These produce diverging municipal resources and capacities, and different degrees of institutionalization of local climate and energy policy. Some municipalities have specialized climate and energy departments within their administrations; others don't. Regional frameworks for local climate and energy policy vary across the *Länder* involved,[95] and as a result, inter-municipal coordination plays a different role in

---

91 Model projects turned out to be difficult to implement because inter-municipal funds, i.e. contributions of member municipalities, would have to be allocated to individual municipalities (interview with J. Eustachi and A. Finger, 3 February 2015).

92 This mission applies in particular to coordinating member municipalities while other, namely economic actors from the metropolitan region are rather addressed by the metropolitan corporation.

93 To the regional consultancy and engineering company *Zentrum für rationelle Energieanwendung und Umwelt* (ZREU).

94 For instance, Rhine-Neckar Region hosts an inter-municipal working group of municipal energy consultants and advise municipalities that are considering re-municipalizing local energy providers or cooperating with neighboring municipalities in wind energy development.

95 Differences include state-level renewable energy policy targets and the landscape of local climate and energy agencies as providers of technical knowledge and practical services for municipal policies. Municipalities from the Rhine-Neckar Region located in Baden-Württemberg have local energy agencies at the district level at their disposal; the Bergstraße district in Hesse also has its own energy agency, while municipalities in Rhineland-Palatinate have to rely on regional offices of the state-level energy agency (*Energieagentur Rheinland-Pfalz*).

the various member municipalities. For instance, smaller municipalities are very interested in regional training and events with clear take-home messages for implementation (Schneider and Eustachi 2014) while interviewees from bigger municipalities found them redundant.

Within the landscape of municipal climate policy in Rhine-Neckar Region, one city stands out as a local pioneer. Heidelberg, rather than being steered by inter-municipal coordination, was an ally of Rhine-Neckar Region with its attempt to establish a regional network of Covenant signatories. Heidelberg proudly highlights that it belongs to the very first signatories of the Covenant, and won further Covenant signatures within the Rhine-Neckar Region (Stadt Heidelberg 2011; 2016a). This leadership role can be explained by comprehensive prior experience in local climate policy and the personal connection of Heidelberg's Lord Mayor to the Covenant.

Heidelberg has a longstanding record in local climate policy that combines exemplary policies with public communication and presents climate mitigation as a means to reduce public spending and obtain additional funds. In 1991, the municipal council decided to adopt the national emission reduction target of 25 % by 2005 (Kern et al. 2005, 51). In the following year, the municipal council approved a climate protection concept with a reduced emission reduction target of 20% and assigned the municipal environmental department (*Umweltamt*) with energy management for municipal buildings (Deutsche Gesellschaft für Internationale Zusammenarbeit 2013, 13). In this way, the municipal council accounted for the limited grasp of the local government on most of the emissions on its territory as asserted in a contract study on municipal climate mitigation potentials by an external research institute, the Institute for Energy and Environmental Research (ifeu), in 1991. Heidelberg aimed for exemplary policies with regard to its own emissions so as to win over actors from other sectors (Kern et al. 2005, 51). The initial climate protection concept was updated in 2004 and 2014 and now aims for climate neutrality by 2050 (ifeu 2014; Stadt Heidelberg 2006). Heidelberg won several international awards for its climate policy and in 2015 was decorated as Global Green City by the United Nations.[96] At the national level, the city finished third in its category in 2006 and 2010 in the competition Climate Protection Capital.

---

96   Including climate policy awards from ICLEI in 1997, from the Climate Alliance in 2002, and
      from the Climate Group in 2005.

From the beginning, transnational municipal climate networks played an important role in Heidelberg's approach to local climate policy. The city was among the first members of the Climate Alliance and Energy Cities, two networks that belong to the consortium running the CoMO on behalf of the European Commission, and is represented at the presidency of Energy Cities since 2004 (Kern et al. 2005, 58). This was also due to personal leadership by Beate Weber, Heidelberg's social-democrat Lord Mayor from 1990 to 2006 (Collier 1997).

A case study on the potentials of local climate policy in Germany found Heidelberg to benefit from network membership not only in terms of exchange with peers, but also with regard to access to the European level and to project acquisition (Kern et al. 2005, 59). Networks connect Heidelberg with partners for project applications, and provide a direct channel for lobbying that enables the city to influence funding programs towards favorable terms. The acquisition of third-party funds is a key argument for the justification of network membership towards the municipal council. Summing up, Heidelberg's approach to local climate policy combines exemplary policies with public communication and perceives both climate mitigation measures and transnational municipal climate networks as economic opportunities.

Eckart Würzner, Heidelberg's Lord Mayor since 2006, engaged in the creation of the Covenant. A non-partisan politician, he was involved in climate and energy policy prior to holding this office, for instance at the Association of Cities of Baden-Württemberg and at the energy efficiency agency Rhine-Neckar. A member of the board of Energy Cities since 1994, he became the network's president in 2006 and participated in the lobbying efforts of networks of local and regional authorities at the European level for the recognition of municipal contributions to climate mitigation, namely the creation of the Covenant (see section 3.2.1.2) (Kehrl 2009; Presse- und Informationsstelle Viernheim 2010). It was a logical step for Würzner to sign the Covenant on behalf of Heidelberg as one of its first signatories.

Würzner also had strong ties within Rhine-Neckar Region. As Lord Mayor of a regional metropolis, he was endowed with several regional offices and functions such as membership at the board of the metropolitan registered society and the patronage of the metropolitan Cluster Energy and Environment. According to official communiques, these offices enable Würzner to connect with the European level (Kehrl 2009; Presse- und Informationsstelle Viernheim 2010). His proximity to climate and energy issues and his offices at the municipal, regional and international levels made Würzner a natural leader for a regional initiative promoting municipal climate action.

5.1.2.2 Disappointed Hopes in the Covenant of Mayors

For Rhine-Neckar Region, the Covenant represented a means to engage the mayors of member municipalities in voluntary climate action so as to back up its energy concept. For Würzner, the inter-municipal Covenant initiative allowed for locally communicating his engagement with Energy Cities, and for further strengthening the Covenant project. From the perspective of Heidelberg's climate department, the inter-municipal Covenant initiative promised to establish a regional network of like-minded municipalities. Indeed, the joint initiative of Rhine-Neckar Region and Heidelberg convinced 14 municipalities from the metropolitan region to follow the example of Heidelberg and become Covenant signatories. The Covenant seemed to be the right tool at an opportune moment and helped to kick off local climate policy coordination.

Five years later, the situation had changed dramatically. The Covenant and related activities were not even mentioned in a progress report on implementation of the regional energy plan although the latter had explicitly referred to it (Verband Region Rhein-Neckar 2015a). How could this be? A majority of actors involved found the Covenant did not live up to their original expectations, and disappointedly turned away from it. Municipalities – especially smaller ones – were overstrained by Covenant reporting obligations. Soon, many were suspended, and others excluded for not submitting the required documents. In response, Rhine-Neckar Region and Heidelberg stopped promoting the Covenant within the metropolitan region and shifted their engagement to other fora.

The possibility for Rhine-Neckar Region to become a supporting structure to the Covenant was first raised in 2009. The officer for European affairs, Jörg Saalbach, had stumbled upon the initiative repeatedly, including in CoR publications, and felt it fit well with the profile and policies of Rhine-Neckar Region (interview with J. Saalbach, 20 January 2015). Arguing that a commitment to the Covenant was in line with regional energy policy in general and the Cluster Energy and Environment in particular, he raised the issue within the administration. A Covenant commitment by Rhine-Neckar Region quickly obtained the support of regional director Stefan Dallinger, a Christian-Democrat, who said that he personally supported inter-municipal climate action as a means of economic development and quality of life (interview with S. Dallinger, 2015, January 29).

Heidelberg had already signed the Covenant and urged Rhine-Neckar Region to take on a coordinating role (interview with J. Eustachi and A. Finger, 3 February 2015). Heidelberg was hoping to achieve a network of neighboring Covenant signatories for enhanced cooperation in local climate policy, including joint projects (interview with J. Saalbach, 20 January 2015). Rhine-Neckar Region, too, aimed to win over as many municipalities as possible for Covenant signature (Kehrl 2009). Dallinger recalled ambitious objectives: "*Of course we lined up to become Europe's strongest region [within the Covenant of Mayors]*" (interview with S. Dallinger, 29 January 2015).[97]

Another recurrent argument for Covenant signature raised by Würzner consisted of additional funding opportunities for local climate policy. Official publications repeatedly cited him emphasizing that Covenant signatories, with the help of Rhine-Neckar Region as a supporting structure, could submit joint project applications so as to gain additional EU funds, to increase their role in European politics, and to establish the metropolitan region at the European level as a pioneer in sustainable energy (Kehrl 2009; Presse- und Informationsstelle Viernheim 2010; Stadt Heidelberg 2011).

Much of these persuasion efforts happened at the regional assembly – where many mayors hold an additional office – on the occasion of discussions on the regional energy plan, and through personal contacts, Dallinger said:

> "At the time, climate mitigation was not on top of the political agenda and one really had to take care to win people of whom one knew that they personally bear upon this issue and not only do it pro forma, but also implement projects at the municipal level afterwards." (interview with S. Dallinger, 29 January 2015).[98]

On the joint initiative of Dallinger and Würzner, mayors and representatives of eight municipalities from Rhine-Neckar Region gathered in Heidelberg on October 2, 2009 for a regional kick-off meeting for concerted climate policy. Würzner called for Rhine-Neckar Region to become a supporting structure to the Cove-

---

97 Author's translation; original quote: "*Wir sind natürlich schon angetreten: Wir wollen quasi die stärkste Region Europas werden.*"

98 Author's translation; original quote: "*Damals stand Klimaschutz nicht so weit oben auf der politischen Agenda und da musste man schon schauen, dass man Menschen gewinnt, wo man – von denen man weiß, die haben einen persönlichen Bezug zu diesem Thema und machen das nicht nur der Form halber, sondern setzen dann auf der kommunalen Ebene auch Projekte um.*"

nant that "would provide strategic, technical and financial guidance for the municipalities of the Metropolitan Region Rhein-Neckar" (Eisermann 2009). Rhine-Neckar Region agreed to do so (Covenant of Mayors Office 2009, 7). Seven municipalities expressed their intention of signing the Covenant (Kehrl 2009; Presse- und Informationsstelle Viernheim 2010).

Würzner and Dallinger convened a local climate conference in Heidelberg on April 16, 2010. About 40 municipalities attended the event (Verband Region Rhein-Neckar 2012b, 40). The regional initiative aimed to support the implementation of the forthcoming regional energy plan, to attract energy businesses, to promote renewable energy development, and to develop tools for GHG emissions monitoring (Presse- und Informationsstelle Viernheim 2010). On the occasion, Rhine-Neckar Region signed a contract to officially become a supporting structure to the Covenant (Verband Region Rhein-Neckar 2012b, 40). Eleven municipalities signed the Covenant and were welcomed on behalf of the European Commission by Pedro Ballesteros Torres, principal administrator at DG TREN (Presse- und Informationsstelle Viernheim 2010).[99] Dallinger stated: "The commitment of municipalities demonstrates that climate protection concerns everyone and that the metropolitan region acts concertedly" (Stadt Mannheim 2010, 1).[100] Sankt Leon-Rot, Ludwigshafen, and Walldorf signed the Covenant successively before the end of 2011 (COOPENERGY 2014, 2).

Covenant signatories expected Rhine Neckar Region to support municipal climate policy by means of technical assistance to SEAP development, advice on implementation, and information and help on the acquisition of third-party funding for the respective measures (Gemeinde Böhl-Iggelheim 2011, 15). Rhine-Neckar Region intended to coordinate local climate policy in strict compliance with the repartition of competences between municipalities and their regional association; i.e. it did not aim to intervene operationally at all (interview with S. Dallinger, 29 January 2015). Its capacities, in particular in terms of human resources, were strictly limited to initial consultancy and events convention (interview with J. Eustachi and A. Finger, 3 February 2015). Covenant coordination was not endowed with a particular budget. Instead, it took

---

99   Mannheim, Viernheim, Heppenheim, Worms, Frankenthal, Wiesloch, Eppelheim, Rimbach, Böhl-Iggelheim, Landau and Limburgerhof.

100 Author's translation; original quote: *"Das Engagement der Kommunen beweist, dass Klimaschutz alle angeht und die Metropolregion gemeinsam handelt."*

"the form of: organising events on the topic of energy transition, setting-up of a working group composed of local climate protection managers, [...] costs of events [being] covered by the participating parties" (COOPENERGY 2014, 1–2).

Up to 2011, Rhine-Neckar Region held a series of events for Covenant signatories and other interested municipalities to inform and exchange on how to establish BEIs, and how to develop SEAPs (interview with J. Eustachi and A. Finger, 3 February 2015). The working group served to support municipal SEAPs through mutual updates on progress made and exchange on experiences.[101]

Once the pioneering spirit of the early days was gone, the Covenant faded from the limelight to a certain extent. Würzner's suggestion on the occasion of the 2010 climate conference to convene yearly meetings of mayors from Rhine-Neckar Region in order to initiate joint climate protection measures was not implemented (Presse- und Informationsstelle Viernheim 2010). Dallinger already knew that he was about to switch offices to become district chief executive of the Rhein-Neckar district. He did not launch any particular projects based on the input from the Covenant signature ceremony but left the field to his successor Ralph Schlusche (interview with S. Dallinger, 29 January 2015). Under Schlusche's directorate, the Covenant continued to figure in official documents such as the regional energy plan but he did not take up Rhine-Neckar Region's Covenant commitment personally.

Local climate policy coordination remained on the inter-municipal agenda because of its institutionalization in the form of the regional climate manager, implementation requirements of the regional energy plan, and ongoing projects. No meetings dealing explicitly with Covenant proceedings were held since 2012 because the requirements of BEIs and SEAPs had been exhaustively discussed (interview with J. Eustachi and A. Finger, 3 February 2015), but new events series and forms of support followed.

In order to learn about the state of local sustainable energy policy and the respective needs for support, the regional climate manager conducted an online survey among member municipalities in summer 2013 (Schneider and Eustachi 2014). Nearly half of the respondents said they did not know the regional energy plan, and had not engaged in developing a local energy and climate concept either. The main reason given for this was a lack of resources in terms of available

---

101 An example of the importance attached to it by participating municipalities consists in the reference made to the inter-municipal working group in the SEAP of Böhl-Iggelheim as an accompanying measure to support implementation (Gemeinde Böhl-Iggelheim 2011).

staff, knowledge, and funding. Asked how Rhine-Neckar Region could support municipalities in developing energy and climate concepts, respondents primarily pointed to the facilitation of knowledge transfer and exchange of experiences through regional networks, information exchange, and events. Rhine-Neckar Region initiated several series of events on municipal climate planning and action for reciprocal exchange and adjustment, including working groups of municipal climate managers and on municipal energy management (*Arbeitskreis Kommunale Klimamanager, Arbeitskreis Kommunales Energiemanagement*). Events address municipal staff and aim to facilitate regular topical knowledge exchange (Verband Region Rhein-Neckar 2016d):

> "It's about [...] preparing topics together and looking for how to implement them at home, coming to the next meeting and saying, ' I had problems here and there, how have you done it', to simply make use of the exchange, to seize synergy effects, in order to make progress within the municipality" (interview with J. Eustachi and A. Finger, 3 February 2015).[102]

Rhine-Neckar Region participated in the EU-project COOPENERGY funded by the Intelligent Energy Europe Program from April 2013 to April 2016. COOPENERGY aimed to

> "support the development of more effective collaborative working between regional and local public authorities to maximise positive energy planning outcomes and accelerate reduction in greenhouse gases" and to avoid that "Sustainable Energy Action Plans at a regional and local level are [.] developed in isolation of each other [.] [which] can lead to duplication, wasted resources and missed opportunities to save time, money and minimise negative impacts" (COOPENERGY 2016b).

The project proposed a set of measures including the promotion of informal or contractual multi-level governance agreements between local and regional authorities and of "key European policies and initiatives (such as the Covenant of Mayors) that can help public authorities plan sustainable energy projects and actions in partnership" (COOPENERGY 2016a). Initially, Rhine-Neckar Region explicitly welcomed this project grant as a means to further develop the regional network of Covenant signatories (Metropolregion Rhein-Neckar GmbH 2013,

---

102 Author's translation; original quote: *"Und da geht's jetzt darum [...] dass man dann gemeinsam auch die Themen erarbeitet und guckt, wie kann ich's daheim umsetzen, wieder zum nächsten Treffen kommt und sagt, ich hab da und da Probleme gehabt, wie habt's Ihr gemacht, einfach den Austausch auch zu nutzen, die Synergieeffekte zu nutzen, um dann auch in der Kommune weiterzukommen."*

4). But the project did not bring about additional Covenant signatories or an enhanced network among existing Covenant signatories.

The promotion of multi-level governance agreements was more successful. In view of achieving blanket coverage with municipal climate plans, Rhine-Neckar Region signed Memoranda of Understanding (*Kooperationsvereinbarungen*) with two rural districts and the near-totality of their affiliated municipalities. The first Memorandum of Understanding was signed between the Rhein-Neckar district, 52 out of 54 affiliated municipalities and the metropolitan corporation in April 2014 (KliBA 2014; Rhein-Neckar-Kreis 2016). It aimed to harmonize regional, district-level and municipal climate and energy concepts and to support cooperation between levels of government. Participating municipalities committed to establishing climate protection concepts until 2020. The Rhein-Neckar district engaged in providing municipalities with baseline energy and emission inventories and consecutive monitoring reports.[103] Two district climate managers took office in January 2015 and, among other tasks, support a network of affiliated municipalities. A second Memorandum of Understanding was signed in September 2015 between the metropolitan corporation, the Rhein-Pfalz district and its affiliated municipalities, and the energy agency of Rhineland-Palatinate. The parties agreed to synergize on monitoring, networking, public communication, and events, and to establish platforms for exchange and knowledge transfer in order to support municipal action (Verband Region Rhein-Neckar 2015).

Although the support provided by the two districts to affiliated municipalities would match the requirements for becoming Covenant Coordinators, neither of them undertook any steps in this direction. The joint submission of a sustainable energy action plan instead of individual municipal documents was not under discussion either, although some affiliated municipalities are remarkably small. The possibility to use the Covenant as an audit for a policy undertaken anyhow was totally neglected. This is all the more surprising as it was Dallinger, former regional director of Rhine-Neckar Region at the time of Covenant adhesion, who signed the Memorandum of Understanding on behalf of the Rhein-Neckar dis-

---

103 Data were compiled by the regional climate and energy agency KliBA with the help of the Heidelberg-based Institute for Energy and Environmental Research (ifeu) and the district-owned AVR Umweltservice GmbH. Energy and emission inventories are made publicly available at http://www.klimaschutz-rnk.de.

trict in his new office as district chief executive. He said he had simply forgotten about the Covenant:

> "To my shame I have to admit I had totally blanked on it. [...] Well, it is so quiet around the Covenant, one does not hear anything" (interview with S. Dallinger, 29 January 2015).[104]

Instead, the Rhein-Neckar district, just as several other municipalities from the region, engaged in an EEA procedure with the help of state-level funding. According to Dallinger, the EEA had the virtue to allow for 1. persuading members of their staff, 2. presenting and positioning the district with its strategic objectives as approved and financially endowed by the district assembly, 3. keeping the issue on the agenda of the district assembly, 4. mobilizing citizens, and 5. benchmarking so as to compare the district's performance with others.

As was repeatedly pointed out by Dallinger and in official publications, no other metropolitan region was better represented within the Covenant than Rhine-Neckar Region with its 15 Covenant signatories (Stadt Mannheim 2010; Verband Region Rhein-Neckar 2012b, 40). But many other Covenant Coordinators gathered larger numbers of Covenant signatories, also compared to the number of member municipalities.[105] Excessively so, Rhine-Neckar Region stopped promoting the Covenant among potential signatories. How could it come to this?

The regional energy plan still affirmed explicitly that Rhine-Neckar Region was committed to its Covenant Coordinator role (Verband Region Rhein-Neckar 2012b, 189). But the Covenant was not mentioned anymore in the shortened executive summary (Verband Region Rhein-Neckar 2012a). This might indicate already that despite the professed intention to further develop the number of, and the support for, Covenant signatories from the Rhine-Neckar Region, the Covenant commitment was no longer a priority.

No municipality from the Rhine-Neckar Region signed the Covenant later than 2011, and the Covenant was no longer dealt with in inter-municipal working group meetings and events after this time. The website profile of Rhine-Neckar Region as a Covenant Coordinator was never used to publish any benchmarks, and was last updated in November 2012. Moreover, the number of recognized Covenant signatories supported by Rhine-Neckar Region decreased from 15 to

---

104   Author's translation; original quote: *"Ich muss zu meiner Schande gestehen, ich hatte ihn dann auch wirklich ausgeblendet. [...] Also, es ist so ruhig um den Konvent, man hört auch nichts mehr."*

105   Examples include Rennes Metropolis (see section 5.2.3) and Val d'Ille (see section 5.2.4).

six. Four municipalities did not maintain the commitment, and five no longer refer to any support from the Rhine-Neckar Region as a Covenant Coordinator (Covenant of Mayors 2016d). These developments were closely linked to the workload of Covenant membership and the fragmented and heterogeneous municipal landscape.

As depicted above, the engagement of Rhine-Neckar Region as a Covenant Coordinator declined from the end of 2011. Interviewees said that Covenant proceedings had turned out to be very challenging, with extensive reporting obligations and the necessity to communicate in English (interview with J. Eustachi and A. Finger, 3 February 2015; interview with J. Saalbach, 20 January 2015). In contrast with former expectations, no EU funding schemes for Covenant signatories were set up to cover these efforts. Interviewees further reported that smaller municipalities in particular were overstrained.

The schedule of Covenant submissions was challenging, too. Municipalities that had done little prior preliminary work before struggled to develop a comprehensive concept within one year after Covenant signature.[106] At the same time, Covenant signatories then waited up to three years for their SEAP to be approved – a delay which seemed inappropriate in comparison to municipal obligations (interview with J. Eustachi and A. Finger, 3 February 2015).

In response to complaints from member municipalities, Rhine-Neckar Region put Covenant promotion on hold. The central questions for inter-municipal staff were:

"What are small municipalities most likely to do, what can we facilitate, and where are our working hours used most effectively?" (interview with J. Eustachi and A. Finger, 3 February 2015).[107]

In the attempt to better match municipal needs and provide more practical support, Rhine-Neckar Region set up new events series. Additionally, other local climate policy schemes came to the fore that were better suited to promote inter-municipal climate policy coordination COOPENERGY 2014, 2. Rhine-Neckar Region now promotes the development of municipal climate protection concepts

---

106 For instance, the Municipality of Böhl-Iggelheim listed the creation of a climate protection concept by an external agency in its SEAP as one of ten actions to be taken (Gemeinde Böhl-Iggelheim 2011, 10).

107 Author's translation; original quote: "*Was machen kleine Kommunen am ehesten, was können wir befördern, und wo ist unsere Arbeitszeit am effektivsten eingesetzt?*"

as funded under the Local Authority Guideline (interview with J. Eustachi and A. Finger, 3 February 2015).

Feedback on the difficulties of municipalities in coping with Covenant reporting obligations was transmitted to the Covenant via the European secretariat of the Climate Alliance in Frankfurt. Other than that, contacts or even exchanges with the Covenant were very much limited. Even when the regional climate manager Julia Eustachi took part in a poll among supporting structures in 2014, she never received any feedback on poll outcomes (interview with J. Eustachi and A. Finger, 3 February 2015). Another example of limited contacts between Rhine-Neckar Region and the Covenant Office consists in the fact that the profile of Rhine-Neckar Region refers to the officer for European affairs as the contact person. In actuality, Covenant activities are handled by the climate and energy officers who, in turn, said they had not been aware of this false information on the website (interview with J. Eustachi and A. Finger, 3 February 2015). All in all, the orchestrator-intermediary relationship between the CoMO and Rhine-Neckar Region remained markedly weak.

The withdrawal of Rhine-Neckar Region explains why no more Covenant signatories could be won after 2011, and mirrors the negative assessment of their Covenant experience by some municipalities that did not live up to their initial commitment and thus left the initiative. At the same time, the end of Covenant-specific offers to member municipalities of course weakened the ties to the remaining Covenant signatories to the point that some no longer refer to support by Rhine-Neckar Region in their Covenant profile.

As the shift to other schemes and the continued Covenant commitment of eleven municipalities from the metropolitan region indicate, the nominal Covenant membership of Rhine-Neckar Region does not imply a withdrawal of the regional association and its member municipalities from local climate policy per se. This also applies to small municipalities, the main complainers in the regional Covenant initiative. While municipal fragmentation turned out to hamper municipal capacities for Covenant reporting, this does not necessarily mean that it impedes local climate action, several interviewees emphasized (interview with S. Dallinger, 29 January 2015; interview with J. Eustachi and A. Finger, 3 February 2015). Much depends on local actors, namely on strong-willed mayors who prioritize climate mitigation– or not (interview with J. Eustachi and A. Finger, 3 February 2015).

Heidelberg, the former municipal leader of the regional Covenant initiative, and Würzner personally had their own unique experiences with the Covenant. With Dallinger's departure from the regional directorate, Würzner lost his high-level ally for Covenant promotion at the metropolitan level. Staff at Rhine-Neckar Region perceived the Covenant primarily as a facilitator for kicking off local climate action. Consequently, they neither used the Covenant as an actual network, nor its methodology for moving forward with local climate planning. Instead, they and their municipal counterparts within the metropolitan region either exchanged ideas among each other, or relied on prior network affiliations with the Climate Alliance or Energy Cities. Quickly, municipal climate protection concepts became the gold standard of municipal climate planning, while the development of SEAPs based on Covenant forms constituted an additional administrative task rather than an actual tool for municipal policy. In sum, a regional network of Covenant signatories as Heidelberg had hoped for did not realize.

From the beginning, Rhine-Neckar Region did not share Würzner's approach to the Covenant as a channel for influence on EU policies or as a venue for lobbying for an increased role of local governments in European governance (interview with S. Dallinger, 29 January 2015; interview with J. Saalbach, 20 January 2015). Consequently, Rhine-Neckar Region and its regional director did not play an active role in the Covenant Club Germany. Würzner, in contrast, acts as one of three spokesmen and actively engaged in the creation phase of the Covenant Club Germany. The latter, according to Heidelberg's website, aims to

"better network German cities and municipalities and increasingly publicize their activities to the public at the national and European level. As a platform, the Club aims to strengthen exchange between cities' Lord Mayors, municipal umbrella organizations, federal ministries, and the EU" (Stadt Heidelberg 2016a).[108]

Over time, the Covenant Club Germany has tended to be sidelined by other national fora. Meetings are scheduled as side-events of larger local climate policy gatherings, for instance the annual Climate Alliance conference. While the first meetings of the Covenant Club Germany were attended by high-level political

---

108   Author's translation; original quote: *"Ziel ist es, die deutschen Städte und Gemeinden besser miteinander zu vernetzen und ihre Aktivitäten auf nationaler und europäischer Ebene verstärkt in der Öffentlichkeit bekannt zu machen. Als Plattform will der Club den Austausch zwischen den Oberbürgermeistern der Städte und den kommunalen Spitzenverbänden, Bundesministerien und der EU verstärken."*

representatives such as mayors, it is now covered mostly by technical staff, and only brings together a minority of German Covenant signatories. Würzner, too, no longer attends in person; Heidelberg is represented by staff from the environmental department instead.

In his function as Energy Cities' president, Würzner continues to tend to the Covenant. He attends the yearly Covenant ceremonies in Brussels sitting in the front row, and participates in strategic discussions over Covenant development at the European level. Repeatedly, he seized the opportunity to publicize Heidelberg's projects for climate mitigation. In contrast, most mayors only participate in the yearly Covenant ceremonies in Brussels once – when they go to sign the Covenant on behalf of their municipality. Even for the signing occasion, some municipalities are represented by administrative staff only, as was the case with Mannheim.

The Covenant remains on the municipal agenda in Heidelberg. Heidelberg's municipal council unanimously decided to commit to the updated targets of the Covenant for 2050 on October 27, 2016 (Stadt Heidelberg 2016b). Heidelberg has not made active use of the Covenant beyond the auditing of existing policies which, due to Würzner's personal involvement and Heidelberg's longstanding record in climate policy, are effectively a matter of course.

5.1.2.3   Case Summary

At the time of Covenant creation, Rhine-Neckar Region was in the process of developing its regional energy plan. The regional association was already on the lookout for tools that could support future implementation. Regional director Dallinger thus seized the opportunity when he came across the possibility of teaming up with Heidelberg's Lord Mayor Würzner in promoting Covenant signature within the metropolitan region. A preparatory meeting in 2009 and a regional climate conference in 2010, both hosted by Heidelberg, became the starting point of the regional Covenant initiative. Rhine-Neckar Region gathered a peak of 15 Covenant signatories. An events series provided signatories with information and opportunities to exchange on BEIs and SEAPs for Covenant reporting.

In the process, participating municipalities complained about the workload of reporting and the absence of corresponding funding schemes. Rhine-Neckar Region therefore stopped promoting further Covenant membership as of 2012, shifted its activities to other schemes and set up new events in order to better meet municipal needs. While a certain level of municipal mobilization was achieved, Rhine-Neckar Region struggled to assert itself as a leader in metropolitan climate and energy governance due to limited resources, the non-mandatory character of its regional energy plan, and a plethora of competing public and private actors within the metropolitan region.

Heidelberg, Rhine-Neckar Region's ally in regional Covenant promotion, had a slightly different, but also mixed experience with the Covenant. Continued and reaffirmed Covenant membership was a matter of course for Heidelberg due to Würzner's personal involvement and the city's ambition to cover all relevant networks, but no tangible value added could be realized.

Summing up, the Covenant was a welcome tool for kicking off inter-municipal climate policy coordination, but could not maintain a lasting role in metropolitan climate and energy governance. Reference to the Covenant helped to synchronize the activities of Rhine-Neckar Region and the regional metropolis Heidelberg and European endorsement further legitimized the inter-municipal initiative for coordinating municipal climate policy. But municipalities as target actors became displeased with their Covenant experience to the point that Rhine-Neckar Region had to stop Covenant promotion in order not to endanger its credibility in metropolitan climate governance. The orchestrator-intermediary relationship was not formally ended, but silently put on hold.

With regard to the orchestrator-intermediary model, the case of Rhine-Neckar Region shows that shared policy goals are sometimes hard to pursue jointly over time by an orchestrator and its intermediary. Quite the contrary, diverging requests of stakeholders can weaken the orchestrator-intermediary relationship to the point that going separate ways seems more promising despite an initially favorable dynamic across multiple levels of governance. At the same time, the case of Rhine-Neckar Region and its regional leader Heidelberg shows that orchestration – for all participating actors to save face – is likely to wither away rather than to be officially ended.

Also, the case of Rhine-Neckar Region points us to the fact that joint policy goals need not be best pursued jointly. A close look at instances of orchestration is needed to determine whether it is, and continues to be, effective. At the same time, the case of Rhine-Neckar Region raises the question of what effective implementation with regard to orchestration really means. A close orchestrator-intermediary relationship and the achievement of the policy goals they share need not coincide. Under certain circumstances, policy goals are better served with a weak orchestrator-intermediary relationship, and other cases could be imagined where an intense orchestrator-intermediary relationship fails to achieve the policy goals it is intended to support.

5.1.3 Stuttgart Region

The case of the Association of the Greater Stuttgart Region (*Verband Region Stuttgart*, in the following referred to as Stuttgart Region) points to the importance of fit between municipal needs, inter-municipal coordination, and superordinate support for successful orchestration. Mere coincidence of policy goals alone is insufficient to foster an intense orchestrator-intermediary relationship.

Stuttgart Region is situated in the heart of the *Land* of Baden-Württemberg within the bigger Stuttgart Metropolitan Region.[109] Stuttgart Region houses a quarter of the population of Baden-Württemberg on 10% of its territory and generates a third of its gross national product (GDP) (Region Stuttgart 2009a). Out of 2.7 million inhabitants of Stuttgart Region, about 600.000 live in the capital city (*Landeshauptstadt*) of Stuttgart, center city of the metropolitan region and the seat of the regional association and of Baden-Württemberg. Other than Stuttgart city, Stuttgart Region comprises another 178 municipalities that belong to five districts (*Landkreise*): the districts of Böblingen, Esslingen, Göppingen, Ludwigsburg, and the Rems-Murr district (see figure 5.3).

Stuttgart Region is one of twelve regional associations (*Regionalverbände*) in Baden-Württemberg that are in charge of spatial planning (see section 5.1.1.3). Stuttgart Region is the only regional association in Baden-Württemberg that is not only responsible for spatial planning, but also for regional transport planning and public transport, landscape planning, and for regional economic development.[110] Economic development is carried out by a subsidiary company, the

---

109 The Stuttgart Metropolitan Region is one of eleven European Metropolitan Regions within Germany as defined by the Conference of Minister for Spatial Planning (*Ministerkonferenz für Raumordnung*) in 1995. Depending on the definition of its territory, the Stuttgart Metropolitan Region is home to up to 5.3 million inhabitants Initiativkreis Europäische Metropolregionen in Deutschland (2016).

110 Stuttgart Region also partly assumes waste management and can engage voluntarily in additional fields such as sports and culture, see Act on the Establishment of the Association of the Greater Stuttgart Region (*Gesetz über die Errichtung des Verbands Region Stuttgart*) as of February 7, 1994.

Stuttgart Region Economic Development Corporation (*Wirtschaftsförderung Region Stuttgart GmbH*, EDC).

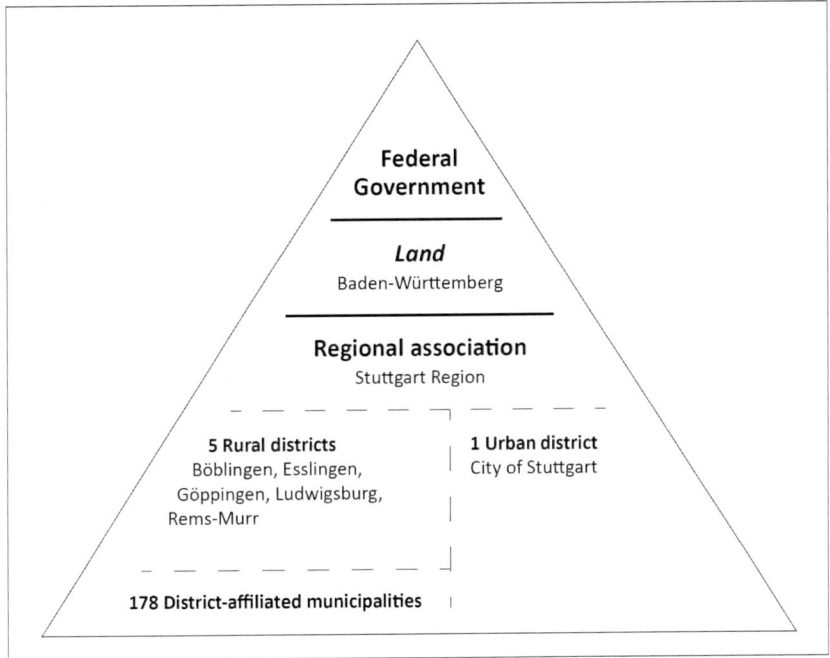

Figure 5.3: Territorial organization of Stuttgart Region
Source: Author.

As an additional particularity, the regional assembly (*Regionalversammlung*) is directly elected, which contributes to politicizing inter-municipal decision-making, but also enhances its legitimacy. Consisting of about 80 members who are elected every five years, the regional assembly chooses an honorary chairman (*Verbandsvorsitzende/r*), an office held by Christian Democrat Thomas S. Bopp (CDU) since 2007.[111] The regional assembly also elects a full-time regional director (*Regionaldirektor/in*) who heads the administration with about 60 employ-

---

111 Thomas S. Bopp (*Christlich Demokratische Union*, CDU) is a member of the regional assembly since 1994. From 2008 to 2011, he was a member of the state parliament of Baden-Württemberg.

ees seated in Stuttgart city, and is responsible for the implementation of the decisions taken by the regional assembly.

Stuttgart Region is both particularly vulnerable to global warming in terms
of heat and flooding due to its geographical setup, and a considerable emitter of
greenhouse gases (GHG) because of its economic structure, notably its important
car industry. Consequently, it has of considerable experience in climate adaptation and actively engages in sustainable mobility in terms of planning, local
transport, and economic development. European leadership, also in climate policy, is part of Stuttgart Region's self-perception. Its benchmarks listed on the
Coordinator profile on the website of the Covenant cover: 1. a regional Climate
Atlas for adaptation and planning purposes, 2. yearly panels on climate mitigation organized by the EDC, and 3. the development of an inventory and scenario
tool for metropolitan regions within $EUCO_2$ 80/50, an EU-project that dealt with
strategies for metropolitan regions to reduce their CO2 emissions by 80% until
2050 (Covenant of Mayors 2015d).

Despite the apparent match of policy goals with the Covenant, out of 179
member municipalities of Stuttgart Region only the three biggest cities became
Covenant signatories: Ludwigsburg and Esslingen, center cities of the districts
of the same name, and the urban district of Stuttgart city. This restricted turnout
raises the question why Stuttgart Region was not able or willing to translate the
shared policy goals into active local coordination among smaller member municipalities on behalf of the Covenant. A closer look at the experiences of Stuttgart
Region as a Covenant Coordinator reveals a limited fit of the support offered
both by the Covenant and Stuttgart Region as its intermediary for member municipalities and by the Covenant for the regional association itself. In sum, these
observations indicate that a coincidence of policy goals alone does not translate
into effective orchestration relationships between municipal targets, inter-municipal intermediaries and superordinate schemes as long as tangible support is lacking (5.1.3.3).

5.1.3.1 Matching Goals of Local Climate Policy

In terms of climate change, Stuttgart Region can be described as a victim and
polluter at the same time. On the one hand, Stuttgart Region is particularly vulnerable to climate change. Situated along the Neckar river valley and surrounded
by mountain ranges, Stuttgart Region belongs to the warmest areas in Germany
and is particularly poor in wind. Climate models predict an above-average rise

of temperatures, more frequent flooding, changes in flora, and increased erosion (Stadt Stuttgart 2013, 10). Climate change induces risks in particular for urban areas, agriculture, forestry, and along river basins, for instance in terms of urban heat, water scarcity, and economic risks such as production downtimes, navigability, and cooling of power plants (BMVBS 2013). On the other hand, Stuttgart Region, home to the automotive manufacturers Mercedes and Porsche, heavily depends on its car industry and exhibits elevated per-capita GHG emissions, also due to commuting within the metropolitan area. Each inhabitant of Stuttgart Region accounts for 16 t of CO_2 emissions per year, compared to about 11 t on German average (Eltrop 2011, 28; European Environmental Agency 2012, 114). Consequently, local priorities oscillate between climate protection and economic development, with sustainable mobility as a potential conciliatory resort. Stuttgart Region describes its strategy for climate and energy policy as consisting of

> "measures to reduce traffic needs (e.g. concentration of settlements along railroad lines), individual traffic behaviour (e.g. e-mobility), measures for infrastructural planning authorities (e.g. a digital clima [sic] atlas) and measures to promote renewable energy technologies" (Covenant of Mayors 2015e).

Stuttgart Region thus seems a natural match for the Covenant in terms of shared policy goals.

Climate policy expertise in Stuttgart Region was developed first and foremost in Stuttgart city. Especially vulnerable to urban heat due to its geographical situation within a caldera, Stuttgart city has a longstanding history in urban climate policy already *"before one knew how to spell climate change"* (interview with U. Reuter, 28 January 2015).[112] The preservation of fresh-air flows was on the agenda of urban policy since the 17th century, and the city disposes of a climatology service (*Abteilung Stadtklimatologie*) which dates back to 1938 (Stadt Stuttgart 2016a). This unit is located within the environmental department (*Umweltamt*) and conducts observations of the urban climate, and establishes models and scenarios in order to inform local planning. According to the department head of urban planning, the involvement of the municipal climatology service is a matter of course for any planning in Stuttgart city due to the service's excellent

---

112 Author's translation, original quote: *"Das hängt damit zusammen, dass diese Arbeitsfelder in [der Stadt] Stuttgart schon lange Thema waren, oder eigentlich schon immer Thema waren, bevor man überhaupt wusste, wie man Klimawandel buchstabiert."*

reputation and its extensive fundamental spadework:[113] *"Hence we as planners need not do anything without having asked the climatologists"* (interview with R. Schulze-Dieckhoff, B. Steinerstauch and R. Kapp, 19 July 2012).[114]

Expertise on the integration of climate considerations into urban planning was summarized in an urban climate reader published in 1999 and updated in 2012. Regional cooperation added to the body of knowledge in Stuttgart city and for neighboring municipalities with fewer resources for climate planning. The municipal climatology service developed a first version of the regional Climate Atlas (*Klimaatlas*) in 1992. The Climate Atlas provides detailed information on local climate conditions for planning purposes by means of a geographical information system not only for Stuttgart city, but also for other municipalities within the region. In this way, the expertise and resources of the center city with its dedicated climatology service are made available to neighboring municipalities. In 2008, an update of the Climate Atlas – established again by Stuttgart city's climatology service, but this time on behalf of Stuttgart Region – was extended so as to encompass the entire territory of Stuttgart Region. This database informed the regional plan adopted in 2009, in particular its targets with regard to green strips (Region Stuttgart 2009a, 10). The regional plan calls for the preservation of free spaces, flood prevention, and the development of so-called green infrastructure within the context of landscape planning.

This is not to say that neighbouring municipalities necessarily share the approach of the centre city Stuttgart, for instance to favour redensification over land consumption. Wolfgang Schuster, Stuttgart's Christian Democrat (CDU) mayor from 1997 to 2013, recalled regular conflicts over land-use planning in the regional assembly:

> "Well, all these measures [for reducing land consumption] naturally only make sense if also the neighbourhood around Stuttgart [city] joins in. Therefore there is this permanent conflict in the regional assembly because some neighbours then ask: When

---

113 Despite well-established working relations on the staff level, an interviewee insinuated that political decisions do not necessarily follow the advice of the municipal climatology service. Conflicts over the weighting of urban climate and other considerations arose for instance in the debates over Stuttgart 21, the reconstruction of Stuttgart's central railway station that was subject to a referendum after considerable public protests (Leggewie 2014; Vatter and Heidelberger 2013). The preservation of a ventilation corridor for fresh-air flow through the city center on the previous track field was part of the arbitration by mediator Heiner Geißler in 2010 (Geißler 2010).

114 Author's translation, original quote: "... *deswegen brauchen wir hier als Planer gar nichts zu machen, ohne die Klimatologen gefragt zu haben.*"

the Stuttgarters tighten their belts in terms of land consumption, then we can be all the more generous – with the logical effect that traffic flows grow accordingly and thus I have won little for the overall ecobalance" (interview with W. Schuster, 16 May 2013).[115]

Also, the above-mentioned documents established by the climatology service of Stuttgart city on behalf of Stuttgart Region should not induce that cooperation in local climate adaptation policy between the city and the regional association was uni-directional, with the experienced city administration as a service provider for Stuttgart Region and its member municipalities. Although the (optimization of the) urban climate by means of considerate planning had been on the agenda of urban policy for centuries, it was no matter of course to establish climate adaptation in a narrower sense within urban policy. Only when climate change arose on the agenda of UN negotiations and local authorities worldwide in the early 1990s did Stuttgart city, by and by, engage in climate adaptation. Initial policies served to adapt to a given climate, while recent policies also account for changes in climate. In this shift, cooperation with Stuttgart Region and inter-municipal planning helped to improve the availability of data and the framework for municipal planning. Noteworthy advances include AMICA (Adaptation and Mitigation – an Integrated Climate Policy Approach), an EU-funded project for an improved integration of climate considerations into regional planning running from 2005 to 2007, and the regional Climate Information System Stuttgart (*Klimainformationssystem Stuttgart*, KISS) developed within KlimaMORO, a pilot project on spatial development strategies in view of climate change funded by the BMVBS from 2009 to 2011 (BMVBS 2014, 62).

In 2012, Stuttgart city adopted a climate adaptation plan (*Klimaanpassungs-konzept Stuttgart*, KLIMAKS). Building on the national adaptation strategy, the climate adaptation plan lists and prioritizes measures to be taken in order to cope with certain climate risks and to enhance the adaptation capacities of the city as a whole (Stadt Stuttgart 2013). It was thus a natural step for Stuttgart city to also join Mayors Adapt in 2014, the climate adaptation spin-off of the Covenant that

---

115   Author's translation, original quote: "*Also all diese Maßnahmen [zur Nachverdichtung und zur Reduzierung des Flächenverbrauchs] machen natürlich nur Sinn, wenn auch die Nachbarschaft rings um [die Stadt] Stuttgart mitmacht. Deshalb gibt es ja auch dort den permanenten Konflikt in dieser Regionalversammlung, weil die Nachbarn zum Teil dann fragen (...?) wenn die Stuttgarter sich den Gürtel so eng schnallen, dass die, was den Flächenverbrauch angeht, dann können wir umso großzügiger sein, mit der logischen Folge, dass damit natürlich die Verkehrsströme wachsen und damit ich eben in der Bilanz, in der Ökobilanz natürlich wenig gewonnen habe.*"

required signatories to develop a comprehensive local adaptation strategy in view of contributing to the EU Adaptation Strategy.[116] In order to further facilitate the integration of climate adaptation considerations into construction planning and with financial support of the Ministry for Environment, Climate and Energy Economy of Baden-Württemberg, the municipal climatology service established the Climate planning pass Stuttgart (*Klimaplanungspass Stuttgart*, KlippS). Published in 2015, this database calculates the importance of selected urban areas for the urban climate and assesses the corresponding need for action in the context of construction planning (Mayer et al. 2015).

Not only do procedural preferences of Stuttgart Region coincide with Covenant proceedings. Policy goals with regard to climate mitigation, too, suggest a considerable overlap between Stuttgart Region and the Covenant. The regional association relies on formal and informal planning in order to ensure the climate-friendly development of regional land-use and transportation; in addition, exemplary projects and communication measures aim to shape the landscape and local transportation both directly and indirectly by targeting member municipalities (Lang 2011, 22). Regional climate policy thus defines a framework for local climate action and provides project opportunities. The responsibilities of Stuttgart Region for regional planning, transportation, and economic development account for two specificities of regional climate policy: its focus on spatial development, and on sustainable mobility.

Planning and communication measures of Stuttgart Region that aim to integrate climate issues into energy, land use, and urban planning are closely interlinked, honorary chairman Bopp said:

> "Very early on, we had technical events with various institutions and institutes and, let me put it like this: educated the regional assembly in terms of climate protection, and at the end it happens in the planning committee" (interview with T. S. Bopp and T. Kiwitt, 28 January 2015).[117]

Training of regional deputies, mayors, and officers serves to sensitize decision-makers to climate issues and educate them about locally feasible solutions. For instance, the EDC has hosted climate protection meetings (*Treffpunkt Klima-*

---

116 Mayors Adapts has been integrated to the Covenant with the 2015 relaunch.
117 Author's translation; original quote: "*Wir haben sehr früh Fachveranstaltungen gehabt mit unterschiedlichen Institutionen und Instituten und haben die Regionalversammlung, ich sag mal: weitergebildet in Sachen Klimaschutz, und am Ende findet das im Planungsausschuss statt.*"

schutz) on behalf of Stuttgart Region once or twice per year since 2008, and invites the entire metropolitan region, i.e. also non-member municipalities, for half-day meetings with practical presentations from experts and practitioners (interview with T. S. Bopp and T. Kiwitt, 28 January 2015).

The paramount formal instrument of climate policy in Stuttgart Region with mandatory specifications for municipal planning consists in the regional plan. The regional plan substantiates provisions from the state level and defines binding targets for the development of settlements, free spaces, and infrastructure. Adopted by the regional assembly in 2009, the current regional plan of Stuttgart Region was designed to guide regional development for 15 years. Updates of selected points are made when required, for example on the allocation of wind energy plants. Quite prominently, the regional plan refers to climate protection as a "central task" in spatial development and planning and to climate adaptation as a means to safeguard the attractiveness of Stuttgart Region as a business location, and calls for local and regional contributions to climate mitigation (Region Stuttgart 2009a, 3).[118]

As regional provisions for transport infrastructure and wind energy significantly restrict municipal autonomy of decision, these decisions have not been taken without conflict in the regional assembly:

> "…advocating such measures is […] not always sunshine and roses and with regard to several questions, we have been controlled in judicial review, we had to weather through court proceedings, that is not always only a bed of roses" (interview with T. S. Bopp and T. Kiwitt, 28 January 2015).[119]

Regional planning within the regional association provides a platform to ensure that regional development accounts for climate mitigation and adaptation as public goods despite diverging interests of individual municipalities. In this way, regional coordination counteracts incentives for non-climate friendly land use by individual municipalities, e.g. greenfield development that would involve in-

---

118 Author's translation; original quote: "*Innovative Ansätze zur Bewältigung nationaler und internationaler Konkurrenz / Sicherung der Standortattraktivität: […] Vorbeugung und Anpassung an die Konsequenzen des Klimawandels […] Regionaler Beitrag zum Klimaschutz: Bei der räumlichen Entwicklung und Ordnung der Region ist der Klimaschutz zentrale Aufgabe. Eine Minderung klimawirksamer Emissionen und Anpassungen an die Folgen der globalen Klimaveränderungen müssen auf lokaler und regionaler Ebene umgesetzt werden.*"

119 Author's translation; original quote: "*… das Einstehen für solche Maßnahmen [ist] […] nicht immer eitel Sonnenschein und wir sind da in verschiedenen Fragen im Rahmen einer Normenkontrolle überprüft worden, wir hatten Gerichtsverfahren durchzustehen, also das ist nicht immer nur Zuckerschlecken.*"

creased traffic within the entire region. Depending on the composition of and politics within the regional assembly, the resulting document provides more or less guidance for climate-friendly regional development.

If the regional development plan identifies climate change as one of the decisive challenges for Stuttgart Region along with demographic change and economic globalization (Region Stuttgart 2009a, 9), this – at least proclaimed – prioritization partly traces back to the convictions of Stuttgart Region's founding regional director Bernd Steinacher (Lang 2008, 13). Due to Steinacher's commitment, honorary chairman Bopp referred to him as the 'Al Gore' of Stuttgart Region (Bopp 2008, 5; interview with T. S. Bopp and T. Kiwitt, 28 January 2015). For instance, Steinacher initiated a transatlantic partnership with the Northern Virginia Regional Commission in 1999 which contributed to the diffusion planning practices that account for climate considerations (OECD 2010, 267).

Steinacher's suicide in September 2008 put an end to his personal leadership for regional climate policy. His successors, Jeanette Wopperer – who retired early for health reasons – and Nicola Schelling, have not shown a particular commitment to regional climate policy, but also struggled to leave their mark in regional policy more generally. Repeatedly, newspapers reported on tensions within the dual leadership with honorary chairman Bopp (Holland 2016; Ikrat 2012; 2015) who had allegedly taken over certain functions of the regional director transitionally while the post was vacant.

While Stuttgart Region perceives the integration of climate considerations into regional planning as an integral part of its responsibilities, the regional association covers a carbon-intensive economy that depends heavily on its car industry. In response, climate mitigation, climate adaptation, and economic development are closely interrelated in inter-municipal governance. This applies in particular to the activities carried out by the EDC:

> "Climate change is also an opportunity for economic development within the region [...] With its clean-tech strategy, Stuttgart Region's economic development activities therefore supports numerous projects, measures, and events in view of positioning the region by 2020 as the world's most energy-efficient region" (Region Stuttgart 02.11.2009, 23).[120]

---

120   Author's translation; original quote: *"Klimawandel ist auch eine Chance für die wirtschaftliche Entwicklung der Region [...] Die Wirtschaftsförderung der Region Stuttgart fördert in ihrer Clean-Tech-Strategie deswegen zahlreiche Projekte, Maßnahmen und Veranstaltungen mit*

Thus, business development through the EDC is an integral part of inter-municipal climate policy. The interlinkages of climate policy and economic development become especially visible in the field of sustainable mobility – a *"very strongly ideologically charged"* issue in local politics according to Stuttgart's longtime mayor Wolfgang Schuster (interview with W. Schuster, 16 May 2013).[121] For instance, regional director Nicola Schelling met with harsh criticism when she opted for a Tesla as an electrified company car rather than a local brand such as Mercedes and Porsche (Faltin 2015).

E-mobility is developed both as a means to promote more sustainable mobility and future economic opportunities alike. From 2009 to 2011, Stuttgart Region was a model region for electric mobility (*Modellregion Elektromobilität*) and implemented pilot projects with hybrid busses, e-vans, e-scooters, and e-bikes within a project of the same name by the BMVBS (BMVBS 2011). Since 2012, Stuttgart Region participates in the follow-up program "Showcase E-Mobility" (*Schaufenster Elektromobilität*) within "LivingLab BWe mobil", a project in Baden-Württemberg as one of four model regions, now funded by the Federal Ministry for Transport and Digital Infrastructure (*Bundesministerium für Verkehr und digitale Infrastruktur*, BMVI) (*BMVI 2016*). Stuttgart Region complements these programs with its own co-funding scheme established in 2012. Run by the EDC, the program "Model region for sustainable mobility" (*Modellregion für nachhaltige Mobilität*) aims to incentivize municipalities to contribute to the development of sustainable transport modes and intermodality (Wirtschaftsförderung Region Stuttgart GmbH 2016a) since 2012. Smaller municipalities in particular benefit from ready-for-use programs such as e-bike rental stations set up at train stations of the regional transport association[122] so as to facilitate intermodality for commuters and tourists (e-mobil BW GmbH 2015).

Other than the promotion of e-mobility, an important part of regional policies for sustainable mobility consists in public transport planning and service provision, including through its local transport provider, the *Stuttgarter Straßenbahnen AG*. The repartition of responsibilities with regard to public transport between the *Land*, the regional association, its districts and member municipalities

---

dem Ziel, die Region bis zum Jahr 2020 als die energieeffizienteste Region der Welt zu etablieren."

121 Author's translation; original quote: *"sehr stark ideologisch besetzt"*.

122 The *Verkehrs- und Tarifverbund Stuttgart* (VVS) covers the territory of Stuttgart Region on the exception of the district of Göppingen, i.e. Stuttgart city and the districts of Böblingen, Ludwigsburg, and Esslingen, and the Rems-Murr district.

has been repeatedly under discussion. Following negotiations over the so-called public local transport compact 2025 (*ÖPNV-Pakt 2025*) in response to congestion of inter-urban trains and disputes over competences, Stuttgart Region obtained new responsibilities for express busses, regional transport management and metropolitan fares in 2015.[123] Chief planner Kiwitt advocated for inter-municipal coordination as a means to cope with functional challenges such as itineraries beyond district borders or the promotion of intermodality.

The regional association aims for a coordinating role among the competing territorial governments of municipalities and districts, and argues that a regional approach rather than bilateral agreements is needed so as to manage public transportation. Naturally, the ideal role of the inter-municipal level looks slightly different from the municipal perspective. The head of the municipal climatology service of Stuttgart city, Ulrich Reuter, highlighted municipal autonomy when he said that responsibilities were clearly assigned for most tasks, with few conflicts, but also few instances of cooperation with the inter-municipal level (interview with U. Reuter, 28 January 2015). In this context of debate over territorial organization and the role of the inter-municipal level, reference to climate targets can serve to establish a need for inter-municipal coordination, and endorsement by an orchestrator such as the Covenant can further legitimize inter-municipal intervention.

Summing up, Stuttgart Region is accustomed to providing data, planning documents, and implementation reports as required by Covenant proceedings. Also, the regional association prominently refers to climate mitigation as a policy goal in the contexts of regional planning, transportation policy, and economic development. This implies a close match, even though this was not originally planned to be the case by local and regional actors, with the policy goals of the Covenant.

---

123 Signed by negotiators of the Land, Stuttgart Region, Stuttgart city and member districts of the regional transport association VVS (see [122]) on February 13, 2014, the public local transport compact was translated into federal state law with the Act on the Development of Public Local Transport in the Stuttgart Region (*Gesetz zur Fortentwicklung des öffentlichen Personennahverkehrs in der Region Stuttgart*) adopted on April 15, 2015, including an update of the 1994 Act on the Establishment of the Association of the Greater Stuttgart Region.

5.1.3.2 Limited Fit of Support to Local Needs

With its record in local climate policy, Stuttgart Region shared the policy goals of the Covenant in form and content. But the Covenant experience of Stuttgart Region demonstrates that policy fit alone is not enough to foster tangible orchestrator-intermediary relationships. For orchestration to succeed, also the support offered by the orchestrator to the intermediary and by the latter to the target actors – here, the municipalities – has to fit actual needs. This point is critical for understanding the case of Stuttgart Region. Since the prerequisite of suitable support was not sufficiently fulfilled, the orchestrator-intermediary relationship remained weak and only a very few joint activities developed.

Stuttgart Region became Germany's first Covenant supporting structure in February 2009.[124] The idea had first been raised by Sylvia Schreiber, head of the European office of Stuttgart Region in Brussels, vis-à-vis honorary chairman Bopp and chief planner Kiwitt on the occasion of a visit in Brussels (interview with T. S. Bopp and T. Kiwitt, 28 January 2015; interview with M. Siehr, 29 November 2012). Covenant signature was justified with prior climate protection activities of Stuttgart Region and its EDC, and the importance attached to supporting member municipalities, in particular smaller ones (Region Stuttgart 2009b, 19). Stuttgart Region pledged to promote the Covenant, to provide financial and technical support as well as networking opportunities for Covenant signatories, and to support the Covenant in terms of lobbying (Covenant of Mayors 2015c). Stuttgart Region aimed to cover this commitment in part with existing activities, such as climate-related networking events for member municipalities, consultancy and cooperation on projects, and with planned cooperation in model projects for climate adaptation, but also intended to develop additional forms of support for municipal climate protection (Region Stuttgart 2009b, 23).

In practice, Stuttgart Region developed very few activities with a clear link to the Covenant. Both the activities of and the number of Covenant signatories from Stuttgart Region as a Covenant Coordinator remained limited (see figure 5.4). After a while, Stuttgart Region silently minimized its Covenant-related activities (interview with M. Siehr, 29 November 2012). This withdrawal was a

---

124 At the time, the Covenant of Mayors did not yet differentiate between Covenant Coordinators (territorial authorities) and Covenant Supporters (networks of local authorities). For an overview of types of Covenant membership, see section 3.2.

reaction to the disappointment with the Covenant both at the municipal and at the inter-municipal level.

29.07.2008	Covenant signature by Heidelberg
02.10.2009	Regional climate policy kick-off meeting in Heidelberg
23.02.2010	Covenant signature by Böhl-Iggelheim
09.03.2010	Covenant signature by St. Leon-Rot
30.03.2010	Covenant signature by Mannheim
16.04.2010	Regional climate policy conference in Heidelberg including Covenant signature by Rhine-Neckar Region
28.11.2012	Latest update of Rhine-Neckar Region's profile on the Covenant website

Figure 5.4: Timeline of Covenant activities in Stuttgart Region
Source: Author.

Interestingly, interviewees from Stuttgart Region still referred to the Covenant commitment of their regional association as being a supporting structure (interview with T. S. Bopp and T. Kiwitt, 28 January 2015; interview with M. Siehr, 29 November 2012): The differentiation in Covenant membership between (territorial) Covenant Coordinators – such as Stuttgart Region – and (organizational) Covenant Supporters had not made it into regional parlance. This further underscores the overall assessment of a weak orchestrator-intermediary-relationship between Stuttgart Region and the Covenant.

In view of the extensive coincidence of policy goals with the Covenant, the limited record of Stuttgart Region as a Covenant Coordinator seems surprising at first sight. It results from limited fit of support in two respects. Firstly, the support that Stuttgart Region was able to offer did not match the assistance that would have been needed to successfully engage member municipalities in the Covenant. Secondly, the expectations of Stuttgart Region in terms of European-level networking opportunities were not met by the offer made by the Covenant.

When Stuttgart Region became a supporting structure to the Covenant in 2009, the regional association aimed to particularly address smaller member municipalities in order to promote Covenant signature and support signatories in the implementation of their commitments (interview with M. Siehr, 29 November 2012). But municipal response remained strictly limited. Only three cities from the territory of Stuttgart Region signed the Covenant. Out of 179 member municipalities of Stuttgart Region, only the three largest ones signed the Covenant,[125] and this despite the fact that Stuttgart Region intended to establish a comprehensive network of climate-active municipalities, aimed to target smaller municipalities in particular, and was able to build on prior experience in climate planning, climate adaptation, and sustainable mobility policy. What motivated the cities of Stuttgart, Ludwigsburg, and Esslingen to sign the Covenant, and what deterred the other municipalities from doing so?

For Stuttgart city with its longstanding record in local climate policy as depicted above, Covenant signature was actually a matter of course – and of prestige. When Stuttgart city signed the Covenant in January 2009 – slightly ahead of Stuttgart Region – it disposed already of a municipal climate plan, the KLIKS. Adopted in 1997, the KLIKS had been evaluated in 2000 and updated in 2007. It entailed three justifications for participating in municipal climate networks: "first, in order to reach the climate-protection targets of the EU, cities need to exchange their experiences; secondly, [the City of] Stuttgart would benefit from the adoption of best practices; and thirdly, municipal networks would enhance effective lobbying in Europe" (Benz et al. 2015, 327). In the words of Ulrich Reuter, head of the municipal climatology service and Covenant contact person for Stuttgart city:

> "The adhesion to the Covenant was linked to the fact that we saw a chance because it actually is the first initiative where the EU communicates with municipalities. [...] And we also saw, well, the possibility, well, that it somehow, I put it crudely: is part of the full panoply, so simply also for reasons of prestige and representation, to say: We are also a member of the Covenant of Mayors. [...] What was appealing about the Covenant was, well, Covenant of Mayors, it sounded relatively high-level, and possibly simply a better access between the EU and municipalities" (interview with U. Reuter, 28 January 2015).[126]

---

125   Out of 2.7 million inhabitants of Stuttgart Region, about 600.000 live in Stuttgart city, 92.000 in Esslingen, and 87.000 in Ludwigsburg Covenant of Mayors (2015e).

126   Author's translation; original quote: "*Der Beitritt zum Konvent hing damit zusammen, dass wir eine Chance sahen, weil es eigentlich so die erste Initiative ist, wo die EU mit Kommunen kommuniziert. [...] Und wir sahen darin auch, ja, die Möglichkeit, ja, dass es irgendwie, ich sag's mal ganz platt: dazugehört, also einfach auch aus Gründen des Prestiges und der Repräsentativität, zu sagen: Wir sind auch Mitglied im Konvent der Bürgermeister. [...] gereizt am Konvent hatte eben, ja Konvent der Bürgermeister, also klang relativ hoch, und vielleicht einfach auch der bessere Zugang zwischen EU und Kommunen.*"

In short, it was important for Stuttgart not to stay behind when new initiatives emerged, and the Covenant promised to provide visibility at and access to the European level.

Building upon the municipal climate plan, achieving the Covenant commitment seemed realistic without major additional efforts. The KLIKS update served as a basis for submitting a SEAP to the CoMO without establishing an entirely new document (Covenant of Mayors 2015c). Although the city had been aware of necessary adaptations of the KLIKS update for meeting Covenant requirements (Benz et al. 2015, 331), Ulrich Reuter reported that SEAP submission turned out unexpectedly laborious because data from the KLIKS, the emerging energy efficiency concept and other sources had to be compiled in the submission form (interview with U. Reuter, 28 January 2015). In a similar vein, Stuttgart city already disposed of a climate adaptation plan (*Klimaanpassungskonzept Stuttgart*, KLIMAKS) from 2012 when it signed Mayors Adapt on October 16, 2014. Consequently, Stuttgart city was able to report on implementation as early as the year following its Mayors Adapt signature (Mayors Adapt 2015).[127]

In return, Covenant signature benefitted Stuttgart city mainly in two ways. Firstly, Stuttgart city made use of the websites of the Covenant and of Mayors Adapt to increase the visibility of its policies and projects (Covenant of Mayors 2013b, 2016; (Climate-ADAPT 2016). Secondly, Covenant signature also served to legitimize further steps of local climate and energy policy. Here, endorsement by the European level became a resource in local politics for those who advocated for updating and supplementing local climate and energy policy, namely within the municipal council. The reduction target of the city's energy plan (*Energiekonzept*), adopted in January 2016, was justified with the commitment to the European 20-20-20 targets (Stadt Stuttgart 2014). The head of the municipal climatology service, Ulrich Reuter, said that his team repeatedly referred to Covenant signature when applying for municipal funds for public communication and citizen participation (interview with U. Reuter, 28 January 2015). As an example, Reuter referred to funds in the lower five-digit range obtained with this strategy which allowed for the development of the city's climate savings book

---

127   Mayors Adapt scheduled a 6-steps procedure: 1. Preparing the ground, 2. Assessing risks and vulnerabilities, 3. Identifying adaptation options, 4. Assessing adaptation options, 5. Implementation, 6. Monitoring and evaluation.

2012, and of an updated version for 2016.[128] This strategy of justifying particular measures or targets with seemingly external constraints has been described as "the immunisation of a city's climate-protection policy against competing claims" (Benz et al. 2015, 329). Notably, a path dependency effect of self-commitment does not imply that the Covenant determined the municipal agenda. Rather, interested actors chose to refer to Covenant signature when it suited their argumentation in internal politics.

In contrast to visibility and legitimation, networking was not a priority for Stuttgart city in its use of the Covenant. In-house competences and other networks such as the Climate Alliance, Energy Cities and Cities for mobility ensure technical and international exchange. The example of Cities for mobility highlights that networks other than the Covenant provide Stuttgart city with distinguished visibility. Building upon the EU-network "Controlling urban mobility" within Urb-Al, a project starting in 2000, Stuttgart city launched Cities for mobility as a permanent international network in 2007 and regularly hosts the network conferences at its city hall (Stadt Stuttgart 2016b).

As Stuttgart city already is a regular participant in projects funded by the EU and national ministries, there also was no particular need for project acquisition, i.e. for identifying potential project partners within the network of the Covenant. Hence, it was not surprising when no projects emerged from its Covenant membership. Instead, Stuttgart city carefully chooses projects so as to maximize support for its existing policies and tasks while at the same time minimizing the workload of project administration. Covenant reporting turned out to require more administrative efforts than expected, but remained acceptable due to prior completed works. Summing up, it was due to Stuttgart city's existing strong record in local climate policy that Covenant signature was an easy step to take. At some points, Covenant membership entailed a certain value added for the city. It benefitted externally with regard to visibility, and internally with regard to a favorable path-dependency.

Ludwigsburg signed the Covenant on March 28, 2012. As in the case of Stuttgart, the city already had an integrated municipal climate plan at the time of Covenant signature. Funded under the Local Authority Guideline, the municipal

---

128 Climate savings books are handbooks with vouchers and information on regional or local low-carbon shopping, food, leisure activities, and initiatives. The format was developed in 2010 and has since been applied to more than 20 cities and regions in Germany (Oekom Verlag 2016).

climate plan had been adopted in 2011 (Fahl et al. 2011). Additional funding for the participatory development of the municipal climate plan had been obtained within Sustainable NOW, an EU-project funded under the IEE scheme. Ludwigsburg had taken part in this project alongside the Climate Alliance which it is a member of (Stadt Ludwigsburg 2011, 12–13). The implementation of international, national and regional funded projects is an important part of the work of Ludwigsburg's sustainable urban development service. Within the service, a dedicated unit Europe and Energy (*Team Europa und Energie*) is in charge of implementing EU-projects and thus familiar with cooperation at the European level (Stadt Ludwigsburg 2014a). As an active participant in regional energy policy, Ludwigsburg participated in the KlimaMORO project guided by Stuttgart Region (see 5.5.1.1), a follow-up project on climate adaptation (*Klimaanpassung Region Stuttgart*, KARS), and in national and regional model projects for electric mobility (Stadt Ludwigsburg 2011, 7, 20, 27-28; 2014b). Competitions and awards ensure the visibility of these activities, including EEA certification in 2011 and 2014. Ludwigsburg also adopted a Strategic Climate Adaptation Concept (*Strategisches Fachkonzept Klimaanpassung*, Klik) in May 2016 – again developed with the help of two funding schemes of the BMU and BMBF – but has not extended its Covenant commitment to also cover adaptation (Stadt Ludwigsburg 2016).

In Esslingen, Covenant signature traces back to prior policies and membership in the Climate Alliance[129] since 2001 rather than to Covenant coordination by Stuttgart Region. Already before the creation of the Covenant, the local council had determined a GHG emission reductions target of 25% by 2020 as compared to 2007, the year of that decision. An integrated climate protection concept (*Integriertes Klimaschutzkonzept*) was approved in the municipal council on June 28, 2010 (Stadt Esslingen am Neckar 2016). Only then and with the decision of the municipal council on December 13, 2010 did Esslingen sign the Covenant. The integrated climate protection concept was submitted as a SEAP, along with a report on climate policy and emission trends from 2007 to 2011 (*Klimabilanzbericht*) (Covenant of Mayors 2014b). Esslingen also won several climate

---

129 The Climate Alliance is part of the consortium running the Covenant of Mayors Office on behalf of the European Commission and promotes the Covenant of Mayors among its members. Several national sections of the Climate Alliance also act as Covenant Supporters. The Climate Alliance Germany supports a total of 89 Covenant signatories, including the cities of Ludwigsburg, Stuttgart, and Esslingen Covenant of Mayors (2016a).

awards. The city already disposed of networks, expertise, and visibility with regard to local climate policy at the time of Covenant signature. Therefore, operative assistance by Stuttgart Region for meeting Covenant requirements was not necessary.

Summing up, the cities of Stuttgart, Ludwigsburg and Esslingen, albeit to varying degrees, already disposed of municipal climate plans and of qualified staff that was familiar with climate policy and European projects prior to Covenant signature. Compliance with Covenant proceedings was thus facilitated, and although the Covenant only provided these cities with moderate benefits, they were well prepared to bear the costs of membership, namely the administrative workload. For them, three cities that perceive of themselves as being actively engaged in climate policy, Covenant signature was primarily a matter of prestige, and staying behind in this European initiative was not an option.

In contrast, smaller municipalities turned out not to be interested in Covenant signature. For them, the calculation of costs and benefits looked different than for the bigger cities. The officer in charge of energy and environment at Stuttgart Region, including Covenant Coordination, said that the main hindrances for Covenant signature by smaller member municipalities consisted in a lack of expertise, resources, and English proficiency (interview with M. Siehr, 29 November 2012). With only one out of 60 employees in charge of energy and environment, Stuttgart Region was not in the position to supplement communal resources to the required extent by funding municipal SEAPs or hiring a translator for Covenant correspondence.

The officer added that the benefits of Covenant signature for German municipalities were limited since they already disposed of networks for exchange among peers, e.g. the Climate Alliance, and could rely on other programs for funding, and increasing the visibility of, their climate policies, e.g. the EEA or the Local Authority Guideline (interview with M. Siehr, 29 November 2012). As he pointed out, the fact that municipalities had often developed municipal climate plans already was not necessarily of help in complying with Covenant procedures. Existing plans were difficult to convert into a SEAP for submission to the Covenant due to divergent baseline years and calculation methodologies. These discrepancies between the benefits and the workload of Covenant membership were also discussed at meetings of the Club Germany in the Covenant and inspired an open letter to the European Commission that called for improving the accessibility of the Covenant for German municipalities (Covenant Club Deutschland 2012).

In response, Stuttgart Region does not actively approach smaller member municipalities for promoting Covenant signature since it would anyway be unable to help overcome their main hindrances (interview with M. Siehr, 29 November 2012). Instead, the regional association focusses on practical assistance to member municipalities (interview with T. S. Bopp and T. Kiwitt, 28 January 2015). Summing up, smaller municipalities from Stuttgart Region refrained from signing the Covenant because they expected they would not to be able to fully engage in the Covenant, and the regional association did not have the resources to support them in the activities that would have been required for implementing a Covenant signature.

These difficulties are also linked to the polycentric structure of Stuttgart Region with nearly half of its inhabitants living in the center city and its belt of district towns, but also with several district towns outside that belt and two thirds of its municipalities having less than 10,000 inhabitants (Region Stuttgart 2009a, 39; Verband Region Stuttgart 2016c).As a result, Stuttgart Region is characterized by heterogeneity between metropolization, polycentricity, and municipal fragmentation. Although Stuttgart city is an important partner for Stuttgart Region in climate policy, e.g. in terms of expertise as in the case of the climate atlas, it cannot impose its approach to, or serve as the engine for, local climate policy in the regional association as a whole.

The difficulties in Covenant promotion have not been publicized. Quite the contrary, Stuttgart Region continues to refer to its Covenant-commitment both internally and externally (e.g. Verband Region Stuttgart 2015, 6; Verband Region Stuttgart 2016b).

Stuttgart Region not only aimed for an – unachieved – regional network of Covenant signatories, but also perceived its Covenant commitment as part of its activities at the European level. Stuttgart Region was the first German region to establish its own representative office in Brussels in 2002 (Wirtschaftsförderung Region Stuttgart GmbH 2016b). The objectives of Stuttgart Region with regard to its activities at the European level read as a list of benefits as proclaimed by the CoMO: access to EU politics, project brokering, learning among peers, self-marketing, and up-to-date information for further dissemination within the region (Wirtschaftsförderung Region Stuttgart GmbH 2016c).

In particular, Stuttgart Region counted on material support from the Covenant in the form of assistance for enhanced visibility and a network. On the one hand, Stuttgart Region was on the lookout for a platform to present its climate policy to a European audience (Verband Region Stuttgart 2016). On the other, the regional association hoped to obtain additional contacts and information through the Covenant. Due to the particularities of Stuttgart Region as compared to other regional associations within Baden-Württemberg, technical exchange with other regional associations from Germany and Europe or within the framework of transatlantic contacts had a longstanding tradition in Stuttgart Region, but the Covenant promised to identify additional partners with relevant experiences (interview with T. S. Bopp and T. Kiwitt, 28 January 2015).

Indeed, Stuttgart Region was able to present its climate policy activities on the website of the Covenant. On the occasion of the signature ceremony of the Covenant in Brussels in 2010, various news posts featured video statements by Regional director Jeanette Wopperer (Covenant of Mayors 2010). Stuttgart Region was among the first nine Covenant members to publish their 'Benchmarks of Excellence', short presentations of local actions, within their membership profile (Covenant of Mayors 2009; 2015d). But neither the Covenant in general nor its Club Germany allowed for new contacts with regard to European-level networking. Since Stuttgart Region already had well-established contacts in Brussels as well as to other regional associations, the Covenant network overlapped largely with existing contacts, e.g. from the fields of planning and economic development (interview with T. S. Bopp and T. Kiwitt, 28 January 2015). Stuttgart Region participates in several European networks, including the European Regions Research and Innovation Network, POLIS – European Cities and Regions Networking for Innovative Transport Solutions, European Metropolitan Transport Authorities, and METREX (Verband Region Stuttgart 2015, 4–7). METREX in particular is better suited to meet the needs of Stuttgart Region because it specifically addresses the metropolitan level, provides distinction and visibility for the regional association and its regional directors, and acts as a partner in funded projects.

In plain terms, the Covenant came in too late and too weak for Stuttgart Region to benefit from its support in terms of lobbying or project acquisition. Stuttgart Region already had established contacts at the European level through its office in Brussels, and with other metropolitan regions through dedicated networks. These networking venues emerged at a more timely moment and better matched the priorities of Stuttgart Region at the European level, namely funding opportunities and the promotion of the role of metropolitan regions. Albeit for different reasons, the result was similar for networking member municipalities; Covenant membership did not provide an actual value added for Stuttgart Region.

5.1.3.3   Case Summary

The case of Stuttgart Region demonstrates that effective orchestration requires a match not only of policy goals, but also a fit between the support offered by the orchestrator and the requirements of the intermediary on the one hand, and the assistance proposed by the intermediary and the needs of the targets on the other. In Stuttgart Region, local climate planning has a long record, especially in terms of climate vulnerability, and sustainable mobility is part of regional strategies for economic development. Policy goals thus match with those of the Covenant both in form and content, and orchestration, at first sight, seems likely to succeed. In practice, it turned out that the Covenant had little value added to offer to Stuttgart Region. Its network did not allow for convening the regional association with new and interesting partners. No privileged access to European actors and partners materialized for Stuttgart Region which was already well-networked at the European level and among metropolitan regions. The agenda set by the Covenant for territorial climate policy coordination, although attractive to Stuttgart Region at first sight as a means to enhance territorial collaboration and its own role in local climate policy, was not functional as it presupposed too much capacity and resources both from the inter-municipal administration and from member municipalities. Notably, the required level of operational support was made impossible by the fragmentation of most member municipalities and the limited size of the intercommunal bodies as compared to district administrations. The CoMO, in turn, was unable to provide tangible assistance to overcome this gap. In view of these difficulties, endorsement alone was not a sufficient motivation for local actors, and coordination with other metropolitan regions through the Covenant remained more than hypothetical.

Instead, Stuttgart Region focused on providing a framework for municipal climate policy by means of informal and formal planning and, in addition, tried to facilitate municipal climate action with the help of inter-municipal competitions, funding schemes, and model projects. While Stuttgart Region maintains its status as a Covenant Coordinator, it has to be assessed as a rather nominal Covenant member for the time being.

5.2 Covenant Coordinators in France

Covenant participation in France remains limited in numbers and compliance. After a peak of more than 110 French Covenant signatories during the study period, the Covenant website now lists 82 that joined between 2008 and 2015, with only 12 commitment made after 2011 (Covenant of Mayors 2014e; 2017a). French Covenant signatories consist of municipalities and agglomerations of all sizes: from 378 inhabitants (Trébédan) to 2.3 million (Paris). The vast majority – 73 of them – is currently put on hold for overdue reporting. Other than the inter-municipal associations examined below, four regional governments engaged as Covenant Coordinators (Covenant of Mayors 2016c).[130] Eight organizations act as Covenant Supporters, but while Energy Cities is affiliated with 51 Covenant signatories, the next favored organization only supports three.[131] French Covenant participants created a national Covenant Club in October 2010 in Paris and expressed a particular satisfaction with this additional forum as compared to other countries (Covenant of Mayors 2014d, 2). It is part of the Alliance for Local Energy (*Alliance pour l'énergie locale*) established in January 2012, an initiative that connects Covenant participants and other actors from France that support a decentralized approach to the energy transition.

The number of Covenant signatories is relatively low when considering the extraordinary number of subnational, especially municipal governments in France, and the proliferation of local climate plans. This section explores the promises and pitfalls of French local climate governance and inter-municipal coordination in particular. Special attention is paid to factors that explain French local governments' short-lived Covenant commitments. First, an introduction to

---

130 The Hérault department and the Territorial Community of Corsica abandoned their commitment (see section 4.2.2); the regions Nord-Pas de Calais and Picardie are still listed separately despite their merger that took effect in January 2016.

131 The other Covenant Supporters are the Association of French Urban Communities, the Association of Mayors of French Large Cities (that have merged in 2015 but still figure separately on the Covenant website), AMORCE, an association of municipalities, their associations and companies dealing with waste management, energy, and district heating, and the association Villes de France (1 – 3 Covenant signatories supported), and the European Association for Local Democracy, the Association of Small Cities of France, and the RURENER network (no individual Covenant signatories supported).

the French framework for inter-municipal climate governance explains political and institutional specificities shaping local scopes of action (5.2.1). Subsequently, case study reports explore inter-municipal climate policy coordination on behalf of the Covenant of Mayors in the merging metropolis of Greater Lyon (5.2.2), the agglomeration Rennes Metropolis (5.2.3), and rural Val d'Ille (5.2.4). They demonstrate that orchestration of local climate policy in France struggles with a lack of municipal capacities, and changeful political agendas.

5.2.1 The French Framework for Inter-municipal Climate Governance

With a mostly reactive and defensive approach to climate policy, French ambitions for a leadership role in European and international climate governance long relied largely on rhetoric and pledge, but France played a more active role as a symbolic leader in recent years (Bocquillon and Evrard 2017). The prospect of hosting the COP 21 in Paris in 2015 encouraged exemplary action at the national and subnational levels. Efforts of the entire diplomatic corps in the run-up to the conference, and Foreign Minister Laurent Fabius' role in brokering of the Paris agreement were praised widely (Kinley 2016).

It would be short-sighted to describe the French framework for local climate policy as being determined by hierarchical government, restricted local capacities, limited records in local climate mitigation, and a dominance of nuclear power in the energy this mix. In actuality, France disposes of a differentiated system of territorial governance and has increasingly engaged in climate policy since the turn of the century. This section first provides an overview of national climate policy with a focus on provisions for subnational action since 2007, including a side note on regional uptake in the regions involved in the case studies (5.2.1.1). It then turns to the local level and subsequently discusses the role of local governments within the decentralized republic (5.2.1.2), the increasing inter-municipal integration within intercommunalities (5.2.1.3), and the landscape of local climate governance (5.2.1.4).

5.2.1.1 National and Regional Climate and Energy Policy

Despite the abundance of natural open space in France outside urban agglomerations as compared to more densely populated EU member states, increased pollution led to the emergence of environmental concerns (Larrue 2002, 201). An environmental ministry was established in 1971. Environmental policies have been driven by health concerns, natural and landscape protection (Chabason and Larrue 1998). Dominated by neo-corporatist arrangements, they underwent incremental change only; in cases of instrumental innovation, European policy was only one of several sources (Halpern 2011).[132] The number of infringements of EU environmental legislation was above average, but in decline (European Commission 2016a). In recent years, increasing priority has been given to green growth policies, including an energy transition (OECD 2016).

France was among the first signatories to the UNFCCC in 1993; today's Inter-ministerial Mission on the Greenhouse Effect (*Mission interministérielle de l'effet de serre*, MIES) traces back to a working group established in 1989. Early GHG emission reductions were mostly a by-product of energy efficiency gains and the promotion of nuclear energy in response to energy security concerns raised by the oil crises (Szarka 2011, 160–61). Accordingly, per-capita emissions have long been below European average, with one of the least carbon-intensive economies in Europe in terms of GHG emissions per unit of GDP (European Environmental Agency 2007, 22–23). (Nuclear) industry actors played a strong role in corporatist climate policy making (Szarka 2000). France stressed that climate policies should not distort economic competitiveness, disapproved of the 1990 baseline under the Kyoto protocol for fear of struggling more than others to realize additional GHG emission reductions, and stressed the importance of equity in terms of per-capita emissions instead (Szarka 2011, 162). Indeed, attempts to tackle energy-related GHG emissions from the demand side with the Climate Plan (*Plan climat*) adopted in 2000 were not very successful (Deroubaix and Lévèque 2006).

French climate policy picked up steam in the early 2000s. In 2003, France committed to a 75% GHG emissions reduction target by 2050, a target often referred to as being 'factor four' in scale (see e.g. Boissieu 2006). The Climate Plan (*Plan Climat*) was adopted in July 2004. Based on the assessment that territorial governments can influence about 12% of the emissions on their territory through their decisions on infrastructure, services, and consumption, the Climate Plan (2004) encouraged territorial governments to develop territorial climate plans (*Plans climat territoriaux*). A corresponding handbook (*Guide Plan Climat Territorial*) delivered in November 2005 emphasized action with regard to

---

132 For an eclectic discussion of the extent to which French environmental policy is subject to Europeanization, see Berny (2011).

transport and housing – sectors with increasing emissions – but also communication (Roussel 2005).

Since then, the central state adopted three major laws to implement the factor four and other climate and energy targets, starting with an act in 2005 that
established climate mitigation as a priority in French energy policy. It also scheduled biannual evaluations and updates of the climate plan. The role of subnational
actors in environmental governance and sustainable development was further
highlighted in the *Grenelle* Act I and II adopted in 2009 and 2010. Named after
a series of public consultations engaged in 2007 by President Nicolas Sarkozy in
the run-up to the French Presidency of the European Council and the final negotiations of the climate and energy package in 2008 (Szarka 2011, 172). They
introduced mandatory tools for regional and local climate planning. Regions
were charged with establishing Regional Climate, Air and Energy Schemes
(*Schémas régionaux climat, air, énergie*, SRCAE) that include emission and energy inventories, provisions for the development of renewable energy, and climate adaptation plans. Subnational governments with at least 50,000 inhabitants
were obliged to develop Territorial Climate and Energy Plans (Plans climat énergie territoriaux, PCET), which will be introduced in more detail in the next section. This legislation was criticized for causing a piling (*empilement*) of PCET
by different levels of government on the same territory (see e.g. Collet 2014).[133]
Although Sarkozy's announcements had raised expectations, further *Grenelle*
recommendations such as a domestic carbon tax were finally abandoned in 2010,
and a Climate Justice Plan presented at the COP 15 in Copenhagen 2009 failed
to have a significant impact on international negotiations for lack of coordination
with EU policy making and UNFCCC processes (Szarka 2011, 173–75).

An even more extensive, decentralized consultation process preceded the
energy transition for green growth act (*Loi relative à la transition énergétique*

---

133   Since SRCAE are not legally binding on the exception of their wind energy section, PCET
      from one territory do not necessarily add up to achieve the SRCAE targets. The ensuing over-
      laps and contradictions have been addressed in the course of the third wave of decentralization.
      In accordance with their responsibilities for spatial planning and, respectively, social policies,
      regions are assigned with leadership in territorial climate and energy policy, while departments
      are charged with measures against energy poverty. The SRCAE will be merged with other
      regional plans within Regional Development, Sustainability and Territorial Equality Schemes
      (*Schéma régional d'aménagement, de développement durable et d'égalité des territoires*,
      SRADDET). Despite the longstanding French taboo for subnational governments to regulate
      over subordinate entities, the SRADDET will provide mandatory long-term orientations for
      sub-regional planning documents such as urban transport plans.

pour la croissance verte). Running from 2012 to 2013, the National Energy Transition Debate (*Débat national transition énergétique*) built on the Grenelle experience in that it involved five groups of actors: trade unions, employers, environmental NGOs, the central state, and subnational governments. The energy transition act, informed by a synthesis report from the debate, was adopted in 2015 after long parliamentary negotiations. It reaffirmed the factor four target for 2050, with an intermediary target of 40% by 2030, as well as an increase of the share of renewable energy in the electricity mix to 23% by 2020, a reduction of the use of fossil fuels by 30% until 2030 as compared to 2012, and a reduction of the share of nuclear energy to 50% by 2025. To these ends, it comprises measures in eight fields including buildings, low-carbon mobility, renewable energy, and territorial governance, i.e. the involvement of non-state actors such as local governments.

Regional uptake of the above provisions varies importantly across France. 13 regional environmental and climate agencies support local action, including RhônAlpEnergie Environnement (RAEE) and Bretagne Environnement; they are organized within the Network of Regional Agencies for Energy and the Environment (*Réseau national des Agences régionales de l'énergie et de l'environnement*) established in 1995. In Rhône-Alpes Region, home to Greater Lyon (see section 5.2.2), the SRCAE was not adopted until 2014 due to discord with the prefect although the region had worked on climate issues before and wind energy planning was the issue causing significant discontent during regional negotiations (Bertrand, Richard, and Larrue 2016). The SRCAE could thus not provide any orientation for the first generation of PCET developed by departments and local governments on its territory. The regional PCET was established as of 2010 in a concertation[134] process that involved territorial governments, state services and agencies, professional associations, the regional climate and energy observatory, and other actors, and took effect as of 2013 (ADEME 2015c). The Rhône

---

134 Concertation refers to the coordination of actors through the exchange of information and deliberative discussion in view of cooperation for a shared policy goal. Concertation aims for a consensus which is often cast into an operational agreement to be signed by all participants. It is thus to be distinguished from other modes of collective decision-making, namely consultation and negotiation: Consultation of stakeholders enables public authorities to obtain information and review positions without engaging in developing consensus. Negotiation, too, does not aim at consensus-building, but refers to the bargaining between interdependent actors with competing objectives in order to settle a difference. For a more detailed delimitation and discussion, see Touzard (2006).

department only launched its PCET in 2012 based on its Local Agenda 21 (ADEME 2015a).

Brittany, where Rennes Metropolis and Val d'Ille are located (see sections 5.2.3 and 5.2.4), adopted its SRCAE in 2013. It prioritized construction, agriculture, adaptation, and governance for the implementation phase up to 2018 (Région Bretagne 2014b). The Brittany Region provides for yearly meetings of the "*Club élus*" (Delegates club), i.e. local councilors who are in charge of PCETs, and for meetings of local officers in the "*Club PCET*" (Région Bretagne 2014a, 17) in order to build capacity for local climate planning, disseminating information on regional climate change impacts, and coordinating PCETs on the territory (Région Bretagne 2011, II-24). Another priority of Breton climate-related policy making consists of transportation;[135] transport infrastructure has played an important role in developing the outermost region (Cole 2013). Brittany's PCET provided no guidance for territorial governments as it only took effect in January 2014 and does not include any particular concertation scheme (ADEME 2015e). The same year, the *département* Ille-et-Vilaine adopted its PCET although the procedure had been launched in 2010; the consultation process only picked up speed after an administrative division on sustainable development and participatory democracy was established for this purpose (ADEME 2015d; Département Ille-et-Vilaine 2016).

5.2.1.2 Local Governments within the Decentralized System

France, by constitution the "one and indivisible republic" (art. 1), is often referred to as a textbook example of a highly centralized state (Wollmann and Bouckaert 2006, 18). Subnational governments consist of 18 regions (*régions*),[136] 101 departments (*départements*),[137] and about 36,600 municipalities (*communes*).[138]

---

135 Examples include an express train connection to Paris (TGV), intermodal mobility, and the metro in Rennes city.
136 The number of regions was reduced from 27 to 18 by territorial reform as of 2016, including the merger of the Auvergne-Rhône-Alpes Region that accommodates Greater Lyon, one of the intercommunalities studied below (see section 5.2.2).
137 Five regions and five departments are located overseas.
138 An additional administrative level consists in the *pays* which correspond roughly to former counties. Introduced in 1995 and strengthened in 1999, they do not have the status of a territorial authority. Under different legal frameworks, they typically cover the territory of several neighboring intercommunalities. The *pays* animate cooperation of public and non-public actors, mainly with regard to land-use planning, and serve as a platform for developing Territorial Coherence Schemes (*Schémas de cohésion territorial*, SCoT). In contrast, two former admin-

An additional layer of the famous 'puff pastry' (*millefeuille*)[139] of French territorial governance (see e.g. Garello 2016; Kuronen and Caillaud 2015) received blanket coverage in 2015 as all municipalities now have to join an intercommunality, an institutionalized form if inter-municipal cooperation that will be introduced in more detail in the next section.

Broadly speaking, regions are in charge of economic development, vocational training, spatial planning, infrastructure, and sustainable development; their influence on policy mainly expresses itself in funding for desired projects. Departments are carrying out social and health policies, and handle culture, sports, and secondary schools. Municipal responsibilities include urbanism, close-range public services such as schools and local infrastructure, waste, and water. Certain municipal missions and services are collectivized at the intercommunal level in order to allow for a pooling of resources and cooperative planning. No level of government has tutelary power over the others (art. 72,5); the constitution stipulates the principle of free administration.

In fact, moves to partly devolve the central state administration, and repeated efforts at decentralization have brought about a more complex system (see e.g. Couzigou 2013; Hoffmann-Martinot and Wollmann 2006).[140] As a result, state services – the prefectures – and territorial institutions (*collectivités territoriales*) coexist at all three territorial layers in an intertwined double structure. The Ministry of Ecology and Sustainable Development (*Ministère de l'Ecologie et du Développement Durable*, MEDD)[141] for example relies on 12 Regional Directorates of Environment, Development and Housing (*Directions régionales de l'environnement, de l'aménagement et du logement*, DREAL),[142] and on agencies of other ministries at the departmental level (see Larrue 2002, 211) for oversee-

---

istrative levels have been abolished: 342 districts (*arrondissements*) and 4,036 cantons (*cantons*).

139 The *millefeuille*, also known as the Napoleon, is a French pastry made up of several layers of puff pastry, alternating with layers of cream. The comparison French territorial organization to this rather heavy, fatty cake often implies that the number of layers – and the costs of public administration – should be reduced. In this context, French interlocutors often enviously refer to German federalism as a layer cake – an inaccurate simplification for it resembles much more a marble cake, as Kropp and Behnke (2016) point out.

140 Decentralization refers to devolution of authority to subnational governments. In contrast, deconcentration of central state administration involves decentralized service provision, but no transfer of authority.

141 From 2007 to 2016: Ministry of Ecology, Sustainable development and Energy (MEDDE).

142 Established in 2009 by merging the Regional Directorates of the environment (DIREN), industry (DRIRE), and infrastructure (DRE).

ing French environmental governance. For instance, the respective DREAL co-authors the SRCAE. In addition, the Environmental and Energy Management Agency (Agence de l'Environnement et de la Maîtrise de l'Energie, ADEME),[143] a largely independent technical agency, and its decentralized directorates, provide subnational governments with expertise and financial support for projects with regard to the energy transition and waste management on behalf of the MEDD and the Ministry of Education. Policy implementation, though, is carried out by regions, departments, and municipalities.

Beyond their distinct responsibilities, municipalities are in charge of pursuing the local interest on behalf of their population based on the general competence to act (*clause générale de compétence*) (Rangeon 2011).[144] Although the latter should not be mistaken for it empowers municipalities to seize any issue of local public interest on the condition they not interfere with another authority's competences.[145] The general competence has been criticized for causing entanglements of policies and increasing public spending, for instance in the case of co-funded projects (Marcou 2011, 60–61; see also Sauviat 2016); it has therefore been abolished for the regional and departmental level in 2015.[146] The general competence to act should not be mistaken for the principle of local self-rule; French municipalities remain an integral part of state administration (Martínez Soria 2007, 1020).

This approach to territorial organization assumes a uniformity of municipal units and does not provide for territorial reorganization in response to municipal fragmentation; instead, the widespread incapacity of small municipalities incites the central state to reduce municipal responsibilities in general to the detriment of more capable municipalities (Martínez Soria 2007, 1021). Le Saout (2011, 80–83) demonstrates that more than a third of all municipalities within the EU stem from France, but their vast majority is very limited in size and resources,

---

143   Created in 1991, the ADEME replaced three existing agencies dealing with waste, air pollution, and energy.

144   French scholars typically refer to subnational governments as territorial governments.

145   The general competence traces back to a traditional understanding of the concept as a principle of liberty since the Constitution of 1791 and has long been codified in jurisdiction only Marcou (2011, 57–60). Originally codified in the municipal act of 1884, it has been extended from municipal to the other subnational governments in 1982.

146   The general competence had partly been abolished already in 2010, but reestablished for regions and departments in 2014.

including municipal budgets and staff.[147] The fragmentation of French municipalities traces back to their institution during the 1789 Revolution based on par-
ishes from the Ancien Régime.[148] Their fragmentation allowed to pacify rural
reclamations and to prevent the emancipation of urban government from the central state (Fougère, Machelon, and Monnier 2002). Until the Third Republic,
mayors were designated by the prefects.[149] During the 20th century, municipalities obtained additional responsibilities, and with the decentralization acts of
1982 and 1983 also increased autonomy – in particular on request of mayors from
big cities (Le Saout 2011, 80–82). Namely, prefects no longer control municipal
acts a priori for their opportunity, but a posteriori for their legality only. Nevertheless, municipalities remain part of deconcentrated state services. For instance,
mayors supervise state law implementation and are hierarchically subordinated
to prefects in this respect. This explains why French municipalities are particularly deeply anchored within the political system and only slowly and recently
gained autonomy from the central state, which they are now reluctant to give up
for the purpose of inter-municipal cooperation (West 2007, 68).

    Despite the assumed uniformity of municipal units, municipalities of differ-
ent sizes organize in distinct umbrella organizations for interest representation
towards the central state, and for exchange on shared challenges. Examples include the Association of Small Cities of France (*Association des petites villes de
France*); Cities of France (*Villes de France*, formerly known as *Fédération des
Villes Moyennes*); and Urban France (France Urbaine), an association of metropolises, agglomerations and large cities that replaced the Association of Mayors
of French Large Cities (*Maires de grandes villes*) and the Associations of Urban
Communities (*Association des Communautés Urbaines de France*). A peak of
municipal lobbying in climate and energy policy occurred during the National
Energy Transition Debate. A key position paper was issued by the Association
of Mayors of French Large Cities together with umbrella organizations from the
intercommunal and regional level to the exclusion of smaller municipalities,

---

147   He further points out that nearly 90% of French municipalities have less than 2,000 inhabitants
       while the other 10% pool three quarters of the French population. Population density and fi-
       nancial resources from local business tax as well as state allocations vary importantly. As a
       result, municipalities with less than 2,000 inhabitants have five staff on average; the five big-
       gest cities have more than 15,000 employees, which obviously involves very different levels
       of political activity, communal services, and project capacities.

148   Decrees of 14 and 22 December 1789.

149   Acts of 28 March 1882, 5 April 1884 and 6 June 1895.

which further illustrates the divergence of interests within the supposedly uniform 'local bloc' (*bloc communal*) (Communautés urbaines de France et al. 2013).

Another form of multi-level interdependencies consists of local politicians' habit to accumulate political offices. The accumulation of offices (*cumul de mandats*) is more prevalent in France than in any other country worldwide and is particularly relevant to local governance, as Marrel (2011) points out. Municipalities and mayors in particular play a strong role in French politics; they embody local democracy and proximity to the citizens. The combination of executive office at the local level and legislative office at the national level facilitates local access to centralized resources with the deputy-mayor directly negotiating with ministries, departments, prefectures, and other local deputies. Local notabilities use the *cumul* to facilitate their reelection and to enhance their parliamentarian career, which entails a professionalization of local politics to the detriment of participation. Despite a modest decrease caused by regulation adopted in 1985 and 2000 which interdicts the accumulation of more than two and the combination of certain kinds of offices, approximately half of all parliamentarians, both at the National assembly and the Senate, still headed a local government in 2010 (Marrel 2011, 115–18). With more strict regulation introduced in 2014, this practice is about to disappear as of 2017.

Local governance in France can only be understood against the background of ongoing rescaling. Territorial organization has been a continuous construction site at least since the 1980s. The first wave of decentralization in 1982 and 1983 put an end to the state's hegemony over subnational governments as it endowed them with specific competences, but provided the prefectures with a coordinating and moderating role between the three legally autonomous and politically competing, but functionally interdependent levels of government (Epstein 2008, 140–41). Ever since, territorial reform has aimed to simplify the generated multi-level system. Several attempts were revoked by traditionalists during the second, mainly consolidating wave of decentralization (*Acte II*) in the 2000s. But municipalities and departments were gradually put into question in view of their suitability to cope with urbanization, globalization, and Europeanization; modernizers call for strengthening the regions, abolishing departments, and introduce direct universal suffrage for intercommunalities (Simoulin 2013, 6; West 2007, 67). The third and ongoing wave of decentralization (*Acte III*) engaged since 2013 has further pushed metropolization to the detriment of municipalities and departments;[150] it also scheduled regional mergers and further strengthened the role of regions (Ghorra-Gobin 2015; Sauviat 2016).[151] With regard to environmental governance, it clarified the repartitioning of responsibilities; for instance, regions take the lead in economic development, public transport, and land-use planning, while municipalities and intercommunalities obtained additional competences for sustainable mobility such as cycling and car-sharing (OECD 2016, 111).

Ongoing reforms and differentiated implementation make French decentralization a moving target hard to appraise. Subnational scopes of action coexist with more hierarchical modes of government and depend on local governments' aptitude to gain the necessary political and financial resources from the state and the EU (Le Galès 2013, 298–300). Local governments rely on business tax and state allocations that have not risen to the same degree as their responsibilities (Béal and Pinson 2013, 261). The succession of territorial reforms undermines planning reliability, also in terms of financial resources which particularly affects infrastructure investments (Simoulin 2013, 7).[152] In his fundamental critique of French decentralization, Epstein (2005; 2008) argues that the emancipation of French territorial governments has been interrupted because the central state became able to "govern at a distance" based on territorial competition. Cities in particular are faced with the disintegration of their local state interlocutors, the proliferation of state agencies, and the generalization of state programs relying on calls for proposals which indirectly increase central control and performance pressure. Thus, any analysis of territorial governance in France needs to carefully account for opposing trends in decentralization as well as for varying impacts across territories and sectors.

5.2.1.3 Inter-Municipal Cooperation within Intercommunalities

Inter-municipal cooperation in France dates back to the introduction of single-purpose associations in 1890.[153] In the beginnings, inter-municipal cooperation

---

150 MAPAM Act on the modernization of territorial public action and metropolises' affirmation (*loi de modernisation de l'action publique territorial et d'affirmation des métropoles*).

151 Act on the delineation of the regions (*loi relative à la délimitation des régions*) and the NOTRe Act on the new territorial organization of the French Republic (*loi portant nouvelle organisation territoriale de la République*).

152 An earlier specificity of French decentralization – significant growth of territorial budgets which further favored coalitions of actors supporting territorial governance – ended with the financial crisis (Le Galès 2013, 299).

153 Loi du 22 mars 1890 sur les syndicats de communes.

mainly served as a means to achieve the required economies of scale in the provision of technical public services such as water, waste, and transport (West 2007, 71). Since the 1960s, institutionalized forms of inter-municipal cooperation within so-called intercommunalities (*intercommunalités*) spread in response to peri-urbanization, i.e. population movements from city centers, but particularly from suburbs towards proximate rural municipalities, and to the resulting segregation (Le Saout 2011, 84–85), and to newly obtained responsibilities in the course of decentralization. As West (2007) demonstrates, these rather bottom-up processes underwent a shift when inter-municipal cooperation became a top-down strategy for rationalizing territorial administration during the 1990s; the central state since sought to facilitate, encourage, and shape inter-municipal cooperation. But municipalities showed considerable power to resist intercommunal integration as introduced in 1992,[154] with "a tendency [...] to avail themselves of the financial incentives for cooperation offered by the State, but to limit the extent of genuine cooperation" (West 2007, 81). In response, the central state pushed for less piecemeal, more federative and integrative forms of intercommunalities. Territorial reforms voted for in 1999[155] and 2004[156] became the starting point of intercommunalities' "spectacular rise" (Epstein 2008, 141), and they were further strengthened in the course of the third wave of French decentralization. Since 2013, intercommunal affiliation is mandatory for all municipalities.

Intercommunalities consist of public establishments for cooperation between local authorities (*Établissements publics de coopération intercommunale*, EPCI). EPCI with their own tax powers include urban communities (*communautés urbaines*) in metropolitan areas, agglomeration communities (*communautés d'agglomération*) in urban areas with less than 50,000 inhabitants, and communities of municipalities (*communautés de communes*) in rural areas.[157] Intercommunalities are governed by community councils (*conseil communautaire*) reflecting the size of member municipalities' populations (see Desage 2011, 283–84). The executive is made up of a president, several vice-presidents, and a number of intercommunal councilors. Since 2014, the latter are directly elected.

---

154   Joxe Act, also called "loi ATR" (Loi du 6 février 1992 relative à l'administration territoriale de la République).

155   Loi du 12 juillet 1999 relative au renforcement et à la simplification de la coopération intercommunale.

156   Loi du 13 août 2004 relative aux libertés et responsabilités locales.

157   EPCI with own tax powers will be referred to in the following as intercommunalities for readability reasons.

For purposes of interest representation, intercommunalities organize within the Assembly of French communities (*Association de communautés de France*).

Municipalities may transfer any responsibility to their intercommunality unless it is explicitly a mayoral duty; transferred functions are then to be exercised exclusively by the intercommunal level (West 2007, 70–71). Intercommunalities may only act with prior agreement of member municipalities with the exception of a limited number of mandatory functions. Climate-related intercommunal responsibilities include public service provision and a coordinating role in local policy. Intercommunalities act as local service providers in the fields of transportation, waste management, and – to a lesser extent – water treatment (OECD 2016, 110). With regard to inter-municipal coordination, they have a leading role with regard to spatial planning and (sustainable) development. Environmental protection is a mandatory task for urban communities and a semi-optional function for agglomeration communities and communities of municipalities.

Metropolises represent a special case of inter-municipal cooperation. The creation of the metropolitan regions Ile-de-France/Paris, Lyon, and Marseille in 1982 represented a renunciation to the French dogma of uniform administration (Martínez Soria 2007, 1026). With territorial reform in 2010, metropolises were introduced as an additional kind of EPCI that was further strengthened in 2014 by the MAPTAM act. The latter extended the metropolitan regime to several intercommunalities, including Rennes Metropolis (see section 5.2.3). Greater Lyon obtained a special status as a territorial government (as opposed to an EPCI) (see section 5.2.2).

French intercommunalities have long seemed to be hardly politicized, mistakenly perceived as administrative units not associated with major political cleavages. In fact, intercommunalities have become major public actors, an important public employer, and their budgets largely exceed those of the regions (Le Saout 2012, 11). Correspondingly, rescaling in favor of intercommunalities, and all the more of metropolises, entails heterogeneous and also competing preferences, with intercommunal leaders and mayors of center-cities typically being in favor of intercommunal coordination while delegates from smaller municipalities show more resistance (Ben Mabrouk 2007, 107).

In this context, it is interesting to note that intercommunalities have so far been exempted from *cumul* regulation. Out of 214 intercommunalities, Marrel (2011) finds 20% to be headed by presidents who hold another mandate at the same time, most often as mayors and as delegates at the department level (*con-*

seillers généraux), sometimes combined with a mandate in national parliament. Intercommunal *cumul*, he argues, stabilizes elitist and clientelist regimes, hampers intercommunal democracy, and obscures delegates' affiliations and loyalties. In 2014, anti-*cumul* regulation was introduced at the intercommunal level along with direct suffrage; intercommunal offices may no longer be held along with other mandates as of 2017.

Competing loyalties also hamper intercommunal capacity building. Négrier, Teillet, and Préau (2008) find that intercommunalities' capacities to seize new responsibilities are very much limited by their own efforts to avoid coming into conflict with municipalities and departments, and by tensions between member municipalities. They point out that local specificities resulted in varying implementation of rescaling, and find that intercommunal capacity building is further decelerated by the inconsistent timing of competence transfers, projects, and elections.

5.2.1.4 Local Climate Governance

While the territorial puff pastry is under attack for its overlap of competences and funding, Béal and Pinson (2013) argue that in response to the resulting need for dealing with these interdependencies in an integrated way, it has actually been a natural seedbed for sustainable governance practices based on procedural provisions and local dynamics of collective action since the 1990s. After a slow start and increasing support from the central state and the regional level since the millennium (Ollitrault 2011, 171–72), there are now nearly 800 municipal and 250 intercommunal Local Agenda 21s throughout France (Comité 21 2017a).[158] Since the mid-2000s, the rise of top-down climate policies based on quantitative and technical objectives weakened these incremental and deliberative practices in favor of a neo-managerial restructuring (Béal and Pinson 2014). But Local Agendas 21 continue to be supported at the national and regional level (Comité 21 2017b), and their climate change sections serve as a basis for local climate plans.

The PCET as a regulated form of local climate planning was first mentioned in the *Grenelle* recommendations in 2007. The *Grenelle I* Act of 2009 encour-

---

158 Committee 21 (Comité 21), a network of nearly 500 members from the realms of subnational governments, businesses, associations, and others established in 1995, provides methodological support and a best practice database Comité 21 (2017c).

aged subnational governments to engage in the procedure; in the following year, it became mandatory for subnational governments with more than 50,000 inhabitants with the *Grenelle II* Act. At the end of the year, the ADEME published a handbook with methodological guidelines for establishing a PCET. As it built on Local Agenda 21 methodology, subnational governments may develop the PCET as one of five Agenda 21 pillars (Klima-Bündnis, Energy Cities, and Klimaschutz- und Energieagentur Baden-Württemberg 2014, 22).

PCET procedures envisage four steps: preparations, inventory and mobilization, planning, and implementation. GHG emission inventories may cover emissions related to public property and regulatory competences only, or include the entirety of GHG emissions from the territory, and may choose between different calculation methods for direct and indirect GHG emissions (*scope 1, 2, 3*). Mandatory fields of local climate action to be covered consist of public property and regulatory competences. In addition, a PCET might also involve contributions from other fields and actors. Since the PCET represents a territorial project rather than a single document, its development takes between 18 months and 3 years (Klima-Bündnis, Energy Cities, and Klimaschutz- und Energieagentur Baden-Württemberg 2014, 18).

Just as the PCET is less common among smaller municipalities, the latter also participate in Cit'ergie, the French counterpart of the EEA program, to a lesser degree. Out of 100 participating municipalities since 2007, only 31 have less than 50,000 inhabitants (European Energy Award 2016a).[159] Cit'ergie is managed by the ADEME that also provides technical assistance and financial support to participating municipalities, and promotes the procedure as an "operational tool for structuring and executing" a PCET, Local Agenda 21, or Covenant commitment (European Energy Award 2016a).

For the implementation of local climate policies, some local governments – including those examined in the three case studies below – can rely on local energy and climate agencies (*agences locales de l'énergie et du climat*, ALEC). These emerged in France as of 1994 in response to a supporting scheme of the European Commission, but have not spread comprehensively. To date, about 35 ALEC with a total of about 300 staff provide consultancy and training for territorial governments, private persons, house owners, and companies with regard to energy and water economies and renewable energies through events, tailored

---

159 Despite the preponderance of small municipalities in France, see section 5.2.1.2.

assistance, and education (Fédération des agences locales de maîtrise de l'énergie et du climat 2017). Since 2004, the ALEC are federated within the Federation of local agencies for energy management and climate (*Fédération des agences locales de maîtrise de l'énergie et du climat*, FLAME).

Local governments rely on several national associations for exchanging experiences, and obtaining information and expertise on climate and energy policy. This includes working groups of local councilors within umbrella organizations such as Urban France, and issue-specific associations that rather address municipal staff such as AMORCE. The transnational municipal climate network most often adhered to by French municipalities is Energy Cities, a member of the CoMO consortium. Founded in 1990 and based in Brussels and Besancon, Energy Cities describes itself as the European association of local authorities in energy transition. It represents over 1000 local authorities from 30 countries; 64 out of 183 individual members stem from France (Energy Cities 2017). Energy Cities' visibility in France is also linked to the co-organization of well-known events such as the annual energy conference (*Assises de l'énergie*), which is held alternately in Dunkirk and Grenoble, two cities known as French sustainability frontrunners. On these occasions, Energy Cities also convenes meetings of the Club France in the Covenant.

5.2.2   Greater Lyon

The Urban Community of Lyon (*Communauté urbaine de Lyon* or *Grand Lyon*, in the following referred to as Greater Lyon) represents France's second-largest agglomeration in terms of inhabitants and economy.[160] Situated in the Rhône-Alpes Region in Western France, the prosperous intercommunality[161] with its significant chemical industry sector comprises 59 municipalities (see figure 5.5).[162] It is structured around the center city Lyon which is also dominating the

---

160   After the Parisian region.
161   In France, inter-municipal associations are institutionalized as so-called intercommunalities, more precisely public establishments for intercommunal cooperation (établissements publics de coopération intercommunale, EPCI); see section 5.2.1.3.
162   Greater Lyon comprises Albigny-sur-Saône, Bron, Cailloux-sur-Fontaines, Caluire-et-Cuire, Champagne-au-Mont-d'Or, Charbonnières-les-Bains, Charly, Chassieu, Collonges-au-Mont-d'Or, Corbas, Couzon-au-Mont-d'Or, Craponne, Curis-au-Mont-d'Or, Dardilly, Décines-Charpieu, Ecully, Feyzin, Fleurieu-sur-Saône, Fontaines-Saint-Martin, Fontaines-sur-Saône, Francheville, Genay, Givors, Grigny, Irigny, Jonage, La Mulatière, La Tour-de-Salvagny, Limonest, Lissieu, Lyon, Marcy-l'Etoile, Meyzieu, Moins, Montanay, Neuville-sur-Saône, Oullins, Pierre-Bénite, Poleymieux-au-Mont-d'Or, Quincieux, Rillieux-la-Pape, Rochetaillée-sur-

intercommunality in political and administrative terms. Out of approx. 1,360,000 inhabitants of Greater Lyon, about 510,000 live in Lyon and another 150,000 in the adjacent city Villeurbanne. Yet, the peri-urban belt is highly urbanized and not as fragmented as in other areas of France; another 21 municipalities have populations of more than 10,000, and no municipality has less than 1,100 inhabitants (Base nationale sur l'intercommunalité 2016b).[163]

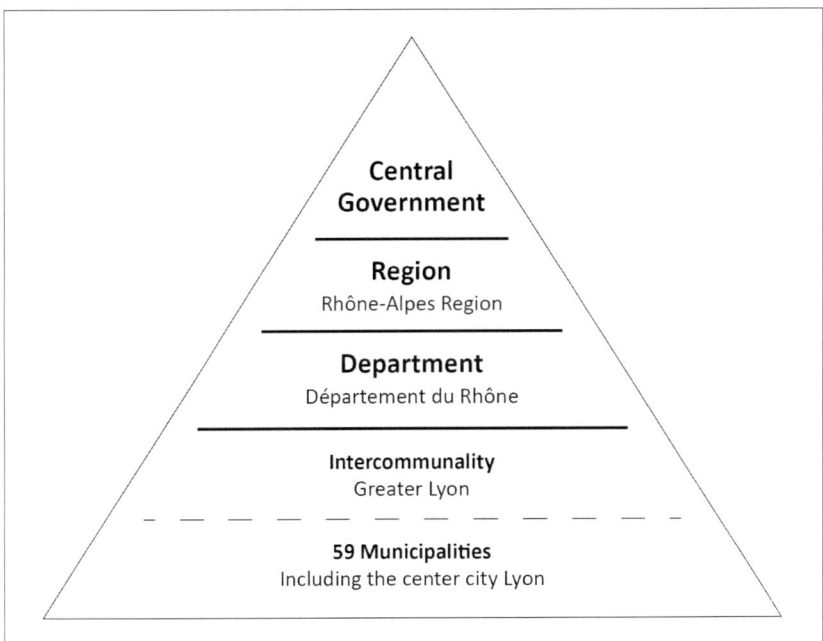

Figure 5.5: Territorial organization of Greater Lyon
Source: Author.

Greater Lyon's legislative body is the community council (*conseil communautaire*) chaired by an intercommunal president. By tradition, this is the Lord

Saône, Saint-Cyr-au-Mont-d'Or, Saint-Didier-au-Mont-d'Or, Sainte-Foy-lès-Lyon, Saint-Fons, Saint-Genis-Laval, Saint-Genis-les-Ollières, Saint-Germain-au-Mont-d'Or, Saint-Priest, Saint-Romain-au-Mont-d'Or, Sathonay-Camp, Sathonay-Village, Solaize, Tassin-la-Demi-Lune, Vaulx-en-Velin, Vénissieux, Vernaison, and Villeurbanne.

163   During the study period, the population was slightly smaller. Out of about 1,330,000 inhabitants, 500,000 lived in Lyon, and 146,000 in Villeurbanne Base nationale sur l'intercommunalité (2014).

Mayor of Lyon city. Since 2001, Greater Lyon is headed by socialist Gérard Collomb, (*Parti socialiste*, PS), who was reelected for a third term at the local elections in March 2014. Community deputies are nominated by their municipal council of origin. Municipalities are represented proportionally but with at least one community deputy. During the study period, the community council consisted of 156 deputies, including 40 vice-presidents, and the intercommunal administration comprised more than 4,700 staff members (Fournier 2012). The budget of Greater Lyon amounted to 1.6 billion Euros in 2009 and has further increased since (Communauté urbaine de Lyon 2009; Métropole de Lyon 2016). Greater Lyon raises a unique business tax (*taxe professionnelle unique*) and a solidarity payment (*dotation de solidarité communautaire*) determined by the intercommunal council which is intended to counterbalance economic inequalities across municipalities. In operation since January 1, 1969, Greater Lyon has obtained a growing range of responsibilities, including in the areas of energy, water, waste, and environment, urban development, economic development, land use planning, transport and roadworks, and housing.[164]

Shortly after the study period, Greater Lyon underwent institutional reform. As of January 1st, 2015, it has become a so-called metropolis with a particular status as introduced by the MAPAM Act of 2014. In contrast to regular metropolises, it is no longer an EPCI, but a fully-fledged subnational government (see section 5.2.1.3). It now unites intercommunal and departmental competences,[165] and holds a general competence for all issues of metropolitan interest (Métropole de Lyon 2015). With this reform, Greater Lyon continues to serve as a laboratory for territorial governance. It has repeatedly obtained additional competences and tested new forms of collaboration and coordination on its territory. Another territorial reform occurred at the regional level. On January 1, 2016, the Rhône-Alpes Region was merged with the Auvergne region.[166] Experimentation with new forms of territorial governance and the prospect of territorial reforms have brought about considerable institutional dynamic. Local climate policy coordination has strengthened the intercommunality Greater Lyon as a collective actor

---

164   For a full list of its 44 competences, see Base nationale sur l'intercommunalité (2016b).
165   The *Département du Rhône* has been reestablished on the territory of its remaining 228 municipalities.
166   The merger was established in the course of territorial reform in 2014 (loi no 2015-29 du 16 janvier 2015 relative à la délimitation des régions, aux élections régionales et départementales et modifiant le calendrier électoral). Provisorily, the Auvergne-Rhône-Alpes Region has its seat in Lyon.

and allowed for its president Collomb to profile as a political leader. The case of Greater Lyon shows how a local government can expand its role in territorial governance by means of climate policy coordination (see section 5.2.2.1), and how a local leader can seize the issue for his political career (see section 5.2.2.2). Continuous activity in local climate policy coordination coincides with only punctual engagement with the Covenant. This calls for a differentiation between the intensity of an orchestration-intermediary relationship and the contribution of an intermediary to shared governance goals, and between the behavior of the intermediary as a collective actor and its internal politics (see 5.2.2.3).

5.2.2.1   Asserting the Metropolitan Level of Government

Environmental issues arose on the agenda of Greater Lyon in the early 1990s during the mandate of Michel Noir, Gaullist mayor of Lyon and intercommunal president. In the absence of legal competences for environmental protection, the intercommunality acted upon its more technical missions such as water, waste, and planning. The department for urban development became home to a newly established Urban Ecology Mission (*Mission écologie urbaine*) which was in charge of drafting the first and second Urban Ecology Charters (1992–1995, 1997–2001) of Greater Lyon. In particular, Greater Lyon has a record in dealing with air pollution, especially since the reinforcement of its competences with the Chevènement Act of 1999 (Pastille Consortium 2002, 31–34).[167] In the same year, Greater Lyon signed the Aalborg Charter, thereby expressing its intention to engage in a Local Agenda 21 process.[168] The Local Agenda 21, including its section on climate policy, was adopted at the community council in March 2005. Starting that same year, a territorial coherence scheme (*schema de coherence territorial*, SCoT) was developed in a concertation process with member municipalities, neighboring intercommunalities, and civil society. Adopted in Decem-

---

167   See section 5.2.1.2.

168   The Aalborg Charter is an urban environment sustainability initiative launched in 1994 at the first European Conference on Sustainable Cities and Towns in Aalborg, Denmark, with a participatory approach to sustainable development. Developed to contribute to the European Union's Environmental Action Program of the time, 'Towards Sustainability', the Aalborg Charter includes a commitment to establishing tangible local targets and a schedule to achieve them, evaluating progress regularly, and reporting to the European Sustainable Cities & Towns Campaign. With more than 3,000 signatories, the European Sustainable Cities & Towns Campaign represents the largest bottom-up movement resulting from UNCED's Agenda 21, and claims to have "prepared the ground" European Sustainable Cities Platform (2015) for other schemes such as the Covenant of Mayors.

ber 2010, it prioritizes land use and urban planning as the main axes of local climate mitigation, namely redensification, the development of railways and public transport, and the preservation of urban green spaces (Barbey 2012, 51).[169] Urban planning had also been at the core of RE-START, an EU-funded project with regard to energy-efficient housing that started in 1997, and of CONCERTO-RENAISSANCE, an EU-funded urban regeneration project aiming to create a sustainable model district in Lyon Confluence as of 2003. Greater Lyon participated as lead partner on land use planning in another EU-project dealing with local climate policy in a narrower sense: Adaptation and Mitigation – an Integrated Climate Policy Approach (AMICA), running from July 2005 to December 2007 (INTERREG IIIC 2009). Greater Lyon is also home to Vélo'v, the first bicycle sharing system in France launched in 2005. To summarize, Greater Lyon repeatedly seized international and national schemes to assert its role as a pioneer in sustainable urban planning, and participated in EU-projects to showcase innovative energy-related projects.

In contrast, the regional climate, air and energy scheme (SRCAE) of the Rhône-Alpes Region was a long time coming due to disagreement between the regional president and the prefect (interview with C. Revol-Buisson, 9 July 2013). Due in July 2011, the SRCAE was approved only in April 2014. In the meantime, Rhône-Alpes Region refrained from a strong coordinating role (interview with M. Athiel, 11 July 2013; interview with L. Bernard, 9 July 2013). Like 34 other local governments within the Rhône-Alpes Region that were obliged to establish a territorial climate and energy plan (PCET), Greater Lyon was left without authoritative guidance. Cooperation between Greater Lyon and the Rhône-Alpes Region consisted mainly in regional funding for pilot projects as opposed to common planning. Also in view of Greater Lyon's considerable financial and human resources, the region focused its means on the support of more rural local governments from its territory (interview with J. Barbaroux, 9 July 2013; interview with C. Revol-Buisson, 9 July 2013).

Thus building on its own experience rather than regional support, Greater Lyon was part of the first wave of Covenant signatories and decided to join the newly-established initiative as early as September 2008. The community council

---

169   Managed by the Syndicat mixte d'études et de programmation de l'agglomération lyonnaise (SEPAL), an agency created in 1985, the SCoT 2030 de l'agglomération lyonnaise applies to Greater Lyon and the adjacent intercommunalities Communauté de Communes de l'Est Lyonnais (8 municipalities) and Communauté de Communes du Pays de l'Ozon (7 municipalities).

estimated that the EU-initiative was *"in total coherence with policies conducted by the urban community"*[170] and expected to meet the commitment with documents and policies already at hand or underway (Communauté urbaine de Lyon 2008). The main goals associated with this step were to visualize Grand Lyon's commitment – also by using the Covenant's logo – and to alter its credibility by this kind of "contract with Europe". Being part of a growing and recognized network at the European level reassured local deputies they were on the right track and clearly showed to the public that local climate action was no amateurish undertaking, but a professional project (interview with P. Crépeaux, 4 July 2013). In addition, the Covenant was expected to allow for exchange and lobbying: *"It's about being recognized, but also about recognizing other active cities"* (interview with F. Moudileno, 10 July 2013).[171]

Greater Lyon's Covenant experience was soon tempered by unexpected difficulties and disappointments. To start with, the intercommunality took some time to submit its BEI. Established in cooperation with Air Rhône-Alpes, the regional air quality agency, it had to be laboriously adapted to the requirements and excel sheets of the Covenant (RhônAlpEnergie Environnement 2012). Once Greater Lyon submitted its SEAP, it pledged to become a Covenant Coordinator. In this respect, Greater Lyon was not particularly successful. Only one more municipality joined the Covenant. Villeurbanne signed the Covenant in November 2009 but was temporarily suspended from 2012 to 2013 due to delayed reporting. Rillieux-la-Pape temporarily figured on the Covenant website but then decided against a formal commitment. At a regional meeting of Covenant signatories, a representative from Greater Lyon deplored a lack of support from the CoMO in implementing its commitment as a Covenant Coordinator:

> "Actually, there is no specific office for supporting structures. That would allow for having a contact person in case of encountering difficulties, but also and mainly at the beginning of signature an accompaniment seems to be in order so as to know how to carry out this mission" (RhônAlpEnergie Environnement 2012, 2–3).[172]

---

170   Author's translation; original quote: *"en totale cohérence avec les politiques poursuivies par la Communauté urbaine"*.

171   Author's translation; original quote: *"C'est être reconnu, mais aussi reconnaître d'autres villes actives."*

172   Author's translation; original quote: *"En effet, il n'existe pas de bureau spécifique pour les structures de soutien. Cela permettrait d'avoir un interlocuteur en cas de difficultés rencontrées, mais c'est aussi et surtout au début de la signature qu'un accompagnement semble être demandé afin de savoir comment mener cette mission."*

Existing networks, and notably Energy Cities, turned out to be more helpful in terms of obtaining information and exchanging ideas with peers:

> "That's the interlocutor that one identifies as being the animator and structure that circulates information here [in France]" (interview with F. Moudileno, 10 July 2013).[173]

In the course of a change in staff, the commitment of Greater Lyon as a Covenant Coordinator sank into oblivion in the inter-municipal climate department. Due to a lack of reporting on Coordinator activities, Greater Lyon was later relabeled a simple signatory by the CoMO.

The unfruitful record of Greater Lyon in terms of promoting Covenant membership among its member municipalities should not be mistaken for a failure in terms of local climate policy coordination. Politically, the development and follow-up of Greater Lyon's climate plan was strongly supported by the ecologist inter-municipal vice-president in charge, Bruno Charles (*Europe Écologie – Les Verts*) (interview with C. Alice, 10 July 2013; interview with G. Ancel, 16 July 2013), who had already advocated Covenant signature (interview with P. Crépeaux, 4 July 2013). Within the administration, Greater Lyon has further institutionalized its commitment to local climate policy. It has a climate department in charge of inter-municipal climate planning that also cultivates contacts to other administrative departments and external networks such as the Covenant and maintains a network of partners for inter-municipal climate action from the territory. In addition, Greater Lyon incrementally established an energy service in order to pool dispersed responsibilities and to improve the integration of efficiency targets in energy-related policy.

Greater Lyon's approach to local climate planning relies on a participatory process and on the concertation of territorial actors beyond member municipalities. A territorial energy and climate conference (*Conférence Énergie Climat*) brings together social housing enterprises, local companies, energy suppliers, academic institutions, and NGOs. Building upon an GHG emissions inventory adopted in 2009 (*Diagnostic Climat*), the territorial climate and energy conference developed several scenarios for Greater Lyon to meet the 20-20-20 targets by 2020, and the factor four target by 2050.[174] Summarized in the 2020 Vision

---

173   Author's translation; original quote: "*C'est l'interlocuteur qu'on identifie comme étant animateur et structure qui fait tourner l'information ici.*"
174   See section 5.2.1.3.

for a low-carbon agglomeration (*Vision 2020 pour une agglomération sobre en carbone*) adopted in 2010, these scenarios served to present all actors with tangible options to contribute to local climate action. Subsequently, participants of the territorial climate and energy conference defined their contributions. All climate mitigation measures, both ongoing and projected, of Greater Lyon and the participants of its territorial climate and energy conference were consolidated in the Cooperative action plan (*Plan d'actions partenarial*). After discussion at the territorial climate and energy conference in November 2011, this compilation was formally adopted as the inter-municipal PCET at the community council in February 2012. It was later submitted to the CoMO as Greater Lyon's SEAP and was formally approved on February 13, 2012. To date, the implementation of Greater Lyon's climate action plan has been evaluated twice. Out of about 150 organizations participating in the territorial climate and energy conference, more than 55 have actively engaged in implementing the cooperative action plan (interview with L. Ponsar, 13 June 2013). Updates on implementation (*Points d'étape*) were published on the occasion of the territorial climate and energy conference meetings in 2013 and 2015. Interestingly, no monitoring is mentioned in Greater Lyon's profile on the Covenant website, i.e. the *points d'étape* have not been translated into Covenant methodology and submitted as implementation reports.

Member municipalities of Greater Lyon are not at the core of concertation within the territorial climate and energy conference; they are helped by the Local energy agency of the Lyon agglomeration (*Agence Locale de l'Energie de l'agglomération lyonnaise*, ALE)[175] mandated by Greater Lyon.

> "The aim of Greater Lyon was to not be the only one having a climate plan, but to nurture climate plans in its surroundings which, if possible, should be as complementary, or as coherent, as possible" (interview with F. Moudileno, 10 July 2013).[176]

---

175   The ALE was established in 2000 following two European projects on local energy. It aims to support energy saving, renewable energy, and climate mitigation by means of information, consultancy, and training. Sectors targeted include housing, companies, public buildings, and local sustainable development including Territorial Climate and Energy Plans. Other than housing companies, businesses, and territorial authorities, it also advises citizens within the scope of its mission as Energy Information Space (*Espace Info Energie*) as mandated by the ADEME and the Rhône-Alpes Region (ALE 2015).

176   Author's translation; original quote: "*Le but du Grand Lyon, c'était de ne pas être seul à avoir un Plan climat, mais de faire pousser autour de lui des Plans climat, et qui soient, si possible, le plus complémentaires possible, ou le plus convergents possible.*"

In response to the lack of policy instruments, the ALE developed a distinct climate planning tool for member municipalities of Greater Lyon in 2009:[177]

"It was a methodological desert. The municipalities really had no clue at all where to start, how to deal with this issue of a climate plan" (interview with F. Moudileno, 10 July 2013).[178]

Also, the ALE conducts emission monitoring on behalf of municipalities, and convenes a working group on climate action plans.[179] Regular meetings every six months for administrative staff serve to accompany and accelerate municipal climate action plans. Other than mutual updates on progress and upcoming projects and technical exchange on challenges and difficulties between participants, the format allows for targeted input and training by invited speakers in response to municipal needs. Staff from Greater Lyon and from deconcentrated state services, the ADEME Rhône-Alpes and the DREAL Rhône-Alpes, joined the meetings to obtain comprehensive updates. Started in 2008, the program gained momentum as of 2010 when a group of 29 municipalities willing to engage in local climate planning was established (interview with F. Moudileno, 10 July 2013). Personal assistance to municipal staff and a blended education scheme for deputies and staff[180] further supplemented the support to municipalities by the ALE (interview with M. Athiel, 11 July 2013). These activities have neither been reported to the CoMO, nor are they communicated as being part of Greater Lyon's Covenant commitment. Other than Greater Lyon itself and Lyon city, 10 other municipalities from the territory developed a territorial climate and energy plan (PCETs)[181] (ADEME 2015b). Only Villeurbanne and Vénissieux were obliged

---

177 The Municipal climate plan toolkit (*Mallette plan climat communal*) is based on European tools, notably the Climate Compass, a Climate Alliance methodology to identify, implement and visualize progress on local climate actions. For the *Mallette*, an ALE collaborator adapted the Climate Compass to the French and the Lyonnais context, notably the repartition of competences. The *Mallette* enables a municipality to evaluate its state in terms of climate action within four or five days, and especially to compare itself to others and identify strengths and tardinesses (Moudileno 2013).

178 Author's translation; original quote: "*C'était un désert méthodologique. Les communes n'avaient vraiment aucun repère sur: Par où commencer, par où prendre ce sujet de plan climat.*"

179 Greater Lyon also animates a similar working group on local Agenda 21 processes.

180 The training carried out in cooperation with Beam21 (Blended Capacity-Building on Sustainable Energy Measures and Action Plans for European Municipalities) comprised 50 hours of free training for several agents and deputies. Still an innovative format within the French landscape of trainings for municipal staff, half of the program consisted in e-learning.

181 For an introduction to the French territorial climate and energy plans (*plans climat énergie territoriaux*), see section 5.2.1.3.

to because their population exceeds 50,000 inhabitants; eight smaller municipalities engaged voluntarily.[182] Additionally, Lyon, Vénissieux and Rillieux-la-Pape engaged in a Cit'ergie[183] procedure under the auspices of Greater Lyon aiming at a more territorial vision of urban policy (interview with L. Bernard, 9 July 2013). In view of establishing an inter-municipal emission monitoring, the ALE also promotes a slimmed-down version of the *Bilan Carbone* as the initial method developed by the ADEME was perceived by municipalities as being too costly (interview with F. Moudileno, 10 July 2013). Covenant signature, in contrast, was perceived as premature by several interviewees but considered possible once municipalities made sufficient progress on their PCET to be sure to be able to meet a Covenant commitment.

Through the territorial climate and energy conference and concertation with member municipalities, Greater Lyon asserts itself as a site for local climate governance. This strategy resembles rescaling efforts towards the inter-municipal level in other policy fields. Academically educated deputies and administrative staff who rely on scientifically-founded instruments to remedy urban dysfunctions introduce and justify new modes of urban governance. These, at the same time, enable local deputies to distinguish themselves, to enhance their legitimacy, and to associate themselves with modernism and pragmatism. Initially developed in the field of urbanism,[184] these new modes of governance in Greater Lyon are marked by consultations with actors, collaboration with relevant professions, and forecasting. During the 1990s, the intercommunality of Lyon still struggled to legitimize itself towards neighboring territorial governments and small, peripheral member municipalities as well as towards the newly emerged networks of actors in urban development (Ben Mabrouk 2007, 109–11). Building upon new modes of governance as applied to local climate policy, metropolization has since been successfully carried forward.

---

182 Saint-Fons, Bron, Corbas, Curis-au-Mont-d'Or, Décines, St Germain-au-Mont-d'Or, La Mulatière, and Rillieux-la-Pape.

183 Cit'ergie represents the French version of the European Energy Award (EEA); see section 5.2.1.3.

184 Namely, these new modes of governance were introduced in the field of urban planning within the urbanism agency. The transformation of the urban space and its understanding started with conferences and workshops held by the urbanism agency. It continued with an increased number of intercommunal projects reflecting the framing of political and administrative leaders of the urbanism agency, and was supplemented by help of symbolic reference to terms such as metropolis, strategy, and comparison to other European agglomerations Ben Mabrouk (2007, 109–11).

The active role of Greater Lyon in local climate governance has been actively supported by its intercommunal president. For example, Collomb's presence at the territorial energy and climate conference demonstrated the importance of this policy for the intercommunality. His support and established contacts to local enterprises were decisive in convincing partners to join (interview with C. Alice, 10 July 2013). A closer look at Collomb's engagement shows how he seized climate governance as an additional arena to distinguish himself.

5.2.2.2 Climate Leadership as a Building Block of a Political Career

Covenant membership of both Greater Lyon and Lyon are closely linked to the engagement of Collomb, their president and mayor. Collomb is a textbook example of the widespread *cumul de mandats* (accumulation of offices) across the French puff pastry of territorial governance (see figure 5.6).

Level of government	Territorial entity	Mandate
National	(Elected within the Rhône department)	Member of the Economic and social council 1994–1999, Senator since 1999
Regional	Rhône-Alpes	Deputy 1992–1999
Departmental	Rhône	Deputy 1981–1988
Intercommunal	Greater Lyon	President since 2001
Municipal	City of Lyon	Deputy since 1977, district mayor since 1995, Mayor since 2001

Figure 5.6: Accumulation of offices by Gérard Collomb
Source: Author.

Collomb started his political career at the municipal level and consecutively obtained additional offices at the departmental and regional levels during the 1980s and the 1990s. Around the turn of the century, he was elected to the offices he holds until today: the French senate and the presidency of Greater Lyon. Since, Collomb seized the issue of local climate policy across all levels of governance he had access to, and actively supported the creation of the Covenant.

The plurality of his offices has endowed Collomb with a vast network of contacts, also in the fields of sustainable development and climate and energy policy. Repeatedly, he made use of his accumulation of offices by hosting national and international events in Lyon. By facilitating the respective conferences, Collomb positioned Lyon as a relevant site and actor of governance and at the same time obtained personal visibility both towards the community gathering at the respective event and towards the local public, his direct constituency. For instance, since 2002, Lyon hosts the *Dialogues en humanité*, a yearly conference on sustainable development.[185] In 2005, Lyon also offered to host the 2006 COP 12. Although the conference was instead given to Nairobi, Kenya, Lyon's candidacy engaged a number of Lyonnais, French, European and international supporters (interview with G. Ancel, 16 July 2013).

Building on his network, Collomb became an instrumental actor in the creation of the Covenant. He had met Gérard Magnin, director of Energy Cities and one of the spin doctors of the Covenant,[186] in the run-up to the UN World Summit on Sustainable Development (WSSD) 2002 in Johannesburg. While Magnin was lobbying the European Commission to cooperate with local governments on energy efficiency, Collomb held the presidency of Eurocities from 2006 to 2008. He wanted to crown his presidency with targets clearly achieved, notably the establishment of the Covenant. In October 2008, he presented the Eurocities Declaration on Climate Change that refers to the Covenant as "an important initiative […] which […] will strengthen the partnership between the European and local levels" and explicitly "supports and complements the aims of the Covenant of Mayors on Energy" (Eurocities 2008, 5). The presentation of the Eurocities Declaration on Climate Change in Lyon in October 2008 took place under the French presidency of the EU, thereby bringing several high-ranking politicians to Collomb's home city: representatives of the European Commission, French Prime Minister François Fillon, and Jean-Louis Borloo, French Minister for ecology, sustainable development, and land use (Communauté urbaine de Lyon 2008). Consequently, Collomb signed the Covenant of Mayors on behalf of Lyon city as well as Greater Lyon, and engaged in convincing other cities from his networks to sign the Covenant, thereby contributing to its early success in membership numbers (interview with G. Ancel, 16 July 2013).

With the strategic use of network offices and the targeted hosting of conferences in Lyon, the issue of climate policy served to enhance Collomb's personal

---

185   In a similar vein, Lyon hosted a conference on information technology solidarity held by the *Fonds de Solidarité Numérique*, a Lyon-based international foundation, on November 24, 2008. Collomb had been elected president of the *Agence mondiale de solidarité numérique*, the foundation's operative agency, in July 2005.

186   See section 3.2.1.2.

visibility. Collomb had considered declaring his candidacy for the presidency of the French Republic in the run-up to the socialist primary elections in autumn 2011, but decided to support François Hollande, the socialist candidate who was elected President of France in 2012 (AFP 2011). In the aftermath of this decision against a candidacy at the national level, Collomb concentrated on preparations for institutional reform in Greater Lyon and the goal of obtaining the status of a metropolis as of 2015. At the municipal elections in 2014, Collomb's withdrawal from national politics went one step further. As electoral trends favored his conservative rival candidate, Collomb distanced himself from the socialist government. For instance, he did not meet with any minister visiting Lyon during the course of his electoral campaign (Mercier 2014). With this strategy, Collomb obtained a third mandate as mayor of Lyon against national trends, and was also reelected to the presidency of Greater Lyon at the community council shortly afterwards.

While climate policy was not a priority during the campaign, Collomb continued to position Greater Lyon and Lyon city as sites and players of climate governance after the end of the core study period in the run-up to the COP 21 in Paris in 2015. For instance, Collomb signed a statement of several European mayors that called for an enhanced role of cities in European climate policy, and included a commitment to reinforce local climate plans (2015). When Lyon hosted the World Summit on Climate and Territories – Societies in Motion on July 1–2, 2015 organized by a group of environmental NGOs and networks of territorial authorities, Collomb appeared with national leaders again as he attended the event alongside François Hollande, President of France, and Annick Girardin and Ségolène Royal, both government ministers.

Summing up, President Collomb was instrumental in institutionalizing local climate policy at Greater Lyon and in establishing the Covenant. He repeatedly seized upon the issue of climate change as an arena for positioning both himself and his metropolis in French politics. While environmental protection had long been an issue sustained by technical staff at Greater Lyon, it lacked political support (Pastille Consortium 2002, 34–36). Collomb, in contrast, endorsed inter-municipal climate planning (interview with C. Alice, 10 July 2013; interview with G. Ancel, 16 July 2013). His engagement in favor of the Covenant and of local climate policy more generally demonstrates how a political leader can take up an issue to advance his career. This assessment does not presume to evaluate or challenge his personal convictions. Rather, it puts Collomb's continuous climate activism in the context of territorial politics.

5.2.2.3 Case Summary

Greater Lyon is actively lobbying for increasing the metropolises' role in French territorial governance. Serving as a laboratory for territorial reform, the intercommunality constantly seeks to assert itself as the appropriate site of governance, including with respect to climate and energy policies. Due to ongoing institutional reform, intergovernmental relations between Greater Lyon and its municipalities as well as superordinate levels of government are determined, to a large extent, by organizational interests. President Collomb availed himself of territorial governance to expand the role of his home metropolis, and his own role within that metropolis and beyond.

In his position as a president of Eurocities, Collomb participated in lobbying efforts for the creation of the Covenant and contributed to the numerical success of the Covenant in its early days. For Greater Lyon, Covenant Coordinator status constituted a European endorsement that legitimized inter-municipal concertation efforts for local climate policy. Once the territorial climate and energy conference had been established, the Covenant, to a certain extent, fell into oblivion. This is not to say, though, that Greater Lyon's commitment to climate action was insincere. An assessment of its achievements has not been the focus of this case study, but the intercommunality undoubtedly has a certain record of sustainable development and climate action since the early 2000s.

What looks like a failure in terms of local climate policy coordination turns out to be a particularly active intercommunality. This finding points to the need to carefully evaluate cases of orchestration. Performance with regard to the orchestrator-intermediary relationship and to their shared governance goals can diverge. The case of Greater Lyon shows how a difficult orchestrator-intermediary relationship coincides with a high level of engagement on the side of the intermediary in the pursuit of shared governance goals. Also, the case of Greater Lyon calls for a differentiation between organizational interests of an intermediary, and the interests of its political leaders. Naturally, these interests tend to overlap, but need not to coincide completely. In order to fully understand the behavior of an intermediary, it is important to consider the intermediary both as a collective actor and in terms of internal processes.

5.2.3 Rennes Metropolis

The case of Rennes Metropolis (*Rennes Métropole*) as a Covenant Coordinator stands out in terms of successful Covenant promotion among member municipalities, but exhibits challenges in the implementation stage. A closer look at the case shows how the intercommunality seizes climate policy coordination as an arena for demonstrating its value added in terms of public policy and services provision by means of support to fragmented member municipalities.

Situated in North-Western France, Rennes Metropolis is an intercommunality[187] that is dominated by its seat and center city Rennes. For instance, urban and metropolitan administrations have been merged to a large extent.[188] The *département* Ille-et-Vilaine and the Brittany Region, too, are seated in Rennes (see figure 5.7).

Apart from Rennes city, most members consist of small municipalities with correspondingly limited administrative resources. Rennes is home to half of the intercommunality's population, namely about 217,000 out of approx. 439,000 inhabitants. Among the other 42 member municipalities, only six have more than 10,000 inhabitants, and ten have less than 2,000 inhabitants (Base nationale sur l'intercommunalité 2016c).

In response to municipal fragmentation, Rennes Metropolis has been endowed with a range of competences including responsibilities in the areas of economic development, land use planning, social housing, urban development, waste, and transport.[189] Rennes Metropolis employs about 1,100 staff and has a budget of more than 490 million Euro (Poppe 2013).[190] Sources of tax revenue for Rennes Metropolis consist of a unique business tax (*fiscalité professionnelle unique*), supplemented by a visitor's tax (Base nationale sur l'intercommunalité 2016c). The intercommunal political bodies consist of a community council of 114 delegates meeting each month in public sessions, and a conference of mayors that brings together municipal councilors (Poppe 2013).

---

187   In France, inter-municipal associations are institutionalized as so-called intercommunalities, more precisely public establishments for intercommunal cooperation (*établissements publics de coopération intercommunale*, EPCI); see section 5.2.1.3.

188   The "*mutualisation de services*" is also expressed by a shared website of both authorities, the "*site de Rennes, Ville et Métropole*".

189   For a full list of its 46 competences, see Base nationale sur l'intercommunalité (2016c).

190   For comparison, the 2013 budget of Rennes Metropolis amounted to 493.15 million Euro while Rennes city had a budget of 566.4 million Euro Poppe (2013).

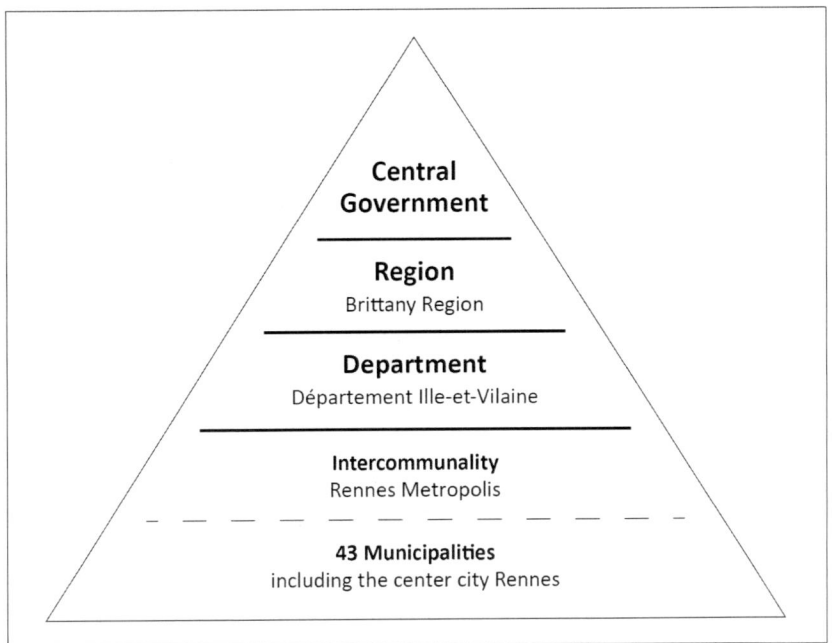

Figure 5.7: Territorial organization of Rennes Metropolis
Source: Author.

Established on December 31, 1999 as a *communauté d'agglomération* replacing a smaller and less powerful *district urbain* created in 1970, Rennes Metropolis has obtained the status of a metropolis[191] on January 1, 2015.[192] Rennes Metropolis only covers the center of the metropolitan area; another approx. 70,000 inhabitants are native to several peri-urban municipalities that have associated in four distinct intercommunalities.[193] Rennes Metropolis holds the responsibility for land-use-planning on its territory, a field with close links to climate planning,

---

191 Just as in the case of Greater Lyon and ten other agglomerations, this transformation was mandatory. Other French intercommunalities may voluntarily adopt metropolitan status in the future under certain conditions.

192 In contrast to Greater Lyon, Rennes Metropolis remains an organization under public law and does not feature the status of a territorial authority. Correspondingly, the Ille-et-Vilaine department remains unaffected.

193 These are the Communauté de Communes du Pays d'Aubigné, the Communauté de Communes du Pays de Châteaugiron, the Communauté de Communes du Pays de Liffré, and the Communauté de Communes du Val d'Ille.

for example in terms of mobility and transport. For purposes of more general spatial planning in the Territorial Coherence Scheme (SCoT), Rennes Metropolis cooperates with the four neighboring intercommunalities under the umbrella of the *Pays de Rennes*.[194] The urban area spans an even larger region with up to 700,000 inhabitants (INSEE 2013). Rennes Metropolis itself continues to grow as additional municipalities join in the intercommunality.[195] In view of repeated territorial reforms and municipal adhesions, territorial organization remains under construction.

Rennes Metropolis joined its center city in signing the Covenant, and achieved a considerable coverage of Covenant signatures among member municipalities. Rennes Metropolis appears to be an outstanding intermediary of the Covenant and has reached out to a large share of it municipalities as target actors of European climate policy, but struggles with implementing its commitment. All but three member municipalities of Rennes Metropolis signed the Covenant.[196]Rennes Metropolis supports nearly half of all (remaining) French Covenant signatories (34 out of 81),[197] but all have been suspended for a lack of reporting (Covenant of Mayors 2016i). Its remarkable record in Covenant promotion among member municipalities raises the questions why it was undertaken, how it was achieved, whether it was then translated into tangible climate action, and what hindrances came up at a later stage. The section starts with an account of how the Covenant came to Rennes Metropolis, and how the intercommunality implemented its Covenant commitment (5.2.3.1). It then depicts how the initial dynamic faded significantly over time (5.2.3.2). The analysis demonstrates that this development was linked to the fact that the intercommunality used its role as a Covenant Coordinator for showcasing the value added of metropolization (5.2.3.3). This highlights the importance of timing for effective orchestration (5.2.3.4).

---

194   Out of these intercommunalities, Val d'Ille also is a Covenant Coordinator (see section 5.2.4).
195   The latest adhesions took effect on January 1st, 2014 when Langan, Romillé, La Chapelle-Chaussée, Miniac-sous-Bécherel, and Bécherel entered Rennes Metropolis after having dissolved their previous intercommunality, the *Communauté de Communes du Pays de Bécherel*.
196   This excludes municipalities that joined the intercommunality at a later point in time.
197   See the introduction to section 5.2.

5.2.3.1   An (Im)Perfect Covenant Dynamic

Covenant signature by Rennes Metropolis had first been suggested in 2008 by Jean-Luc Daubaire, ecologist deputy mayor (*Les Verts*) at Rennes city in charge of urban ecology (interview with B. Catherine, 3 November 2014; interview with J.-L. Daubaire, 26 November 2014; interview with D. Guillotin, 1 December 2014). In this function, Daubaire also headed the local energy agency (*Agence Locale de l'Energie et du Climat du Pays de Rennes*, ALEC), and he had established ties to Energy Cities, one of the networks of local and regional authorities that run the CoMO. At the time, Rennes city already had an evaluation of its first climate plan – a voluntary precursor of the PCET – at hand, and projections gave reason to believe that the city could meet the Covenant commitment (interview with J.-L. Daubaire, 26 November 2014). The interviewees further pointed out that the idea was quickly taken up by Daubaire's intercommunal counterpart, socialist Bernard Poirier[198] (*Parti socialiste*). In his function as first vice-president of Rennes Metropolis in charge of foresight and sustainable development, Poirier suggested a collective engagement:

> "The center city sets out on its own again and we, the municipalities, are also interested. We also want to join the Covenant of Mayors" (interview with J.-L. Daubaire, 26 November 2014).[199]

A closer look at Covenant proceedings revealed methodological challenges for the member municipalities of Rennes Metropolis. Since the intercommunality was in charge of several policy fields to be covered by SEAPs, namely transport and housing, municipalities from Rennes Metropolis were unable to schedule the required actions in these fields themselves. Daubaire said that after consultation with the European Commission on this issue, an integrated planning approach was agreed:

> "We chose to designate Rennes Metropolis as main coordinator [...] and each municipality adds its derivate individual action plan with complementary actions [...]. Thus, it's a hybrid plan. There is one part of Rennes Metropolis and that's why Rennes Metropolis is the bearer of the project, and the parts that remained municipal responsibilities and are going to constitute the content of municipal action plans. Thus, we, Rennes Metropolis, at the Covenant of Mayors, we present an intercom-

---

198   Bernard Poirier also held the office of mayor of Mordelles from 1995 to 2014.
199   Author's translation; original quote: "*La ville centre part encore toute seule et nous, les communes, on est intéressé aussi. On veut intégrer la Convention des maires aussi.*"

munal action plan plus the plan for each municipality as being our commitment" (interview with J.-L. Daubaire, 26 November 2014).[200]

In short, the idea of the center city signing the Covenant was taken up at the intercommunal level so as to initiate a collective project, and in response to challenges in coping with Covenant methodology that were caused by interlinkages between municipal and intercommunal responsibilities with regard to sustainable energy policy, Rennes Metropolis took on the role of Covenant Coordinator.

The idea to collectively sign the Covenant was then presented and discussed at several meetings with member municipalities, a municipal councilor recalled, and a territorial *"dynamic"* ensued (interview with D. Guillotin, 1 December 2014). The community council of Rennes Metropolis approved adhesion to the Covenant on December 18, 2008. A first group of member municipalities signed the Covenant around the turn of the year 2008/2009; others followed gradually until November 2009. A state agency highlighted the comprehensive "mobilization of municipalities of Rennes Metropolis through the collective signature of the Covenant of Mayors" as a key success factor of the metropolitan climate policy (ADEME 2015f).

Asked how the high rate of participation among member municipalities was achieved, interviewees estimated that while first-movers were genuinely convinced of the endeavor, latecomers were also affected by the territorial dynamic that made it difficult to stay behind (interview with B. Catherine, 3 November 2014; interview with J.-L. Daubaire, 26 November 2014; 26 November 2014; interview with J.-P. de Nayer, 25 November 2014). Also, Rennes Metropolis had undertaken several efforts to demonstrate that the Covenant commitment, namely a GHG emission reduction of 20% until 2020, was actually feasible. Firstly, the metropolitan climate diagnostic with a 2006 baseline provided the data required for any calculation of GHG emission trends. Secondly, Rennes Metropolis had achieved a special understanding with the European Commission that municipal GHG emission reduction commitments would only cover the sectors under their

---

200   Author's translation; original quote: *"On a choisi de faire rentrer la métropole de Rennes comme coordinateur principal [...] et chaque commune ajoute son plan d'actions typique en déclinaison avec des actions complémentaires [...]. Donc, c'est un plan hybride. Il y a une partie de Rennes Métropole et c'est pour ça que Rennes Métropole est porteur du projet, et des parties qui sont restées des compétences des communes et qui vont constituer le contenu des plans d'actions communaux. Donc, nous, Rennes Métropole, à la Covenantion des maires, on présente un plan d'actions intercommunal plus le plan pour chacune des communes comme étant notre engagement."*

competence, and that GHG emission reduction targets could be formulated per capita (Conseil municipal de Thorigné-Fouillard 2009). This dispelled doubts concerning the impact of municipal policies on sectors governed by superordinate levels, and the capacity to reduce total emissions in view of continuous population growth. Thirdly, municipal and inter-municipal officers agreed that later signatories were also motivated by the intercommunal support for municipal climate policy that was beginning to take form (interview with B. Catherine, 3 November 2014; interview with J.-P. de Nayer, 25 November 2014).

Summing up, the idea of joining the Covenant was first raised by a convinced ecologist at the center city and subsequently taken up by the intercommunality that then adopted the role of a Covenant Coordinator. A territorial dynamic also carried along skeptical municipalities, and an outstanding share of member municipalities signed the Covenant. This was also due to the efforts of Rennes Metropolis to demonstrate the feasibility of the Covenant commitment by providing data, achieving methodological concessions from the CoMO, and offering support for municipal climate policy. By means of this support, Rennes metropolis further implemented its commitment as a Covenant Coordinator beyond mere Covenant promotion.

Rennes Metropolis conceives of its climate and energy policy as an integral part of its sustainable development strategy. Correspondingly, Rennes Metropolis presents its territorial climate action plan (PCET) as an element of its Agenda 21 (Rennes Métropole 2012). The PCET of Rennes Metropolis was launched in February 2007 and entered into effect in January 2010. It builds on a 2005 emissions inventory according to which the transport sector accounts for nearly half of all energy-related GHG emissions on the territory, and aims for a 20% GHG emissions reduction per inhabitant.[201] For the construction of its PCET, Rennes Metropolis cooperated with the *Agence locale de l'énergie et du climat du Pays de Rennes* (ALEC), the local energy agency, which is also a key partner for implementing the PCET (ADEME 2015f). Established in 1997 under the name of CLé in a joint endeavor by Rennes Metropolis, Rennes city and the *Agence de l'Environnement et de la Maitrise de l'Energie* (ADEME) Bretagne, the regional environmental agency of the central state, the ALEC provides energy-related consultancy to public and private clients from the territory of the Pays de Rennes

---

201   Rennes Metropolis experiences continuous population growth.

with a team of about 15 staff. In addition, Rennes Metropolis engaged in the Cit'ergie audit, the French EEA procedure (ADEME 2015f).

Rennes Metropolis is dominated by its center city in terms of population, politics, and climate governance. Climate policy of Rennes Metropolis and Rennes city are closely interlinked due to shared administrative resources and urban infrastructure, and Rennes city supports the intercommunality in local co-ordination. The very fact that Rennes has more resources and administrative ca-pacities than other member municipalities of the intercommunality results in a local leadership role. Holder of a Cit'ergie audit since 2011, Rennes city built its PCET on an emissions inventory with a 2006 baseline (Barbier 2012). Launched in September 2009, the PCET was developed in a series of conferences. After preparatory workshops in September and October 2009, an open forum was held in December that same year. The consultation process resulted in a white paper officially delivered in May 2010 (ADEME 2012). About half of the measures included in its PCET trace back to this white paper (Schilken et al. 2013, 11). The state agency ADEME describes a shared commitment by deputies and ad-ministrative staff as well as an extensive baseline study as the main strengths of this PCET (ADEME 2012). It further highlights the mobilization of territorial actors together with Rennes Metropolis as a challenge for successful implemen-tation.

With the signature of the Covenant, member municipalities of Rennes Me-tropolis committed to establishing a sustainable energy action plan (SEAP) alt-hough with the exception of Rennes city, due to their limited size,[202] they are not obliged by national law to develop territorial climate and energy plans (PCETs). Here, the Covenant helped to set the agenda, and to establish instruments for the implementation of local climate policy since PCET methodology was neither mandatory nor suitable for small municipalities. Once the Covenant had been successfully promoted among member municipalities, Rennes Metropolis was faced with the challenge of assisting member municipalities in translating their commitments into action. At this stage, municipalities expected tangible assis-tance, an officer of the local energy agency of the time explained:

"What was absolutely demanded was to have this accompaniment [...] here by an external organization that can help them with implementation. Not to act on their

---

202   Only Rennes city and the metropolis itself exceed the legal threshold of 50,000 inhabitants.

behalf, not at all, but to help them with implementation, to help, help to follow-up, to report progress and evaluate" (interview with D. Guillotin, 1 December 2014).[203]

Thus, operational support was the next step for Rennes Metropolis as a Covenant Coordinator. To begin with, the intercommunal climate department suggested a common framework consisting in mandatory and optional issue areas to be covered by each municipal SEAP, an intercommunal officer recalled (interview with B. Catherine, 3 November 2014). Mandatory issue areas included municipal buildings, street lighting, citizen sensitization, housing, and local transport (Conseil municipal de Thorigné-Fouillard 2010). In addition to assistance by more or less in-house services such as its climate department and the ALEC, Rennes Metropolis commissioned a development office as of July 2009 to accompany the entirety of municipalities in establishing their SEAPs (Conseil municipal de Thorigné-Fouillard 2009).

The local energy agency ALEC was assigned to animate a working group on local climate planning to accompany municipalities during 2011 and 2012. Open to deputies and officers from all member municipalities, the working group was attended by Covenant signatories only, an intercommunal officer recalled (interview with B. Catherine, 3 November 2014). The ALEC officer in charge at that time reported that two to three meetings were held per year. In preparation of the meetings, he said, he took stock of progress on the SEAP of each municipality. At the meetings, priority was given to the presentation of actions by participant municipalities in order to provide inspiration for others, e.g. with regard to municipal buildings, street lighting, and other municipal responsibilities (interview with D. Guillotin, 1 December 2014). In a similar vein, an intercommunal officer summarized these meetings as occasions where experiences and best practices were exchanged, difficulties were discussed, and progress on the metropolitan PCET was presented (interview with B. Catherine, 3 November 2014). This PCET was also submitted as a SEAP to the Covenant.

In order to prevent laggard municipalities from spoiling the common project, Rennes Metropolis focused its support on those municipalities that showed the least initiative or required the most assistance (interview with J.-L. Daubaire, 26 November 2014). Despite the guidance of Rennes Metropolis, the verification

---

203   Author's translation; original quote: *"Ce qui était demandé absolument c'était d'avoir cet accompagnement [...] à côté par un organisme extérieur qui puisse les aider dans la mise en œuvre. Pas faire à leur place, surtout pas, mais les aider sur la mise en œuvre, les aider, les aider à suivre, à faire un bilan de suivi et d'évaluation, quoi."*

of action plan drafts by the development office, and accompaniment in the ALEC working group, some municipalities struggled to establish their SEAP. As a former ALEC officer put it, the smaller a municipality, the more it requires operational support because it lacks the necessary human resources for local climate planning (interview with D. Guillotin, 1 December 2014). In response, Rennes Metropolis commissioned the ALEC to help out (interview with B. Catherine, 3 November 2014).

Rennes Metropolis advocated for a comprehensive set of SEAPs on its territory which were developed in coherence with the orientations of its own PCET (Barbier 2012; Ville de Rennes 2014). The Covenant continued to facilitate this process as it entailed clear targets to aim for, and legitimized the coordinating role of Rennes Metropolis and its intercommunal PCET. A municipal councilor noted:

> "The Covenant really was a tool for mobilizing local authorities from the agglomeration. I am not sure whether they would have been equally dynamic if engaged simply in a climate and energy plan at the municipal level rather than the Covenant of Mayors, a commitment taken with certain objectives, namely the 20-20-20 targets. Without that, I am not sure whether we would have seen as many municipalities mobilized through the energy-climate plan of the Rennes agglomeration" (interview with D. Guillotin, 1 December 2014).[204]

As a result of the local climate policy coordination activities of Rennes Metropolis, all 34 participant municipalities adopted a climate plan in the course of 2010 (Barbier 2012). Covenant signatories gathered once a year for exchange supported by the local energy agency ALEC which also assisted them in updating their climate plans throughout 2011 and 2012 (ALEC 2013, 2). In November 2013, municipal climate plans were taken stock of at a meeting animated again by the ALEC (ALEC 2014, 10). A former ALEC officer said that a questionnaire and visits in member municipalities were used to obtain a comprehensive assessment of implementation across the metropolis (interview with D. Guillotin, 1 December 2014). It turned out that municipalities focused their activities on re-

---

204   Author's translation; original quote: *"La Convention [des maires] était vraiment un outil de mobilisation pour les collectivités locales de l'agglomération. Je ne suis pas sûr que si l'on simplement engage un plan énergie et climat à l'échelle des communes, elles soient aussi moteur que de s'engager autour de la Convention des maires, alors d'un acte qu'on pose avec des objectifs, notamment les trois fois vingt. Je ne suis pas sûr que si il n'y avait pas eu cela, on aurait vu autant de communes se mobiliser au travers du plan énergie-climat de l'agglomération rennaise."*

ducing the energy use by municipal buildings, and had implemented or begun to implement the vast majority of actions scheduled in local climate plans (ALEC 2014, 10).

5.2.3.2 A Loss in Momentum

Once the momentum had been created, difficulties with Covenant proceedings came up, and ideational support from the Covenant became less important. Consequentially, the orchestrator-intermediary relationship tapered off, although a revival remains conceivable.

Rennes Metropolis had committed to submitting municipal climate plans to the CoMO in order to reduce the administrative burden for individual signatories (Conseil municipal de Thorigné-Fouillard 2010). It turned out that a lack of disaggregated data on territorial GHG emissions complicated the submission of municipal SEAPs. Also, the translation into English and the entry of municipal action plans into the Covenant tables entailed a significant workload (interview with B. Catherine, 3 November 2014; interview with J.-L. Daubaire, 26 November 2014). The intercommunal officer thus agreed with the CoMO that Rennes Metropolis would, in the future, resort to joint SEAP submission:

> "The demand of the Covenant of Mayors is the following: For your next action plan, if you could not do one document for Rennes Metropolis, one document for Rennes city, and one document for each municipality, but a single document within which there are also the actions of municipalities, that would be better. And thus, we consider, for the revision of our territorial climate and energy plan, to make a single document which maybe the municipalities could also sign or vote in the municipal council. We have not yet determined the way it will happen, but that's it: We will switch from thirty different documents with an excessive workload behind that does not necessarily have any added value to a much more integrated work" (interview with B. Catherine, 3 November 2014).[205]

---

205 Author's translation; original quote: *"La demande de la Convention des maires, c'est: Pour votre prochain plan d'actions, si vous pouviez ne pas faire un document Rennes Métropole, un document Ville de Rennes, et un document pour chacune des communes, mais un seul document dans lequel il y ait les actions aussi des communes, ce serait mieux. Et donc, on réfléchit, pour la révision de notre PCET, à faire un seul document que peut-être les communes pourraient, elles aussi, signer ou voter en conseil municipal. On n'a pas encore arrêté la façon dont ça va se passer, mais voilà: On passerait d'une trentaine de documents différents avec un travail trop important qui n'a pas forcément de plus-value, en plus, derrière, à un travail beaucoup plus intégré."*

Although individual municipal proceedings had the strength to facilitate appropriation of the intercommunal climate policy project, he added, such an integrated approach would also facilitate his task to follow-up and to monitor climate policy implementation, a task of his complicated by the diversity and the lack of precision of the first generation of local climate plans. In the meantime, the interviewee explained, he was focusing his limited working hours on the revision of the intercommunal action plan rather than reporting to the Covenant on the implementation of plans that had already turned out to require revision (interview with B. Catherine, 3 November 2014).

As a result, only the SEAP of Rennes Metropolis is registered on the website of the Covenant to date, and Rennes Metropolis has not yet submitted an implementation report. Since municipal and intercommunal action plans have not been updated yet, a joint SEAP has not been submitted either. Covenant signatory status of all member municipalities, including Rennes city, has been put on hold for the delay in SEAP submission; the profile of Rennes Metropolis on the Covenant website has not been updated since November 28, 2012; benchmark reports date back to 2010 (Covenant of Mayors 2016g).

Also, neither the network nor the methodology of the Covenant had much practical relevance for the implementation and monitoring of municipal SEAPs. Councilors and officers from Rennes Metropolis exchange information with peers in their Breton network or with – mainly French – contacts from Energy Cities rather than searching for other Covenant signatories across Europe or for benchmarks from the Covenant website to learn from (interview with B. Catherine, 3 November 2014; interview with J.-L. Daubaire, 26 November 2014). Rennes Metropolis and its municipalities use the Covenant as a seal of approval or an audit rather than as an actual network. For monitoring the implementation of the intercommunal PCET, Rennes Metropolis relies on internal tools and has not engaged in any systematic evaluation (ADEME 2015f).

The provision of intercommunal assistance for municipal climate policy declined at the end of the electoral term. The last meeting of Covenant signatories supported by the ALEC on behalf of Rennes Metropolis took place at the end of 2013 (interview with B. Catherine, 3 November 2014). Consequently, no initiative was taken to promote Covenant signatures among municipalities that joined the intercommunality in January 2014 (interview with D. Guillotin, 1 December 2014).

Local elections in March 2014 resulted in a change of political personnel that potentially entails a weakening of ties to the Covenant. Some delegates who had participated in two visits in Brussels for the launch and the anniversary of Covenant signature "*to get our picture taken*" (interview with J.-L. Daubaire, 26 November 2014)[206] are no longer in office. An intermunicipal officer estimated that the celebratory signature ceremonies in Brussels played an important role in committing municipal delegates to the Covenant, and wondered how to create a similar experience for new delegates:

> "If one does not want this engagement to decrease during the 2014–2020 term, it could be important to find a means to either bring Europe to Rennes, or a way to resolemnize, to make this a solemn commitment again. I do not quite know yet how to do this" (interview with B. Catherine, 3 November 2014).[207]

A municipal councilor emphasized that after electoral change, the perpetuation of the Covenant commitment of a municipality depends not only on the new mayor who might have other political priorities and certainly lacks time to cover all issues herself (interview with J.-L. Daubaire, 26 November 2014). Instead, the interviewee highlighted the importance of other local actors, namely the continuity of administrative staff and the engagement of the new intercommunal councilor in charge. He also called for a supportive framework at superordinate levels of government:

> "That is what the Covenant needs today: a relay not for the state to take over, it is still a local movement, but for the Commission to hold states responsible just to avoid that changes in local politics at every election stop long-term projects that go in the right direction" (interview with J.-L. Daubaire, 26 November 2014).

Staff from the climate department might be able to raise the issue to new councilors in charge when reminders of required reporting on SEAP implementation come from the CoMO. In this way, orchestration by the Covenant could again help to set the agenda for local climate policy. Also, a certain staff continuity is assured as the new councilor in charge of urban ecology (i.e., of the Covenant) in succession of Daubaire at the municipal council of Rennes, Daniel Guillotin,

---

206   Author's translation, original quote: "*Nous, on est allé se prendre en photo.*"
207   Author's translation, original quote: "*Si on ne veut pas que cet engagement s'affaiblisse sur ce mandat 2014–2020, il pourrait être important de trouver soit un moyen d'amener l'Europe à Rennes, soit une façon de resolenniser, de rerendre solennelle cet engagement. Voilà, je ne sais pas encore comment on pourrait faire.*"

worked at the ALEC prior to this mandate and was in charge of networking Covenant signatories.

As demonstrated above, the Covenant had an attractive offer to make for Rennes Metropolis at a particular point in time, but then became less relevant. Why was this the case? Has Rennes Metropolis abandoned the policy goals once shared with the Covenant? Has Covenant support become unalluring? Or, has Rennes Metropolis simply reached its goals related to Covenant signature? A closer look at the situation of territorial governance at the time reveals that the commitment of Rennes Metropolis as a Covenant Coordinator needs to be understood as part of an intercommunal rescaling strategy.

5.2.3.3 Rescaling Climate Policy in a Metropolis Under Way

Intercommunal vice-president Poirier embraced Covenant membership, more precisely the coordinator status, as an opportunity to make an additional offer to member municipalities, thereby enhancing the attractiveness and legitimacy of the emerging metropolis. As first vice-president of Rennes Metropolis, Poirier was able to advocate a joint commitment to local climate action as a means to strengthen the ties between member municipalities:

> "Bernard Poirier had the intuition that the Covenant of Mayors could be a tool for making municipalities from the territory cooperate. And thus, he wanted Rennes Metropolis to mobilize the municipalities from the territory to also commit, along with Rennes Metropolis and Rennes city, to the Covenant of Mayors. So it was really something political that was done, and very much something on the issue of the climate and energy question, that is something that unites us, and we have to be able to create a common dynamic on this subject. Like that, it was a link, after all, a tool to make contact between the municipalities from the territory" (interview with B. Catherine, 3 November 2014).[208]

In this process, Covenant signature provided a framework for municipalities to solemnly join in the intercommunal initiative to tackle climate change.

---

208   Author's translation; original quote: "*Bernard Poirier a eu l'intuition que la Convention des maires pouvait être un outil pour faire travailler ensemble les communes du territoire. Et donc, il a souhaité que Rennes Métropole mobilise les communes du territoire pour qu'elles s'engagent, elles aussi, avec Rennes Métropole et avec la Ville de Rennes dans la Convention des maires. Donc c'est vraiment quelque chose de politique qui s'est fait, et beaucoup quelque chose sur le fait de la question énergie-climat, c'est quelque chose qui nous réunit, et on doit pouvoir créer une dynamique commune sur ce sujet-là. Voilà, ça a été un lien, fin, un outil pour faire du lien entre les communes du territoire.*"

Close cooperation was also in order because of interdependencies between the intercommunal and the municipal levels, namely in the transport and housing sectors where some important municipal competences were transferred to the intercommunality, and between the center city and the surrounding municipalities, for instance with regard to shared funding of infrastructure. Thus, semi-yearly meetings of Covenant signatories also served for concertation on the intercommunal PCET (ADEME 2015f).

Intercommunal assistance was crucial for municipalities to participate in the endeavor. In order to enhance the capacities of the intercommunal administration for climate planning and for the accompaniment of member municipalities, the climate departments of Rennes Metropolis and Rennes city were merged in 2010 (*"mutualisation des services"*). Additionally, Rennes Metropolis provides co-funding for energy monitoring contracts between municipalities and the local energy agency ALEC and commissioned the latter with the accompaniment of municipalities, including the working group as an opportunity to exchange experiences with peers, to obtain expert input tailored to local needs, and to participate in technical training where necessary.

Experiences with the campaign '*Je change, ça change tout*' ('I change, it changes everything') illustrate the intercommunal attempt to provide tailored assistance and, at the same time, highlight the limitations of this strategy for territorial mobilization. An evaluation of municipal SEAPs, an intercommunal officer said, had shown a universal call for citizen involvement (interview with B. Catherine, 3 November 2014). This focus, a municipal councilor explained, was due to the limited share of territorial emissions that municipalities had direct control over (interview with J.-L. Daubaire, 26 November 2014).[209] In response, the interviewees further pointed out, Rennes Metropolis developed '*Je change, ça change tout*' as part of ENGAGE, a European project coordinated by Energy Cities. Municipalities had been very responsive to joining the Covenant as a European initiative for local sustainable energy. Now that it came to implementing this commitment, a municipal councilor reported, the framework of another European project provided additional ideational support:

> "It was interesting at the time and for us reassuring to see that we were actually very much in line with the cities of Europe and at the moment that we asked ourselves

---

209 Municipalities' direct impact was limited, as mentioned earlier, by the fact that crucial sectors of public policy like transport and housing, which account for important shares of GHG emissions on the territory of Rennes Metropolis, have been allocated at the intercommunal level.

how to do it, how to be efficient, how to make sure that things happen on the territory, other countries in Europe asked the same question" (interview with J.-L. Daubaire, 26 November 2014).[210]

The campaign aimed for mobilizing citizens by picturing inhabitants and their personal contribution to the Covenant commitment on posters and Facebook posts. This ready-made tool-kit was available for municipalities on request and included an event booth with photographer and animator provided by the metropolis. During the course of the campaign from 2011 until 2013, 32 photo sessions resulted in about 1,000 posters (Schilken et al. 2013).

Other than facilitating citizen mobilization in member municipalities and establishing the intercommunality as the provider of attractive services for municipal policies, a municipal councilor mentioned another benefit of the campaign: increased pressure on municipalities as target actors of local climate policy. The campaign allowed for addressing citizens even in municipalities that lacked either political will or human resources to actively work on their climate plan, he said, and explained that the campaign created public attention that increased the pressure on mayors to have some municipal measure to showcase (interview with J.-L. Daubaire, 26 November 2014). But municipal response did not meet the expectations of an intercommunal officer who traced back this reluctance to tensions between the municipal and the intercommunal levels:

"One could have wished for them [municipalities] to take this up a little more. Thus, this is interesting because it is the difficulty between the level of the intercommunality and the municipalities. It's different. It's two distinct things, in the minds. Anyway, it's how I feel it. Anyway, it's very personal. But I have felt that in the minds of municipal councilors, Rennes Metropolis, that was a different stratum and thus, it is not because Rennes Metropolis does something for the municipalities that it's good, one is going to take it up. There is still a little distance between the two levels" (interview with B. Catherine, 3 November 2014).[211]

210　Author's translation; original quote: "*C'était intéressant à l'époque pour nous et rassurant de voir qu'en fait, on était très parallèle avec les villes d'Europe et qu'au moment où nous, on se posait la question comment faire, comment être efficace, comment faire que les choses existent sur le terrain, les autres pays d'Europe se posaient la même.*"

211　Author's translation; original quote: "*On aurait pu souhaiter qu'elles [les communes] s'en emparent un petit peu plus. Donc, c'est intéressant parce que c'est la difficulté entre l'échelle de l'intercommunalité et les communes. C'est différent. C'est deux choses différentes, c'est dans les esprits. De toute façon, c'est la façon de laquelle je le ressenti. De toute façon, c'est très personnel. Mais j'ai ressenti que dans l'esprit des élus des communes, Rennes Métropole c'était une strate différente et donc, c'est pas parce que Rennes Métropole fait quelque chose*"

In response, municipalities – in particular those with divergent goals for energy policy and territorial organization – remained reluctant to take up the ENGAGE campaign. In other words, service provision by Rennes Metropolis was not innocent or, to put it more neutrally, purely functional, but had a political dimension. It served to advance the intercommunal climate policy project. This project, in turn, served two political goals, namely climate mitigation – as advocated by Daubaire – and metropolization as pursued by Poirier.

Other than such instances of misfit between intercommunal and municipal policy goals, two additional factors slowed down implementation of municipal climate plans and the intercommunal climate policy project as a whole. Firstly, several interviewees referred to financial constraints deriving from the economic crisis as a major hindrance for local climate policy implementation, especially with regard to measures requiring substantial investment (interview with J.-L. Daubaire, 26 November 2014; interview with J.-P. de Nayer, 25 November 2014). Secondly, two major events occupied political attention: In the run-up to municipal elections in March 2014 and the nationally administered transformation of the intercommunality into a metropolis as of January 2015, both climate mitigation and the legitimization of metropolization towards municipalities and citizens paled in comparison to the need to urgently reorganize local politics which, according to several interviewees, tied up a lot of resources (e.g. interview with O. Dehaese, 12 January 2015). Thus, the intercommunal climate department first had to get to know the new intercommunal and municipal councilors and wait for the reorganization of metropolitan politics to settle before being able to potentially revitalize the intercommunal climate policy project (interview with B. Catherine, 3 November 2014). In the meantime, a municipal officer reported, Rennes Metropolis concentrated on monitoring rather than actively supporting municipal climate policy to an even larger extent than it did before the elections (interview with J.-P. de Nayer, 25 November 2014).

An update of the metropolitan climate plan had turned out to be necessary in order to specify targets, on the one hand, and to adopt a more integrative planning approach together with municipalities, on the other. Member municipalities will be invited to join in this process regardless of whether they are Covenant signatories or not, the new president of the local energy agency ALEC said (in-

---

pour les communes que c'est bien, on va le prendre. Voilà, il y a quand même une petite distance entre les deux échelles."

terview with O. Dehaese, 12 January 2015). He added that Covenant adhesion was going to be suggested to non-signatories on this occasion. At the community council on November 19, 2015, intercommunal president Emmanuel Couet[212] situated the revision of the metropolitan climate plan within an overarching strategy with "the objective to become an actual eco metropolis" (Gandon 2015). At this meeting, the Covenant recurred as an audit for intercommunal climate policy when socialist vice-president André Crocq[213] announced the launch of a new territorial climate, air and energy plan during the course of the year 2016 and referred to Rennes Metropolis as "a pioneer territory in terms of energy transition and climate issues. The mobilization of municipalities on our territory was unique[214]on the European level with, since 2008, the quasi-totality of municipalities signing the Covenant of Mayors" (Gandon 2015).[215] According to Crocq, the new plan is going to maintain housing and transport as prioritized sectors, but a new focus is to be put on the integration of climate and energy targets in related metropolitan documents.[216] Crocq further signaled the creation of an inter-municipal working group for fostering exchange and a local energy transition conference in order to mobilize territorial actors. Interestingly, no reference to the previous working group promoted by the ALEC was made. In a similar vein, Rennes Metropolis neither joined Mayors Adapt, nor did it commit to the 2050 targets after the relaunch of the Covenant. Instead, Couet signed Eurocities' Green Digital Charter in the run-up to COP 21, thereby committing to reducing

---

212   Emmanuel Couet, socialist mayor of Saint-Jacques-de-la-Lande, took over intercommunal presidency in April 2014 from Daniel Delaveau, socialist mayor of Rennes, who retired from politics.

213   As vice-president of Rennes Metropolis, André Crocq is in charge of sustainable development and territorial animation and the metropolitan go-between with the territorial coherence scheme. He also presides over the Pays de Rennes and, as a regional councilor, is in charge of sustainable development. Due to this accumulation of offices, he resigned as mayor of Chavagne in April 2016, but he maintained a seat as municipal councilor (Le Guillou 2016).

214   Author's translation; original quote: "... *un territoire précurseur en matière de transition énergétique et d'enjeux climatiques. La mobilisation des communes de notre territoire a été unique à l'échelle européenne avec, depuis 2008, la quasi-totalité des communes signataires de la Convention des maires.*"

215   Crocq's acclamation is at least immoderate. In fact, among French Covenant Coordinators, Val d'Ille was the only one where all member municipalities sign the Covenant. Also, and in contrast to Rennes Metropolis, Val d'Ille successfully submitted a joint SEAP on behalf of its municipalities. In other countries, several Covenant Coordinators achieved similar degrees of mobilization and successfully supported municipalities in submitting their SEAPs.

216   Namely, the local housing plan, the urban transport plan and the local plan of intercommunal urbanism are also about to be established.

emissions through the use of information and communication technologies (Eurocities 2015). This might indicate that new political personnel were searching for ways to take possession of local climate policy. A new project was required to allow for the perpetuation of the Covenant commitment.

5.2.3.4   Case Summary

Orchestration by the Covenant was an attractive offer for Rennes Metropolis at the beginning of its activities in local climate policy coordination. The agenda set by Covenant procedures, namely the commitment of signatories to establish a BEI and to develop a SEAP, helped to guide and instrument local climate policy in its early stages. Backed by a European initiative, Rennes Metropolis succeeded to establish itself as a legitimate coordinator and to kick off a regional dynamic of considerable coverage. Ideational support by the Covenant, in the form of official endorsement of Rennes Metropolis, enhanced the intercommunal authority and increased the pressure on laggard municipalities to join in the effort.

The commitment of Rennes Metropolis as a Covenant Coordinator was also driven by territorial considerations. The intercommunality aimed to grow in terms of the number of member municipalities as well as in responsibilities and resources. For this, Poirier was on the lookout for projects that could enable Rennes Metropolis to network with member municipalities behind a shared policy goal, thereby creating a sense of community and at the same time establishing the metropolis as a legitimate and benevolent coordinator. Once the momentum had been created, the Covenant was no longer able to offer tangible value added to Rennes Metropolis, and thus became less important in local communication of climate and energy policy. In parallel, local politics began to focus on local elections and the implementation of territorial reform. After a period of self-involvement, Rennes Metropolis is now undertaking an update of its climate plan. In this context, the remarkable Covenant dynamic of the past has become a reference in public communication again, and a window might open for either local actors or the CoMO to revitalize the orchestrator-intermediary relationship.

The case of Rennes Metropolis as a Covenant Coordinator shows that agendas of an orchestrator and its intermediary need to coincide for effective orchestration to take place. In itself, this is not a surprising finding since the O-I-T-model assumes that joint policy goals are a prerequisite for the establishment of an orchestrator-intermediary relationship. What the case of Rennes Metropolis sheds light on more precisely is the fact that such a coincidence is bound by the moment, i.e. it is not a given. Once the priorities of the intermediary change, it might abandon the common policy project. Another insight of the case consists in the necessity to perpetuate the benefits of orchestration for actors within the intermediary organization across legislative periods. The Covenant provides a solemn signature ceremony – and the corresponding visibility – for the political personnel that initiates the commitment, but does not offer any such opportunity for successors. Thus, it is up to local actors to appropriate the Covenant commitment of the previous government, e.g. through a relaunch of local climate policy. Here, the Covenant could provide more support, be it in terms of agenda setting for updating local climate policy, or in terms of resolemnization, for instance by means of facilitating conventions for existing signatories or some kind of audit for local climate policy that would go beyond mere mention on the website profile.

In order to adapt the orchestration offer to local circumstances, namely territorial organization and population growth, the CoMO agreed to municipal SEAPs covering only those fields of public policy not transferred to the intercommunal level, and to formulate per-capita rather than total emission reduction targets. This understanding demonstrates that the orchestrator-intermediary relationship is not a one-way road, but allows for mutual adjustments for the benefit of the joint pursuit of the shared policy goal.

5.2.4 Val d'Ille

In order to understand the case of the Community of Communes of Val d'Ille (*Communauté de Communes du Val d'Ille*, in the following referred to as Val d'Ille), it is important to envision that this is a rural intercommunality and by far the smallest intercommunality studied here.[217] Situated within the département Ille-et-Vilaine of the Brittany region in North-Western France, Val d'Ille consists of 10 municipalities (see figure 5.8).[218] After decades of population growth since 1970, Val d'Ille now has about 19,500 inhabitants (INSEE 2012).

---

217 In France, inter-municipal associations are institutionalized as so-called intercommunalities, more precisely public establishments for intercommunal cooperation (établissements publics de coopération intercommunale, EPCI); see section 5.2.1.3.

218 Other than its seat town Montreuil-le-Gast, Val d'Ille comprises Guipel, La Mézière, Langouet, Melesse, Saint-Germain-sur-Ille, Saint-Gondran, Saint-Médard-sur-Ille, Saint-Symphorien, and Vignoc. Except for Melesse and La Mézière, all member municipalities have less than 2,000 inhabitants (Base nationale sur l'intercommunalité 2016a).

The municipal councils elect 35 delegates to the intercommunal council seated in Montreuil-le-Gast that, in turn, choses the intercommunal president. Established on December 31, 1993, the intercommunality raises a unique business tax (*fiscalité professionnelle unique*) (Base nationale sur l'intercommunalité 2016a). Val d'Ille is in charge of economic development, social housing, roadworks, social services, facilities for culture and sports, environment, tourism, sustainable transport, communication, and employment (ADEME 2014; François 2013b). The transport is particularly relevant for local climate policy as Val d'Ille "shows an historically hedged farmland, barely hilly and wooded landscape, which can be considered as a peri-urban territory. [...] [I]ts economic activity mainly consists of agriculture (breeding) and services (commercial area), with significant daily transport flows towards Rennes" (François 2013b). Situated in the hinterland of Rennes Metropolis, Val d'Ille benefits from the metropolitan economy and infrastructures that are within easy reach for commuters and excursionists.

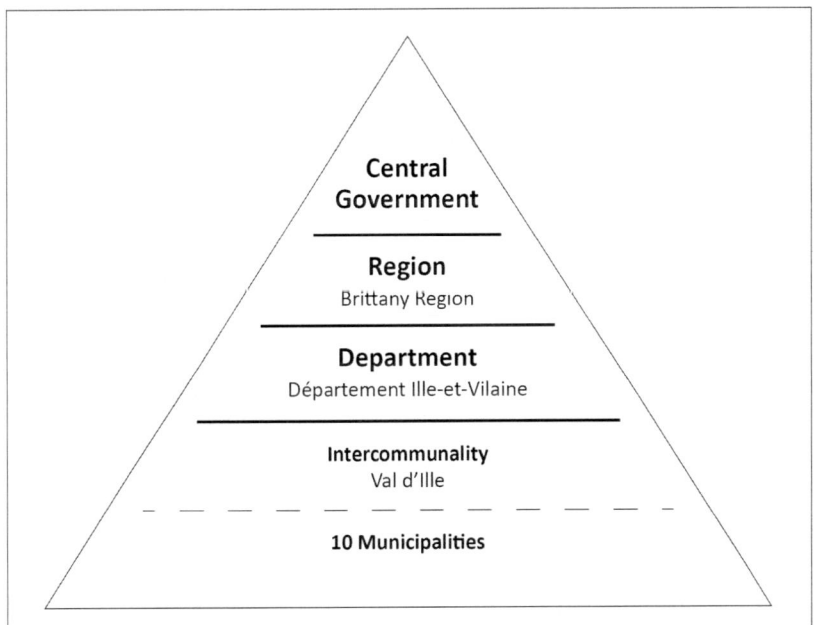

Figure 5.8: Territorial organization of Val d'Ille
Source: Author.

The size, rural character, and peri-urban location of Val d'Ille also shape its climate policy (see 5.2.4.1). The case of Val d'Ille shows how rural inter-municipal associations can overcome limited resources for local climate action provided they have dedicated political leaders and technical staff. Leading the way, the intercommunal president set off a dynamic among territorial actors and administrative staff (see 5.2.4.2). Despite initial difficulties in the communication between Val d'Ille and the CoMO, the intercommunality became instrumental in developing a new procedure within the Covenant, the joint SEAP for rural communities (see 5.2.4.3). As a closer look at the case will demonstrate, the experience of Val d'Ille as a Covenant Coordinator highlights the bidirectional character of the orchestrator-intermediary relationship (see 5.2.4.4).

5.2.4.1   Climate Policy Making in a Rural Community

Climate policy at Val d'Ille is marked by its rural character and the limited size of the intercommunality along with an outstanding political will to act on climate mitigation. From 2008 to 2014, Val d'Ille was headed by Daniel Cueff, an ecologist. Prior to his intercommunal presidency, the then non-partisan politician had already given proof of his decidedly environmental agenda as mayor of Langouët since 1999.[219] With a strong interest in ecology in general and climate policy in particular, he made this small and very rural municipality a laboratory for experimentation (interview with M. Janssens, 11 December 2014). At the municipal elections in 2008, Cueff ran for office via the list of the newly-established association *Bretagne écologie*. As soon as he was elected to the intercommunal presidency on April 10, 2008, he prioritized environmental action within all intercommunal policies and activities.

With his environmentalist agenda, Cueff was able to build upon prior experience at Val d'Ille. Sustainable development had been inscribed to the intercommunal statute on the creation of the intercommunality, a local Agenda 21 was initiated in 2007, and several municipalities started climate policies in 2006 (François 2013b). Although the intercommunal administration only comprises about 40 staff in total, two officers are assigned to address climate and energy issues. This illustrates the importance given to climate and energy issues because

---

219   Intercommunal presidency is typically assumed by a mayor from a member municipality which automatically leads to an accumulation of offices. In addition, Cueff was also a member of Brittany's regional council from 2010 to 2015.

most often, one or two officers are in charge of a whole area of intercommunal public action (Communauté de Communes du Val d'Ille 2016). An interviewee from Rennes Metropolis acknowledged the choice to dedicate two staff to the issue and concluded that although size matters in local climate policy, priorities are decisive:

> "A small structure that wants to tackle this issue and which make it a priority, can. Val d'Ille is not a large territory at all. Nevertheless, they demonstrate a lot of determination" (interview with B. Catherine, 3 November 2014).[220]

Due to Cueff's strong political will, Val d'Ille also joined the Covenant (ADEME 2014; interview with M. Janssens, 11 December 2014) and thereby committed to establishing a sustainable energy action plan (SEAP) although, due to its limited size of less than 50,000 inhabitants, it was not obliged by French law to establish a territorial climate and energy plan (PCET). When the local energy agency ALEC suggested signing the Covenant, Cueff instantly seized the opportunity (interview with M. Janssens, 11 December 2014). At the ceremony in Brussels on February 10, 2009, he signed the Covenant on behalf of the intercommunality within the first year of his presidency.

Following Cueff's signature of the Covenant in February 2009, a working group was established within the administration in order to establish a SEAP. Building on the agenda of the Covenant, these works were "an opportunity to broaden and structure a lot of initiatives, started sometimes before the CoM accession" (François 2013a) and resulted in the development of the *Diagnostique de la politique Energie-climat 2010*, a voluntary territorial climate and energy plan (PCET).[221] "The operational action plan document has slowly evolved to a 10 pages document with a one page summary table. A simple and convenient way to share the progresses made and still to come between elected representatives and staff" (Covenant of Mayors 2012). The intercommunal council formally approved the climate-energy diagnosis without a dissenting voice on March 1, 2011. It identifies the transport and agriculture sectors as the main sources of GHG emissions in Val d'Ille and provides an inventory of existing climate-related policies. 143 climate-related policies of the intercommunality (24) and its member municipalities (119) are identified and focus on energy is-

---

220   Author's translation; original quote: "...*une petite structure qui voudrait travailler ce sujet-là et qui en ferait un sujet d'importance, peut. Le Val d'Ille n'est pas un gros territoire. Et malgré tout, ils font preuve de beaucoup de volontarisme.*"

221   With less than 50,000 inhabitants, Val d'Ille was not obliged to develop a PCET.

sues to a large extent (67%). Examples include energetic retrofitting of public buildings and the promotion of photovoltaics. The climate-energy diagnosis concludes with a call for the mobilization of citizens and socio-economic actors from the territory and for more consideration of the transport and agriculture sectors in local climate policy (Communauté de Communes du Val d'Ille 2011).

The challenge to ensure climate-friendly mobility in a sparsely populated, peri-urban area makes transport policy a natural intercommunal priority. Val d'Ille aims to provide attractive services to citizens and counteract mobility poverty on the one hand, and to implementing sustainable energy in the transport sector on the other. Public transport – mainly with busses – is a core activity of the intercommunality. An alteration under Cueff's presidency consisted in the development of a sustainable mobility strategy during 2009 and 2010, including measures promoting intermodality and the launch of an electric bike hiring program as an incentive to bicycle to work even with longer distances (Clark 2015). For its mobility policy, Val d'Ille was awarded with the Brittany regional sustainable development trophy in 2012 (François 2013b).

With regard to agriculture, Val d'Ille has had a record in supporting organic farming since the 1990s because of its community kitchens in schools and support for short distance trade. Also, Val d'Ille has now begun to produce hedgerow pellets – one reason it aimed for the replanting of hedgerows within the *Breizh Bocage* program (Régnier 2012).[222] Biomass from hedgerow and other pellets is an essential component of sustainable energy policy for Val d'Ille because a photovoltaic project could not be realized and the territory does not allow for large-scale wind turbines. The promotion of non-fossil fuels is all the more important as the intercommunality also committed in 2011 to becoming a positive energy territory by the year 2030 (100% RES Communities 2015). Several measures aim to reduce energy consumption and to develop renewable energy. In 2012, Val d'Ille engaged in two energy projects. Local energy loop (*Boucle Energétique Locale*), a program of the Brittany Region, aims to reduce energy consumption, promote renewable energy, and to harmonize local energy consumption and production. Pilot projects for reduced energy use in the housing sector addressed private households. In the agricultural sector, professionals were addressed by information events. In the business sector, the *GreenFit* project aimed at promoting sustainable energy use in business parks with regard to mobility, buildings,

---

222   Breizh is the Breton name of Brittany.

behavior and certification (ALEC 2016; Communauté de Communes du Val d'Ille 2016). A particularly ambitious intercommunal policy under Cueff's presidency consisted in the promotion of energy-efficient commercial areas at the price of rejecting the introduction of climate-detrimental businesses (interview with C. Barais, 14 January 2015; interview with M. Janssens, 11 December 2014).

The climate-energy diagnosis was recognized by the Covenant, after some methodological challenges during the submission process (see 5.3.3), as a SEAP envisaging an overall CO_2 emission reduction target of 20%. An interviewee of the Brittany region credited the climate-energy diagnosis with containing interesting measures, and as being both ambitious and functional (interview with C. Barais, 2015, January 14). As factors of success of this PCET, the ADEME highlights strong political will, exemplarity, and interdisciplinarity (Agence de l'Environnement et de la Maîtrise de l'Energie 2014). Cueff intended to not only push forward climate and energy policy at the intercommunal level, but to also win over member municipalities and his fellow mayors. In this undertaking, the Covenant became a valuable source of support.

5.2.4.2 A Militant Intermediary for Local Climate and Energy Policy

Following Cueff's signature of the Covenant, Val d'Ille engaged in a double struggle with both member municipalities and the CoMO (see 5.2.4.3). In view of member municipalities, the Covenant was helpful with regard to persuasion efforts, and for structuring the joint climate policy:

> "The goal was that he [Cueff] knew very well that the ten municipalities were not ready at all [...] The nine other mayors were not as militant as himself and some were even rather reluctant towards the Covenant of Mayors. [...] We really had to convince them of the necessity to work on [...] climate and to not only [...] rely on classic competences: urbanism, waste management. They had no political vision at all for working on climate and energy. [...] The president very much committed himself for each commune to sign on its respective competences. [...] For us, it was a beautiful victory. The advantage of the Covenant of Mayors was rather a role that allowed for uniting many deputies, but also associations from the territory for a common goal. We have mainly taken the Covenant of Mayors like that: as something that was a trigger, it was a commitment that was formal [...] which has shown the

political will and has allowed uniting" (interview with M. Janssens, 11 December 2014).[223]

The Covenant served Val d'Ille and its president as a means to legitimize inter-communal climate policy coordination primarily towards member municipalities and municipal representatives at the intercommunal council. The Covenant provided European endorsement and helped setting the agenda on steps and measures to undertake. Leading the way, Cueff increased the pressure on his fellow mayors to join in the effort and cooperate in inter-municipal climate and energy policy.

With regard to the implementation of its Covenant commitment, Val d'Ille had a preference for tangible action rather than cumbersome planning. For instance, its PCET does not include a detailed list and schedule of actions to be taken as is the case in the climate action plans of other inter-municipal associations studied here, and Val d'Ille does not participate in the Club France in the Covenant:

> "Don't spend too much time in meetings, be hands-on! […] It is demanded from us to not spend too much time participating in reflections. But that's because we are a small authority. Deputies want the means available to be dedicated to actions because that is something visible. Besides a deputy, like all deputies, needs to be reelected, and thus needs something visible" (interview with M. Janssens, 11 December 2014).[224]

In line with this hands-on approach, detailed evaluation of measures from the climate action plan is limited to the ones that are actually discussed at the yearly

---

223   Author's translation; original quote: "*Le but c'était qu'il savait très bien que les dix communes n'étaient pas prêtes du tout […] Les neuf autres maires n'étaient pas aussi militants que lui et quelques-uns étaient même plutôt réticents à cette Convention des maires. […] Il a vraiment fallu convaincre de la nécessité de travailler sur […] le climat et de ne pas […] travailler uniquement sur les compétences classiques: l'urbanisme, les déchets. Ils n'avaient pas du tout une vision politique pour travailler sur le climat, sur l'énergie. […] Le président s'est beaucoup impliqué à ce que chacune commune signe sur ses compétences propres. […] Pour nous, c'était une belle victoire. L'avantage de la Convention des maires a été plus un rôle qui a permis de fédérer pas mal d'élus, mais aussi pas mal d'associations du territoire autour d'un but commun. Nous, on a surtout pris la Convention comme ça: comme quelque chose qui était un déclic, c'était un engagement qui était formel, […] ça a montré la volonté politique et ça a permis de fédérer.*"

224   Author's translation; original quote: "*Passez pas trop de temps en réunions, soyez dans l'action! […] On nous demande de ne pas passer trop de temps à participer à des réflexions. Mais ça, c'est aussi parce qu'on est une petite collectivité. Les élus veulent que les moyens qu'il y a, qu'ils soient consacrés à des actions parce que c'est des choses qui se voient. Et puis un élu, comme tous les élus, il a besoin d'être réélu et donc il a besoin de choses qui se voient.*"

monitoring meetings of the executive: "Soazig Rouillard, in charge of SEAP co-ordination, notes that [...] carrying out the evaluation of all actions would be time demanding and finally useless compared to the decisions made" (Covenant of Mayors 2012).

Intercommunal measures targeting municipalities need to strictly respect legal competences as member municipalities vigilantly guard their sphere of responsibility. At the same time, municipalities also happen to demand intercommunal intervention. For example, Val d'Ille co-funds a contract on communal property management with the *Agence locale de l'énergie du Pays de Rennes* (ALEC), the local energy agency. The intercommunality also provides data for municipal actions or for legitimizing its own intervention, for instance with a feasibility study on wood-based district heating. Another strategy of Val d'Ille consists in finding partner municipalities to implement exemplary measures it can then refer to: When municipalities engage in climate- or energy-relevant actions, staff from Val d'Ille contacts them, if need be, to suggest better solutions, for example more energy-efficient heating for a new building, or alternative construction materials. This consultancy is all the more important as two thirds of member municipalities have no technical staff. Deputies also rely on intercommunal staff for preliminary and free advice (interview with M. Janssens, 11 December 2014).

The mobilization of partners is a crucial challenge for Val d'Ille (ADEME 2014). In order to present citizens with the intercommunal climate policy and the climate-energy diagnosis, and to gather their feedback, citizen meetings were organized. Here, a decentralized approach was chosen because experience had shown that citizens rarely attend central, intercommunal events (interview with M. Janssens, 11 December 2014). In order to mobilize citizens and actors from the territory, a working group of deputies on energy autonomy was established. Although open to civil society, it is primarily attended by environmental associations (ADEME 2014).

Other than municipalities, citizens and associations from the territory, partners of Val d'Ille in climate and energy policy include several engineering offices and companies, the Brittany Region, and the local energy agency ALEC (interview with M. Janssens, 11 December 2014). Val d'Ille maintains a long-term partnership with ALEC. It relies on the ALEC to improve the energy performance of intercommunal buildings, and provides 50% co-funding for this consultancy service for member municipalities. Further cooperation projects include a funding scheme to support energy-efficient social housing, a series of public information events, the development of best practices for companies, the examination of construction plans of new businesses with regard to energy consumption and renewables, and reporting documents to the Covenant (ALEC 2016; Communauté de Communes du Val d'Ille 2016).

Representatives from Val d'Ille actively engage in the regional network, animated by the ADEME Bretagne, by attending meetings, presenting their actions, and taking part in the development of tools and in trainings (interview with C. Barais, 14 January 2015). Val d'Ille also cooperates with the ADEME in funding particular measures. The intercommunality is also active in the network of positive energy territories (TEPOS) animated by CLER – Réseau pour la transition énergétique[225] and the Taranis network for local and citizen energy in Brittany created in 2011, both co-founded by Val d'Ille. In contrast, the Département d'Ille-et-Vilaine is not very involved in climate and energy issues (interview with M. Janssens, 11 December 2014).

In the endeavor of implementing Val d'Ille's commitment to the Covenant, Cueff found committed allies among intercommunal administrative staff. This allowed for an anchorage of Val d'Ille's climate commitment within the intercommunal administration despite its limited human resources. With regard to the integration of climate and energy issues in intercommunal policies and activities, the relative weakness of intercommunal resources is balanced by the fact that the administrative structure is organized in a rather horizontal and non-hierarchical manner. Referees interact directly with each other instead of having to communicate via some superior. They are constantly brought up to date on projects in the making, also when those are handled by fellow referees. The resulting interdisciplinary approach of Val d'Ille's PCET approach has been highlighted by the ADEME (ADEME 2014). Climate-sensitive referees can introduce climate considerations at an early stage of project development in direct communication with the respective referees and deputies in charge. For example, higher energy efficiency standards were successfully introduced for a communal building. For climate-friendly suggestions to succeed, the presentation of co-benefits turned out to be essential, such as savings on energy costs in the case of construction, new economic opportunities in cases such as the allocation of climate-related busi-

---

225   The CLER was formerly known as Renewable Energies Liaison Committee (Comité de Liaison Énergies Renouvelables).

nesses or the installation of renewable power plants on the territory (interview with M. Janssens, 11 December 2014).

The limited institutionalization of Val d'Ille's climate commitment makes leadership particularly important. As there is no particular climate department or the like which could maintain sustainable energy policy on the inter-municipal agenda, political will becomes decisive: *"The secret is to have motivated deputies. To have tangible political support, that is really the first thing"* (interview with C. Barais, 14 January 2015).[226] Individual dedicated administrative staff alone simply cannot make this difference. Thus, the change in presidency resulting from local elections in March 2014 poses a particular challenge to the perpetuation of intercommunal climate policy coordination. Under the new intercommunal president Philippe Chevrel (*Parti Socialiste*, PS), priority is given to economic development and employment although climate adaptation and climate mitigation still figure among five priorities established by Chevrel (Communauté de Communes du Val d'Ille 2014). Compared to the Cueff's presidency, environment and energy policy are *"not at all abandoned [...] but there is not at all the same impulsion anymore"* (interview with M. Janssens, 11 December 2014).[227]

It is up to the remaining climate-committed staff within the administration to keep up their agenda and make corresponding suggestions to their fellow referees as well as to their deputies and president. In this endeavor, they will likely rely on their horizontal cooperation style and their experience in advertising climate action via its co-benefits. As both the PCET and the Covenant foresee an evaluation of the implementation of the climate action plan, occasions for raising the issue towards deputies might arise. At the moment, the evaluation of Val d'Ille's PCET process is still under construction (ADEME 2014). The Covenant might again become instrumental in terms of agenda-setting for evaluation as its proceedings require the submission of a monitoring report. The latter was due – after several extensions granted by the CoMO – at the end of 2015 (interview with M. Janssens, 11 December 2014), and still has not been registered on the website of the Covenant, which suggests there have been hindrances in the process.

---

226 Author's translation; original quote: *"Le secret est d'avoir des élus motivés. Avoir un vrai portage politique, c'est vraiment le premier point."*

227 Author's translation; original quote: *"Le volet environnement, énergie est un peu mis à côté. Il n'est pas du tout abandonné [...] mais il n'y a plus du tout la même impulsion."*

Despite the possibility for Val d'Ille to not fully meet its objectives as de-termined in the SEAP, an interviewee from the Val d'Ille administration refers to the Covenant as a remarkable success. He argues that the support from the Covenant was decisive for kicking off a unifying process around a common goal among deputies despite their different political and municipal affiliations. Also with regard to interactions between the intercommunality as an intermediary and the CoMO as its orchestrator, he gave a positive summary. After initial difficul-ties, a new format for reporting was developed to better suit the local circum-stances of smaller governments such as Val d'Ille (interview with M. Janssens, 11 December 2014).

5.2.4.3 Teaching the Orchestrator: Joint SEAP Submission

Val d'Ille aimed for all member municipalities to also sign the Covenant. The small size of Val d'Ille's member municipalities and their respective administra-tions made Covenant reporting obligations a considerable hurdle for municipal staff. Smaller member municipalities only have a secretary and one technical staff. Both the additional workload and a lack of expertise for reporting turned out to be challenging.

> "Once committed to the Covenant of Mayors, there is administration to do, who does it? And how to make sure that administration is as little a burden as possible, because if it is too heavy, the small municipalities, just for the administrative burden, won't commit to the Covenant of Mayors. Which is a pity because sometimes, they under-take really very precise actions that absolutely match the Covenant of Mayors" (in-terview with M. Janssens, 11 December 2014).[228]

In order to overcome this hurdle, Val d'Ille assisted its member municipalities in drafting and submitting the required documents. But even at the intercommunal administration of Val d'Ille itself, only two staff are in charge of climate and energy issues. As their time budget did not allow for spending too many working hours on filling in Covenant forms for member municipalities, they asked for instructions as they struggled to understand the first forms. It was not easy to obtain the necessary information from the CoMO:

---

228   Author's translation; original quote: *"S'engager à la Convention, il y a de l'administratif à faire, qui le fait? Et comment on fait en sorte que l'administratif soit le moins lourd possible, parce que si il est trop lourd, les petites collectivités, rien que pour la lourdeur de l'adminis-tratif, ne vont ne pas s'engager à la Convention des maires. Ce qui est dommage parce que quelque fois, elles font des actions vraiment très orientées et qui peuvent tout à fait s'inscrire dans la Convention des maires."*

"We are a small territorial authority with the means that correspond to a small territorial authority. […] They did not understand why we did not manage to cope on our own. But we do not have the means" (interview with M. Janssens, 11 December 2014).[229]

In such a way, early contacts gave the impression that staff at the CoMO was not used to communicating with small territorial governments and was not fully aware of their specific difficulties. Itself overloaded with the supervision of Covenant signatories due to the numerical success of the Covenant, the CoMO had little idea, in the beginning, of the limited resources in a small, rural community.

Val d'Ille perceived these difficulties as *"no insurmountable concern at all. It simply took a little more time to get things going"* (interview with M. Janssens, 11 December 2014).[230] Instead of giving up due to seemingly prohibitory Covenant reporting obligations, a new procedure was created inspired by exchange between Val d'Ille and the CoMO: the joint SEAP submission. Especially designed to suit the needs of rural areas, this new format introduced in 2011 allows handing in an overarching SEAP accompanied by a list of selected communal measures of individual member municipalities. Compared to each commune submitting a separate SEAP, this implies a significant reduction of the workload associated with Covenant membership for small, rural municipalities. The submission of a joint SEAP is an option for instance for French *intercommunalités* or German *Landkreise*. Val d'Ille then became a Covenant Coordinator that same year, and submitted its climate-energy diagnosis along with selected actions planned by member municipalities. (François 2013a; interview with M. Janssens, 11 December 2014) The joint SEAP submission has since been refined. As a result, Val d'Ille no longer figures as a Covenant Coordinator, but a signatory with associated local governments (Covenant of Mayors 2014f).

With the establishment of a more feasible reporting procedure, the Val d'Ille experience shows that even a small member, if engaged, can make a difference in terms of the Covenant's evolution. However, this does not imply a close working relationship between Val d'Ille and the CoMO. Contacts between Val d'Ille and the CoMO are very limited since the submission of the joint SEAP. On the one hand, the offers made by the CoMO do not meet the needs of Val d'Ille.

---

229 Author's translation; original quote: *"Nous, on est une petite collectivité avec les moyens qui vont avec une petite collectivité. […] Ils ne comprenaient pas comment on n'arrivait pas à se débrouiller tout seul. Mais nous, on n'a pas les moyens."*

230 Author's translation; original quote: *"Ce n'était pas du tout un souci insurmontable. Simplement on mettait un peu plus de temps pour mettre les choses en route."*

"According to Soazig Rouillard, in charge of mobility and Territorial Climate and Energy Action Plan [...], the CoM has been [...] barely used as technical support mean or help network. CoM webinars for example look more suitable for large communities. Likewise, financing solutions offered through the CoM cannot either be employed [sic] because they are dedicated to large-scale investment programmes" (François 2013b). The validation of the joint SEAP by the JRC took over a year, while – to the frustration of Val d'Ille officers – "direct interlocutors [at the CoMO] only take care of procedural aspects and do not provide any support on the substance" (François 2013a). On the other hand, Val d'Ille was able to rely on alternative resources that were better suited to local needs and priorities. For instance, Val d'Ille has not used the Covenant as a network for exchanging with peers but rather relies on contacts with, and with Covenant signatories from, Rennes Metropolis. Moreover, Val d'Ille does not refer to the Covenant when implementing climate and energy policies. Rather than relying on European endorsement, the intercommunality promotes its climate and energy measures with reference to their co-benefits (interview with M. Janssens, 11 December 2014)

5.2.4.4   Case Summary

The case of Val d'Ille demonstrates that limited resources alone do not impede inter-municipal climate policy coordination as long as dedicated politicians and administrators sustain the issue.[231] Endorsement by the Covenant as an orchestrator was used to legitimize the decidedly environmental intercommunal policy, and Covenant proceedings were used to set the agenda for the development of municipal climate action plans. Lean administration in a small, rural intercommunality necessarily entails limited resources for institutionalizing climate action. Thus, local climate policy remains vulnerable to changes in political personnel. Coordination by the Covenant in the form of reporting requirements might support remaining proponents of local climate policy as they represent a welcome path dependency.

The case of Val d'Ille highlights the fact that orchestration is not a unilateral relationship. Agency of the intermediary is of course included in the O-I-T-model

---

231   In terms of research methodology, limited human resources and corresponding working habits hampered data collection with regard to the case of Val d'Ille. For a discussion of potentially resulting limitations of the scope of conclusions from this case, see section 4.4.3.

which assumes voluntary cooperation in the pursuit of a joint policy goal (see section 3.1.3). Still, the model suggests a sequence of events where the orchestrator defines the goals to be pursued collaboratively, and its offer towards potential intermediaries, which can then be accepted or refused by the latter. The development of better-suited proceedings for rural local governments by the CoMO shows how the orchestrator adapts to the needs and demands of intermediaries in the process of joint governance. Also, the case of Val d'Ille confirms indication found in prior applications of the O-I-T-model that intermediaries strategically seek orchestration as a means to enhance their capacities. Cueff had a record in decidedly environmental policy prior to his commitment to the Covenant, and Val d'Ille had already engaged in sustainability and climate policy, for example with its launch of a local Agenda 21. Orchestration by the CoMO was not an occasion to adopt a new policy but rather one to strengthen local actors in the pursuit of policy goals they were already committed to.

5.3 Summary of Findings

Five inter-municipal associations from Germany and France engaged as Covenant Coordinators. Greater Lyon, Rennes Metropolis, Val d'Ille, Rhine-Neckar Region and Stuttgart Region voluntarily became intermediaries of the Covenant. This EU initiative aims to orchestrate local climate and energy policy in view of the European climate and energy targets. The following sub-questions are examined with regard to each case (see section 1.3): Why have inter-municipal associations been attracted by the Covenant? How have they assumed their role as intermediaries, and which modes of governance emerged? An overview of the five cases demonstrates that despite strong variances in local circumstances, drivers, and forms of inter-municipal climate policy coordination on behalf of the Covenant, there are some overarching patterns.

The inter-municipal associations under study varied in important ways (see figure 5.9). The number of inhabitants ranged from approximately 20,000 (Val d'Ille) to about 2,700,000 (Stuttgart Region). The number of territorial governments comprised by an inter-municipal association spanned from 10 municipalities (Val d'Ille) to 295 municipalities and districts (Rhine-Neckar Region). While French inter-municipal associations consisted exclusively in municipalities, German inter-municipal associations also comprised districts as an additional, intermediate level of government. The German Covenant Coordinators under study had significantly fewer staff than their French counterparts although they were larger in terms of population and member governments. As French inter-municipal associations also are in charge of additional, personnel-intensive tasks, the size of teams available for local climate policy hardly differs.

By definition, inter-municipal associations group neighboring municipalities, but the structure of membership differs in terms of settlement patterns. While Greater Lyon and Rennes Metropolis are clearly structured around a center city, Stuttgart Region and Rhine-Neckar Region are polycentric with several competing regional metropolises. This impacts the roles of center cities as local pioneers in climate policy. Where inter-municipal associations consist of a single regional metropolis with a peri-urban belt surrounded by rural municipalities, center cities can act as local leaders. They can lead by example, cooperate closely with the inter-municipal association, and also shape its climate policy. Where inter-municipal associations are polycentric, no individual city can shape inter-municipal policy as it pleases. Cooperation with the inter-municipal association and leadership by example remain possible, but not to the same extent. Consequently, French and German inter-municipal associations meet different challenges in developing a shared vision for regional development. In short, French agglomerations have to reconcile the interests of the center and the periphery, while German metropolitan regions work on establishing a common identity. In the case of Val d'Ille, no municipality plays a clear center role. The inter-municipal association did not rely on the influence of an individual member municipality, but on personal leadership and persuasion efforts by the inter-municipal president.

One might assume that these differences translate into Covenant coordination in an easily predictable way, i.e. that larger inter-municipal associations with more resources at their disposal find it easier to promote the Covenant among member municipalities. Another possible assumption would be that German inter-municipal associations benefitted from a favorable national framework, namely ambitious climate policies and high degrees of local autonomy. Case studies showed that this was not the case (see again figure 5.9). The number of Covenant signatories supported ranged from 3 (Greater Lyon, Stuttgart Region) to 33 (Rennes Metropolis).

	Population	Municipalities[a]	Structure	Covenant signatories	Climate plan	Status
Greater Lyon	1,360,000	59	Center city	3	Territorial climate and energy plan; SEAP	Covenant signatory
Rennes Metropolis	439,000	43	Center city	33	Territorial climate and energy plan; SEAP	Nominal Covenant Coordinator
Val d'Ille	20,000	10	Rural	10	Territorial climate and energy plan; Joint SEAP	Active Covenant Coordinator
Rhine-Neckar Region	2,350,000	295[b]	Polycentric	15	Regional energy plan	Nominal Covenant Coordinator
Stuttgart Region	2,700,000	184[c]	Polycentric with center city	3	–	Nominal Covenant Coordinator

[a] more precisely: number of territorial governments comprised (municipalities and other)
[b] 8 urban districts, 7 rural districts, 280 district-affiliated municipalities
[c] 1 urban district, 5 rural districts, 178 district-affiliated municipalities

Figure 5.9: Overview of case studies
Source: Author.

In contrast to their German counterparts, French Covenant Coordinators all submitted their own Sustainable energy action plan (SEAP) to the Covenant. They translated their climate action plan as established by national law along Covenant methodology.[232] Here, the case of Val d'Ille stood out in that its Territorial climate and energy plan was voluntary, and blended into a joint SEAP that also comprised individual municipal contributions. This option was also envisaged, but not implemented by Rennes Metropolis. German Covenant Coordinators developed regional energy plans of a more strategic, less operational nature. With

---

232   In a similar vein, Covenant signatories typically translate their local climate plan based on national methodology – the German municipal climate concept or the French PCET - into the forms of the Covenant of Mayors for submission. Thus, the local climate plan and the SEAP of a given municipality are no distinct plans, but rather two formats of the same planning document. If no local climate plan is developed (as in the case of Val d'Ille's member municipalities), SEAP development is accordingly more laborious.

regard to the current status of Covenant Coordination, the case of Val d'Ille again stood out as it was the only inter-municipal association that was fully up to date with its Covenant reporting obligations and thus could be considered as an active intermediary. Greater Lyon, because of its failure to report on coordination activities, was relabeled a simple Covenant signatory. The other three inter-municipal associations were still listed as Covenant Coordinators on the Covenant's website, but were only nominal intermediaries.

The comparison between Germany and France highlights a difference in their inter-municipal associations' legal status that impacts the perceived necessity for self-legitimation. German regional associations have statutory independence since they are established by *Länder*-level legislation, i.e. their very existence and mandates are not locally determined. This is not to say that the legitimacy afforded by regionalization was irrelevant to local governance, but that their legal status restricts the necessity for legitimacy-seeking behavior. In contrast, French intercommunalities have been described as a form of voluntary cooperation

> "despite the fact that French law introduces […] a series of mandatory tasks […] because the establishment of a particular communauté urbaine depends on the joint decision of the local governments to do so. Moreover, the communes can resign their membership or jointly decide to dissolve the communauté" (Hulst and van Montfort 2007b, 11).

Hence, they have to demonstrate the benefits of intercommunal affiliation to present and potential member municipalities. As a result, legitimacy-seeking behavior more strongly motivates inter-municipal associations' Covenant participation in France than in Germany.

At the same time, differences in responsibilities impact the options available for inter-municipal climate policy coordination. German regional associations lack operational capacities because they only intervene in public service provision at selected instances such as public transport. This significantly hampers their ability to govern by provision. But they have extensive competences for mandatory planning with regard to regional development, land use, energy, and transportation. These enable them to govern by authority and enforce the consideration of climate considerations at the municipal level. French intercommunalities are involved more closely in urban development and housing and have a closer relationship with member municipalities. This facilitates governance through enabling by operational assistance, including working groups for municipal politicians and staff.

Correspondingly, differences in available modes of governance shape inter-municipal associations' behavior as Covenant Coordinators. Unlike German regional associations, French intercommunalities assist member municipalities in Covenant reporting and in compiling joint SEAPs on their behalf. In Germany, this approach would better match the profile of rural districts. The latter participate in multi-level cooperation agreements, for example with the Rhine-Neckar Region, and in the National Climate Protection Initiative, but to date not a single *Landkreis* has become a Covenant Coordinator.

The mixed picture of Covenant Coordination in Germany and France is a surprising finding against the background of enthusiastic accounts of the Covenant and its ever growing number of signatories. This raises the question why the orchestrator-intermediary relationship between the Covenant and inter-municipal associations in Germany and France seems to be such a difficult one. Challenges derive from the factors motivating inter-municipal associations to engage in orchestration in the first place and from their experiences in implementation which caused all five Covenant Coordinators under study to temper their engagement over time.

6 Analyzing Orchestration in the Covenant of Mayors

The Covenant can be conceptualized as an orchestration arrangement where the CoMO acts as an orchestrator on behalf of the European Commission. Inter-municipal associations that engage as Covenant Coordinators are understood as intermediaries in addressing the actual target actors, namely municipalities. This does not exclude the possibility that participating actors might have a genuine interest in contributing to climate mitigation – especially when their actions are driven by engaged individuals (Hooghe and Marks 2001a, 70). It builds on existing literature on voluntary non-state climate action when it assumes that the participation in a particular scheme or program is most likely motivated by some kind of co-benefits (see 3.1.1). In this respect, the O-I-T-model points to the importance of support from the orchestrator for motivating the intermediary to engage voluntarily in the pursuit of shared policy goals. In a similar vein, support from Covenant Coordinators can be expected to motivate municipal action. The latter addresses collective actors and citizens whose behavior constitutes the ultimate target of climate and energy policies.

The case descriptions in chapter 5 have shown that the roles of inter-municipal associations in local climate governance vary largely depending on how coordination is put into practice. They have also demonstrated that despite differences between individual Covenant Coordinators, similar challenges to orchestration arise. A strong orchestrator-intermediary relationship between the Covenant and inter-municipal associations is just as far from self-evident as is being able to successfully address municipal target actors. This chapter aims to capture and explain variations and patterns seen in the case studies building on a cross-case analysis. It focuses on inter-municipal associations as coordinators of local climate policy but includes insights on municipal behavior where applicable.

The analysis begins by discussing what kinds of co-benefits inter-municipal associations aim for when engaging in local climate policy coordination (6.1) and how they build upon Covenant support in practice (6.2). The motivation that

© Springer Fachmedien Wiesbaden GmbH, part of Springer Nature 2020
L. Bendlin, *Orchestrating Local Climate Policy in the European Union*,
Energiepolitik und Klimaschutz. Energy Policy and Climate Protection,
https://doi.org/10.1007/978-3-658-26506-9_6

intermediaries derive from the Covenant of Mayors as well as their use of the support it provides are examined in detail. Subsequently, the analysis proceeds with a typology of municipal target actors' benefits and attempts to further improve the latter so as to explain target actors' response to the offer made by the CoMO and their respective Covenant Coordinator (6.3). Disappointments about co-benefits in general and Covenant support in particular weaken the relationships between participating actors and, as a result, the Covenant's ability to orchestrate local climate policy effectively (6.4). The chapter concludes with a summary of its key arguments (6.5).

6.1 Intermediaries' Motivations: Why Become a Covenant Coordinator

A key question in examining orchestration arrangements consists in why intermediaries agree to participate. When asking why inter-municipal associations act as Covenant Coordinators, it is important to differentiate between motivations to engage in local climate policy coordination in the first place, and to commit to the Covenant. The Covenant commitment typically amends rather than initiates inter-municipal local climate policy coordination. Metaphorically speaking, tinder needs a spark to ignite. This section first establishes a typology of inter-municipal motivations for local climate policy coordination. Secondly, it displays how the motivational background of an inter-municipal association impacts the attractiveness of the support offered by the Covenant.

Inter-municipal climate policy coordination is not self-evident. In the absence of legal obligations, spending local resources on a global common good requires consistent justification (Benz et al. 2015, 319). We know from the literature that municipalities engage in local climate policy in order to advance climate governance and to assert their role therein, but also to obtain co-benefits "such as cost savings, clean air, regional economic development and job creation, alleviation of energy poverty, accessibility and livability of the city" (Alber 2009, 8). Case studies have shown that the motivations of inter-municipal associations are similar to those of municipalities in certain respects, but go beyond them in others. This will be explained by their slightly different position within the multilevel system, which entails specific challenges and opportunities. Inter-municipal motivations for engaging in local climate policy relate to opportunities for agency in the sense of genuine choice (see 3.1.3) and to local governments' ability to link climate action to their self-conception and development project. Three main drivers deriving from an individual, horizontal, or vertical perspective can be identified: local leadership, territorial competition, and rescaling (see figure 6.1). Reasons for becoming a Covenant Coordinator differ along these lines.

	Perspective		
	Individual	**Horizontal**	**Vertical**
Motivations	**Local leadership** • Genuine conviction • Career building	**Territorial competition** • Economic opportunities • Quality of life • Organizational growth	**Rescaling** • Capacity gains from below • Capacity gains from above • Self-legitimation

Figure 6.1: Why inter-municipal associations engage in local climate policy coordination
Source: Author.

The local leadership of individual politicians is a recurrent motivation for inter-municipal associations to engage in climate policy coordination. This finding fits well with accounts of local politics being driven by personalities, i.e. local politicians, rather than party politics as observed at superordinate levels of government (Ryan 2015). Local politicians might pursue different interests in taking on a leadership role for local climate policy coordination, namely conviction as well as career-building.

From the perspective of local leaders, participation in the orchestration arrangement of the Covenant is attractive when its policy goals match their political priorities or when Covenant support strengthens their political position. Some politicians become local leaders for climate policy coordination because they hold the genuine conviction that climate change needs to be mitigated, that their territorial government has a contribution to make in this respect, and that it is their role to push for corresponding policies. Other local leaders show an instrumental use of their engagement in climate policy as a means of career-building. This does not necessarily imply a lack of conviction; career-building opportunities might as well constitute a welcome co-benefit of local climate policy engagement.[233]

---

233   The research design of this inquiry does not allow assessing whether career-building local leaders have initially engaged in climate policy out of calculation or conviction. To that end, narrative interviews with the respective leaders about their personal motivations would have been in order rather than expert interviews with their co-workers about political proceedings.

Politicians who play a local leadership role are either mayors or inter-municipal presidents, or function as both at the same time. Mayors such as Eckart Würzner from Heidelberg make climate mitigation a priority for the development of their municipality and attempt to involve other municipalities from their inter-municipal association in this endeavor. Sometimes, local leaders also are or become inter-municipal presidents; examples include Gérard Collomb from Lyon/Greater Lyon and Daniel Cueff from Langouet/Val d'Ille.

Mayors have a direct grasp on municipal priorities, policies, and services. They are often well networked with fellow mayors (not only) from their inter-municipal association, and garner legitimacy from their proximity to citizens (and voters). This enables them to directly engage their municipalities, and to mobilize local stakeholders. Inter-municipal presidents have high public visibility within the inter-municipal association and institutionalized ties to member municipalities and their mayors. They can use these resources in order to create a common project summoning municipal and non-governmental actors from the entire inter-municipal association to take part. Their individual legitimacy and influence on inter-municipal priorities, policies, services and agencies such as local and regional energy agencies depend to a large extent on the institutional architecture of the respective inter-municipal association, including provisions for the accumulation of offices. Inter-municipal presidents in Germany are weakened by the fact that their function is an honorary one, and their mandate may not be combined with parliamentary office. The inter-municipal administration is led by a full-time director; the leadership role in climate action of an administration can therefore be more decisive than the preferences of the president her- or himself.

Two other driving factors of inter-municipal associations engaging in local climate policy coordination are territorial competition and rescaling, i.e. horizontal vs. vertical intergovernmental relations. Territorial competition pertains to horizontal relations with other inter-municipal associations. Comprising similar co-benefits as described for municipal climate policy (Alber 2009; Benz et al. 2015); voluntary climate action by subnational governments can be understood as an instrument of economic development (Hsueh and Prakash 2012). Low-carbon technologies, renewable energies and 'green' businesses can provide economic growth opportunities, increased tax revenues and job creation. Quality of life and the livability of a region contribute to demographic growth and attract the residents needed to ensure there is a sufficient workforce for local companies.

In view of organizational growth, inter-municipal services, such as assistance for municipal climate policy, can attract additional municipalities to join an inter-municipal association. This motivational factor linked to local climate policy co-ordination is most important where municipalities are free to choose an inter-municipal association for affiliation. Rennes Metropolis for example recently af-filiated new municipalities. Economic and demographic growth was frequently referred to in the Stuttgart and Rhine-Neckar Regions as reasons to engage in climate action.

Rescaling regards vertical relations with subordinate and superordinate lev-els of government, more precisely activities aiming at changing the architecture of multi-level climate or territorial governance more broadly. Here, the key ques-tions are 'who gets what' and 'who gets to decide' (Reed and Bruyneel 2010). Local climate policy coordination represents one more opportunity to showcase how an increased role for the inter-municipal level could improve public policy performance. In order to gain additional responsibilities and resources from af-filiated municipalities, urban and rural districts, inter-municipal associations have to demonstrate their ability to govern effectively. They must show their superior problem-solving abilities, or offer efficiency gains through economies of scale. Pooling municipal capacities at the inter-municipal level has to be pre-sented as being in a municipality's own interest, in particular in fields of public policy where municipalities determine the inter-municipal association's scope of action. Inter-municipal associations also need to demonstrate their ability to pro-vide governance services at the regional level and, most importantly, at the na-tional level. They do this, for example, by lobbying for territorial reform (i.e. decentralization or metropolization) that will give them new capacities. Within the struggle between levels of government over competences and resources, le-gitimacy is a valuable resource in relations with citizens and other territorial ac-tors. Contributing to the public good by local climate policy coordination be-comes a means to this end.

One particularly ambitious (and successful) proponent of this approach was Greater Lyon. President Collomb lobbied for territorial reform and aimed to po-sition Greater Lyon as a laboratory for metropolitan regions' enhanced role in French territorial governance. The intercommunality sought to establish itself as a platform for convening all relevant actors from the territory for developing a joint climate plan. Eventually, Greater Lyon obtained additional competences and corresponding funds both from the municipal and the – locally abolished –

departmental level. Once the reform was under way, this perspective became less important, and so did the Covenant commitment. In contrast, interviewees from Rennes Metropolis referred to metropolization as an obligation: to fulfil additional tasks, and to reorganize inter-municipal governance. It was important to demonstrate that the intercommunality rendered valuable services to affiliated, often considerably fragmented municipalities. Thus, Rennes Metropolis aimed to provide municipalities with tangible services in accordance with their needs. These slightly different motivations resulted in correspondingly different approaches to implementing the commitment as a Covenant Coordinator.

Whether inter-municipal associations were motivated primarily by local leadership, territorial competition, or rescaling, all cases had one thing in common. The inter-municipal associations under study, or at least actors within them, had engaged in local climate policy coordination already before they became Covenant Coordinators. Interviewees frequently referred to the Covenant commitment as the next logical step, a continuation or expression of prior activities, or as a means to enhance their visibility. They insisted that the Covenant had not sparked, but complemented local climate action. Put differently, they emphasized not adopting external policy goals, but instead finding an ally for activities they were undertaking anyhow – an ally that offered ideational and material support. For the intermediary, it is important to demonstrate that participation in the orchestration arrangement is in its own interest as well as that of its stakeholders. If motivated by local leadership, Covenant participation has to correspond to local leaders' own visions. If driven by territorial competition, it has to involve more economic benefits than costs, and if used for rescaling, it has to enhance rather than restrict policy-making autonomy.

Different dominating factors inspired local climate policy coordination by inter-municipal associations. Correspondingly, different aspects of Covenant support were most attractive to them. Local leaders primarily benefit from the Covenant in terms of its endorsement of their climate policy project. Formal recognition by the European Commission enhances the legitimacy of local leadership and increases the pressure on the inter-municipal assembly and reluctant mayors to support the initiative. Covenant events and publications also provide personal visibility for inter-municipal presidents and mayors who take the lead. Depending on the availability of alternative schemes and supporters, local leaders can also rely on the Covenant for agenda-setting and assistance. Ideational Covenant support in convening territorial actors to engage in economic development projects and legitimizing the inter-municipal agenda is an asset for territorial competition. Assistance in public communication via Covenant media enhances Covenant Coordinators' visibility and supports place branding. For instance, Stuttgart Region added three so-called benchmarks of excellence to its website profile. Covenant support in third-party funds acquisition is another important value added sought by Covenant Coordinators motivated by territorial competition. An illustration of this competitive approach consists in the frequent references made to the leading role played by Rhine-Neckar Region in the Covenant as compared to other German regional associations. When Covenant Coordinators engage in rescaling, endorsement is obviously beneficial as they can build their argument on the European level for why they need to play a role in climate governance. In addition, coordination by the Covenant can further legitimize lobbying efforts as it documents European-level expectations and thereby highlights individual gaps in inter-municipal associations' responsibilities and resources in view of fully participating in the Covenant.

In short, the Covenant's offer of material and in particular ideational support for local climate policy coordination motivated inter-municipal associations to participate in orchestration. Inter-municipal associations can use these additional resources for local purposes, namely for local leadership, territorial competition, and rescaling. These motivations can be distinguished as being primarily individual, horizontal, or vertical, but overlap in practice. Therefore, the same form of Covenant support may attract inter-municipal associations driven by different motivations. Notably, the importance of Covenant support for joining does not necessarily imply that inter-municipal associations were able to effectively apply these resources.

6.2 Intermediaries' Approach to Implementing their Covenant Commitment

Once an inter-municipal association has engaged as a Covenant Coordinator, the question remains how to proceed. This section discusses how intermediaries implement their commitment, and how they can build upon support by the CoMO in this endeavor. Covenant support has been discussed as a means to motivate inter-municipal associations' participation. This section examines how inter-municipal associations use and benefit from Covenant support in practice. Inter-municipal associations have responsibilities for matters like land-use planning, housing, and transportation. But they only have limited regulatory power over member municipalities and restricted resources at their disposal. Which means and strategies do they deploy in order to implement the commitments they have made, to mobilize member municipalities and other actors, and to play a coordinating role in local climate governance? Inter-municipal proceedings can be clustered along four non-exclusive approaches based on their corresponding target groups: 1. A political approach to local climate policy coordination targets municipal politicians, 2. a technical approach focuses on municipal administrations, 3. a concerted approach strives to involve non-governmental actors from the territory, and 4. a practical approach concentrates on one's own behavior (see figure 6.2). Hence, modes of inter-municipal climate policy coordination differ slightly from those of municipal climate policy as described in the literature (Alber and Kern; Bulkeley et al. 2011; Bulkeley and Kern 2006; OECD 2010). Political, technical and concerted approaches to local climate policy coordination rely on enabling modes of governance to a large extent. Other modes of local climate governance, i.e. self-governing, governing by provision (of services, programs, and plans) and by authority (over member municipalities and other actors), relate primarily to the practical approach to inter-municipal climate policy coordination.

Approach	Political	Technical	Concerted	Practical
Target group	Municipal politicians	Municipal administrations	Non-governmental actors	Inter-municipal association
Means and strategies	• Information and networking • Obligation through public commitment • Peer pressure	• Information and networking • Obligation through public commitment • Training • Technical support • Financial support	• Information and networking • Obligation through public commitment • Financial and promotional incentives	• Obligation through public commitment • Integration into inter-municipal planning • Integration into inter-municipal services and policies

Figure 6.2: Means and strategies of Covenant Coordinators based on target groups
Source: Author.

Each inter-municipal association chooses a specific combination of means and strategies for the implementation of its engagement in local climate policy coordination. These decisions are taken against the background of locally, regionally, and nationally determined scopes of action, and inter-municipal activities interact with the means and strategies deployed by other municipal or non-governmental actors in their territory.

An inter-municipal association might conceptualize its engagement for local climate policy coordination as a political task, i.e. as an issue of convincing and engaging political decision-makers. It will then target municipal politicians, primarily the mayors of its member municipalities and, if these do not coincide, the representatives of its municipalities at the inter-municipal assembly. The first step then consists of informing these municipalities about local climate policy and possible targets and undertakings, including Covenant signature. Networking meetings among interested local politicians can help create and maintain a dynamic among participating municipalities. Meetings ensure that the issue remains on the agenda, and networking opportunities help to reassure politicians about the feasibility of local climate action. This dynamic can be enhanced and secured through public commitment, for example by means of an inter-municipal climate conference where municipalities announce their targets or action plans. Examples include the local climate conference convened by Rhine-Neckar Region along with Heidelberg city as a platform for a collective Covenant commitment. Covenant endorsement further legitimizes the inter-municipal initiative in the eyes of municipal politicians and mayors who might be particularly attracted by the Covenant's promise for EU-level visibility. When such a dynamic captures a majority of the member municipalities of an inter-municipal association, it can result in peer pressure on resistant politicians to join despite initial disinterest or refusal, as was the case in Rennes Metropolis.

A more technical approach to local climate policy coordination concentrates on involving the administrations, or more precisely the staff of member municipalities who are in charge of climate and energy management and municipal action plans. Working groups bring together municipal staff and serve as a platform for disseminating information on local climate and energy policies. Issues covered typically range from technical aspects of GHG emission monitoring to suggestions for implementation and advice on funding schemes. Covenant signature and proceedings can help to set the agenda and the instruments to be used in the inter-municipal initiative. Regular meetings facilitate networking between municipal staff and provide an opportunity to exchange best practices, difficulties in implementation, and tested remedies. Municipal staff value the proximity to their every-day work both in terms of travel (time and funds for official journeys are often scarce) and content that is tailor-made to their needs. For more detailed technical support, inter-municipal associations typically rely on local energy agencies, which assist municipal staff in establishing local climate plans and monitoring energy use or GHG emissions. In some cases, local energy agencies and inter-municipal staff cooperate in developing guidelines and templates for local climate planning and monitoring. This facilitates local climate and energy planning in general and Covenant reporting in particular, and has inspired the 'grouped SEAP approach' in which the Joint Research Centre reduces its work-load as it "only analyses a representative sample of SEAPs from municipalities supported by the same Coordinator and entrusts the latter with the responsibility of communicating one single feedback report to the larger group of municipali-ties" (Covenant of Mayors 2013c, 7). Other Coordinators go one step further in that they submit a joint SEAP, an option introduced in 2012 in response to feed-back from local governments, including Val d'Ille. A joint SEAP either compiles individual or shared municipal commitments and lists both individual and col-lective measures of the participating municipalities (Covenant of Mayors 2014c). Financial support from the inter-municipal level will remain limited for a lack of available funds for this voluntary activity, and the political necessity to evenly support member municipalities. Exceptions include co-funding of services from the local energy agency to member municipalities, and funding for ready-made project toolkits, such as the Energy Caravane in Rhine-Neckar Region and the ENGAGE campaign by Rennes Metropolis.

When an inter-municipal association adopts a concerted approach to local climate policy coordination, it focuses on the involvement of non-governmental actors. These may include local businesses, housing companies, energy and transport providers, civil society, universities and other research institutes, etc. Inter-municipal climate conferences serve to disseminate information on possi-ble measures and co-benefits such as economic opportunities and quality of life. On such occasions, reference to the Covenant commitment serves to provide Eu-ropean-level endorsement, i.e. legitimacy and increased pressure on target actors to participate. Examples include yearly networking events in Rhine-Neckar Re-gion, but Greater Lyon went a significant step further as it established the climate conference as a platform for the elaboration of a territorial climate plan that included inter-municipal as well as non-governmental contributions. Inter-municipal associations are unable to provide broad financial support for all non-governmental actors addressed. Typically, they resort to competitive schemes such as calls for proposals instead.

Finally, a practical approach to local climate policy coordination primarily targets inter-municipal behavior as a consumer, provider, and regulator. Although these modes of governance have been labeled as self-governance, they should not be mistaken for being self-involved. Exemplary inter-municipal consumption is a prerequisite for leadership by example. Governance by provision and by authority necessarily involves target actors outside the organization, including users of inter-municipal services such as public transport or waste removal, and actors affected by inter-municipal regulations such as land-use plans and energetic standards for construction. Here, the Covenant commitment serves to engage the entire organization. Multiple administrative departments have to be mobilized in order to integrate climate and energy objectives into inter-municipal planning, and to ensure that climate and energy considerations are taken into account during the development of inter-municipal services and policies. A Covenant Coordinator focusing on self-governance is likely to develop and submit its own SEAP, although this is no mandatory requirement of the CoMO. In such a case, the inter-municipal association can build upon Covenant support in similar ways as municipalities (see 6.3).

Summing up, Covenant Coordinators do not necessarily focus on Covenant promotion in a narrower sense but interpret their commitment to local climate policy coordination more broadly. Aiming to stimulate and mainstream territorial climate action within or outside the Covenant, inter-municipal associations address municipal politicians and staff as well as non-governmental actors or their own organization. These approaches can be categorized as political, technical, concerted, and practical, and entail corresponding emphases in inter-municipal means and strategies for coordinating local climate policy. Notwithstanding variations in the kinds of Covenant support that are most valuable to Covenant Coordinators, Covenant support does not suffice to foster behavior that would not have occurred in the absence of the Covenant. Rather, Covenant support facilitates and complements existing initiatives and policies. A similar assessment applies to municipalities, the target actors in the orchestration arrangement of the Covenant.

6.3   Target Actors' Response: Municipal Covenant Participation

The Covenant involves actors from multiple levels of governance, but ultimately aims at stimulating climate action by local governments, namely by municipalities. Hence, municipal participation determines whether the orchestration arrangement works – or remains an empty shell. In order to understand municipal response to the Covenant, it is important to consider that institutional structures of local climate policy are not exogenous, but result from the accumulated decisions of local actors over time. The benefits of Covenant signature from a municipal perspective thus depend on the level of development of local climate policies. Beginners in local climate action are likely to lack the knowledge, visibility, authority, and networks required for effective climate mitigation. At a more mature stage of local climate policy development, municipalities will have set up energy and emission monitoring schemes and action plans; they will have started to implement and update them. The municipal administration might have developed partnerships with relevant actors and been provided with financial and human resources for local climate action by the municipal council. Whether a municipality is at an early or mature stage of local climate policy, local actors perceive Covenant benefits from an internal as well as an external perspective (see figure 6.3).

	Internal benefits Municipal politics	**External benefits** Citizens, peers, and superordinate levels of government
Early stage of climate policy	• Establish authoritative constraints for taking action • Benchmark for setting targets • Methodology for climate and energy planning	• Visibility for local leaders • Enhanced legitimacy of local climate action • Facilitate funding acquisition
Mature stage of climate policy	• Secure local climate action	• Audit for existing policies • Advocacy platform

Figure 6.3: Covenant benefits for municipalities based on stage of policy
Source: Author.

At an early stage of local climate policy formulation, Covenant signature can facilitate internal political and administrative processes. Proponents of climate action within local governments and administrations often have a marginal position. As put forward by Aylett (2015), they must first win basic support from other departments and actors. Then, they need to seize the planning process as an opportunity to translate the global problem into the local context and forge coalitions. As Covenant signature is legally non-binding, it is easy to achieve consensual support for Covenant participation in the municipal council. Nevertheless, the Covenant establishes an authoritative outside constraint on reluctant factions or administrative departments. Once a municipality has committed to the Covenant with its signature, it is now bound by its honor to take action. Covenant rules provide orientation on how to proceed with targets, monitoring, and planning, especially if superordinate levels of government do not assist in agenda-setting. When local climate and energy plans were introduced in France on a voluntary basis, local governments relied on Covenant methodology until the ADEME came up with more nationally adapted tools. Covenant signature also entails a certain level of visibility for local leaders already at the outset of local climate action. European endorsement enhances local governments' legitimacy and authority to act on climate change towards external stakeholders such as citizens or superordinate levels of government. Early Covenant signatories expected facilitated access to European funding through specific funding schemes or preferential access to other schemes. Early considerations on how to design the Covenant made explicit reference to addressing the question of financing (European Commission 2006a, 28–29). Later signatories observed that the anticipated funding opportunities for Covenant members did not materialize. Few European and national funding schemes explicitly refer to Covenant signature,[234] but it can be mentioned in funding applications to enhance their credibility.

At a more mature stage of climate policy, municipalities benefit from Covenant signature in different ways. Internally, accounting and reporting routines are already in place and there is no need to rely on Covenant methodology; existing documents will simply be translated into Covenant forms. But the Covenant's schedule can have an agenda-setting function, setting the basis for evaluations or updates of a local climate plan. Since municipalities are not monolithic actors, Covenant signature can help interested actors overcome hindrances to pursuing and updating local policies or perpetuating the institutionalization of local climate policy such as dedicated administrative departments and their re-

---

234 Exceptions include the Rural Web Energy Learning Network for Action (eReNet).

sources. The official Covenant commitment, adopted by the local council and signed by the mayor, combined with the threat of being shamefully suspended in case of non-compliance, compels local actors to continue local climate action. Municipal staff frequently reported having referred to the municipality's Covenant signature in budgetary negotiations, or in order to engage newly elected councilors in a local climate project. Similar to the authoritative constraint at earlier stages of local climate policy, Covenant signature helps to establish a favorable path dependency (Levin et al. 2012).

Externally, endorsement by the Covenant is also used as a quality audit for existing policies, i.e. as an authoritative recognition of their ambition and legitimacy. Numerous interviewees referred to Covenant signature as a logical continuation of what they had already been doing rather than an incentive to undertake more or different actions. In these instances, local climate action cannot be traced back to the Covenant; rather, the audit becomes an additional resource for political communication. Adjacent literature has described similar behavior of cities as using EU-projects as a platform for profiling and identity-building (Wolffhardt et al. 2005, 94–95), and as branding (Dinnie 2011), with TMCNs representing both brands and forums for branding (Busch 2015). Covenant signatories use the Covenant as an audit or brand when they include its logo in publications and refer to EU-level recognition when communicating local policies. And they use it as a forum for branding when they seize opportunities for self-promotion, such as the upload of best practices to their website profile.[235] Conceptualizing this behavior as place branding or as green city branding would imply a somewhat coherent branding strategy (Busch and Anderberg 2015). This could not be affirmed. Instead, case studies indicated that auditing either motivated only some actors within a given municipality (or inter-municipal association), or served genuinely climate-related objectives – such as the creation of favorable path-dependencies or lobbying for additional responsibilities and resources – rather than mere territorial competition.

Auditing and branding behaviors illustrate a fundamental tension within the Covenant. Originally conceived as a network of mayors from frontrunner me-

---

235 Where local leaders hold a prominent position within the Covenant of Mayors, such as the presidency of a network from the CoMO consortium, the brand is not only attached to the municipality, but also to the individual. Participating in or hosting Covenant events can back political leaders' claims that her or his city counts, thereby supporting their electoral campaigns or career ambitions for higher levels of government. The same behavior has been encountered at the inter-municipal level.

tropolises, it developed into a broader platform that also addresses followers and claims to be a "mainstream movement" (Covenant of Mayors 2015a). Early Covenant signatories mostly consisted of larger cities with considerable prior records in local climate policy formation. They continued to highlight that they were among the first to join the Covenant as a kind of proof for their being part of a climate avant-garde. In the meantime, the Covenant's numerical success has reduced its value added for city marketing, because branding is based on distinction. At the same time, the Covenant still helps legitimize voluntary climate action as a common behavior. This can be important in contexts where local governments must struggle to obtain the necessary authority to engage in GHG emission reducing activities. Covenant membership may over time become less attractive to pioneers, and more attractive to late comers. Later Covenant signatories more often consist of smaller municipalities with limited prior experience in local climate policy. This dynamic partly explains the limited number of Covenant signatories from EU member states with extensive experience in local and national climate policy as compared to higher numbers in some EU member states with lower levels of local climate action.

Last but not least, the Covenant can provide an additional platform and channel for municipal advocacy and lobbying. Covenant signatories benefit from European endorsement in debates over national climate governance in terms of ambition and rescaling. This was the case during the National Energy Transition Debate in France (Réseau action climat (RAC) France). Moreover, the Covenant provides a channel for local governments to lobby the European Commission. Examples include an open letter from the Covenant Club Germany that called for revisions in Covenant proceedings so as to reduce the workload for German municipalities in complying with Covenant rules (Covenant Club Deutschland 2012). This attempt to improve the cost-benefit ratio of Covenant signature shows that target actors do not condone the Covenant as set up by the European Commission, but try to influence its design.

Notably, case studies found that municipalities did not rely on the Covenant for learning and networking. Interviewees frequently broached the issue of a perceived overlap between the Covenant and TMCNs. With their respective municipalities already members of Energy Cities and/or the Climate Alliance, they said they had well-established ties with network staff and other municipalities from the network through regular network conferences and personal contacts. Hence, they would rely on these contacts for learning and exchange rather than browsing through best practices on the Covenant's website or attending Covenant events in remote places in foreign languages. In addition, they expressed a preference for national or regional contacts and events, explaining this was because they were more likely to share challenges and opportunities. This, they explained, facilitated the transferability of experiences. Some openly criticized the redundancy of being a member of a given TMCN and the Covenant at the same time. This critique highlights a certain misunderstanding, as the Covenant is not designed as an additional TMCN, but as an overarching platform. In particular, its standardized reporting methodology allows for documenting local governments' climate action beyond TMCN affiliation and national borders. These data help to demonstrate the breadth and depth of local climate action, which, in turn, can serve as an argument in debates at superordinate levels. On the one hand, it helps the CoMO when calling for additional EU schemes and programs in support of this voluntary groundswell. On the other hand, it supports the European Commission as it enhances its credibility in global climate governance and can help to dispel member state objections to EU climate policies. But this does not benefit individual Covenant signatories and is of little significance to local actors. They rather see the additional workload of mandatory reporting and see little value in the development of comparable local documents and in the control of SEAPs by the JRC. Hence, superordinate levels have to offer excludable benefits to motivate municipal participation in the orchestration arrangement.

Interestingly, not a single interviewee referred to party politics as a relevant factor for the decision to make or sustain a Covenant commitment or not. Participation in the Covenant is typically voted for unanimously or with at least a broad majority of the factions in a council. Municipalities accepting inter-municipal support for their climate policy are not necessarily governed by the same party as the respective Covenant Coordinator or other participating municipalities. However, electoral change was repeatedly referred to as a crossroads for local climate policy (see section 6.4).

As shown throughout chapter 5, municipalities frequently struggle to realize the benefits they aimed for when they signed the Covenant. Two kinds of factors burden their cost-benefit calculation of Covenant participation: elevated costs of program participation due to the workload of Covenant proceedings, and reduced benefits from program participation due to program deficiencies and overlaps. Attempts to alter the program accordingly have not been conclusive (see figure 6.4).

	Challenges	Remedy attempts
Cost-increasing factors	Lack of technical assistance	Mission specifications for Covenant Coordinators & local and regional energy agencies
	Workload of reporting	Native-language documents Joint SEAP
	Costs of network usage	National events & contact persons Webinars
Benefit-reducing factors	CoMO's overload	Differentiated helpdesk Handbooks accompanying projects
	JRC's overload	Grouped SEAP approach
	Network overlap	National Covenant clubs
	Competing schemes	Methodological alignment

Figure 6.4: Attempts to improve Covenant signatories' cost-benefit calculation
Source: Author.

That Covenant signature is often voted for unanimously by local councils is linked to the fact that no legal obligations and no obvious implications for the municipal budget ensue from the decision. In actuality, Covenant signatories experience significant costs in the form of program participation and have explicitly complained about this to Covenant Coordinators and the CoMO. The development and implementation of BEIs, SEAPs, and monitoring requires considerable levels of expertise and dedicated person hours. Indeed, the mission of Covenant Coordinators has been specified in this regard. It is now distinguished from less operational assistance by Covenant Supporters, and supplemented by more technical assistance from local and regional energy agencies. Covenant Coordinators do not fully live up to this mission statement in practice, as will be discussed in more detail in the next section. For instance, they do not necessarily assist Covenant signatories in Covenant reporting, the workload of the latter being a recurrent source of discontent. Few actually make use of the possibility of submitting a joint SEAP on behalf of their member municipalities. At least, native-language Covenant documents such as guidelines and forms have reduced the language barrier for municipal staff as compared to solely English communication. Language proficiency also plays a role with regard to active participation in Covenant events for networking and training; other cost drivers include the time and funds required for international travel. In response, national contact persons often attend to Covenant signatories, and webinars supplement face-to-face events, especially when it comes to more technical issues.

In addition to factors that increase the costs of program participation, Covenant signatories experience various hindrances and thus do not fully make use of program benefits. Firstly, some difficulties derive from program deficiencies in a narrower sense, such as the Covenant's inability to scale up the capacities of the CoMO and the JRC for attending to Covenant signatories' requests and processing their documents in correspondence with its relatively large number of participants. The CoMO was overwhelmed by Covenant signatories' inquiries and tried to reduce their need for personal attendance by publishing numerous handbooks and guidelines in national languages. Assistance to Covenant signatories was reorganized in collaboration with the JRC and with country-specific helpdesks. Still, Covenant helpdesks cannot provide tailored advice or extensive support but only answer punctiform questions, mainly with regard to Covenant methodology such as reporting forms. In addition to the reorganization of the CoMO, specific challenges identified in exchange with Covenant stakeholders have since been addressed by accompanying EU-projects that aim to promote Covenant participation, provide technical assistance in SEAP development, and support capacity-building for SEAP implementation. Some target a particular group of members, others network among Covenant participants including Co-ordinators, Supporters, and signatories (European Commission 2014a; 2016b).[236] Secondly, Covenant signatories' difficulties in realizing program benefits derive from overlaps with trans-local or national programs. Since the Covenant is not designed as a substitute for TMCNs, it is no surprise that many Covenant signatories from Germany and France are longstanding members of the Climate Alli-

---

236   Examples of to date 38 projects accompanying the Covenant's implementation include 'Sharing urban sustainable energy strategies - promoting the Covenant of Mayors' (COME2COM, 01/05/2010 -30/04/2012), 'Local authorities communicating to engage stakeholders and citizens '(ENGAGE, 01/06/2010 - 30/11/2012), 'Networking the Covenant of Mayors' (NET-COM, 01/06/2011 - 30/11/2013), 'Capacity Building of Local Governments to Advance Local Climate and Energy Action – from Planning to Action to Monitoring' (COVENANT CAPAC-ITY, 01/06/2011 - 31/05/2014), 'Cities Exchanging on Local Energy Leadership' (CASCADE, 01/06/2011 - 31/05/2014), 'Adding to SEAP – more participants, more content across Europe (SEAP-PLUS, 01/04/2012 - 30/09/2014), and 'Empowering Supporting Structures of the Covenant of Mayors to assist Local Authorities in implementing and monitoring their Sustainable Energy Action Plan (MAYORS IN ACTION, 01/03/2014 - 28/02/2017).

ance or Energy Cities. These are the networks they rely on for information, exchange of experiences, and networking with project partners. From their perspective, the Covenant is largely redundant. In order to emphasize the value added of an overarching platform, national Covenant clubs have been established so as to animate national networks of Covenant signatories and provide them with a venue for collective lobbying. But participation in national Covenant clubs has decreased in number and rank after the initial project period, as few exclusive channels towards the national or European level opened up. An even more fundamental challenge consists in competing national schemes. Municipalities are not likely to accept the Covenant as an additional framework if the development of a climate action plan based on a specific methodology is mandatory because of national legislation, as for certain local governments in France. The same applies if a voluntary program is accompanied by specific funding schemes, as in the case of the EEA or the German Municipal Directive for Climate Protection. Admittedly, this does not imply a refusal to act on climate change at all, but prevents the documentation of municipal action through Covenant reporting.

Summing up, municipalities from Germany and France have engaged in the Covenant in view of benefits with regard to internal politics and external relations. Their actual use of Covenant signature depends on the stage of local climate policy but has been hampered consistently by factors that increased the costs of program participation while at the same time reducing its benefits. Consequently, reluctant municipal response has thwarted Covenant Coordinators' initiatives.

6.4   The Limits of Orchestration: Durability, Intensity, and Causality

The number of Covenant Coordinators only grows slowly, and some have been suspended or reclassified as simple signatories because of their failure to submit reports (Covenant of Mayors 2013c, 1). This development points to the significant resources required to live up to the role of an operational Covenant facilitator (Covenant of Mayors 2014a, 2–4). Cross-case analysis allows for a more grounded assessment. The case studies have shown that all five of the Covenant Coordinators reduced their corresponding activities over time, a withdrawal caused by six mutually non-exclusive factors (see figure 6.5). In the beginning, inter-municipal associations actively promoted Covenant signature. For this purpose, they relied on inter-municipal fora such as their assembly, and on high-publicity events, e.g. the Rhine-Neckar Region's climate conference. Rennes Metropolis, Val d'Ille, and, to a certain extent, Rhine-Neckar Region succeeded in creating a dynamic that convinced numerous member municipalities to sign the Covenant; Greater Lyon and Stuttgart Region did not. No Covenant Coordinator managed to collect additional Covenant signatures after their starting periods.

	Municipal feedback	Competing schemes	Absence of funding	Electoral change	Institutional change	Change in staff
Greater Lyon	✔	✔			✔	✔
Rennes Metropolis	✔				✔	
Val d'Ille				✔		
Rhine-Neckar Region	✔	✔	✔	✔		
Stuttgart Region	✔	✔	✔	✔		

Figure 6.5: Why Covenant Coordinators reduced their activities
Source: Author.

Subsequently, all five Covenant Coordinators tried to support Covenant signatories in establishing the required documents for Covenant reporting as depicted above. During the submission process, municipalities found that Covenant proceedings entailed a significant workload – another instance of negative feedback towards Covenant Coordinators. Only Val d'Ille really succeeded in overcoming this obstacle by means of a joint SEAP submission on behalf of member municipalities. Against the background of these negative experiences with Covenant proceedings, competing European and national schemes became more important. At the national level, new schemes emerged that had not been available at the time of Covenant creation. In France, the ADEME developed dedicated tools for establishing territorial climate and energy plans. Needless to say, these better matched the situation of French municipalities than Covenant methodology possibly could. In Germany, municipal climate protection concepts were funded under the Local Authority Guideline within the NKI. Consequently, a SEAP became an add-on to be developed – or not – based on a prior climate protection concept. In both countries, the EEA procedure was more attractive than a Covenant commitment because it provided more guidance, more possibilities to adapt the procedure to local needs, and was supported with national or state-level funding. While early Covenant signatories and supporting structures had hoped for dedicated funding schemes, such programs did not come about. A range of accompanying projects aim to support Covenant implementation, but are limited in scope so that German and French municipalities cannot take part. Some target specific countries or kinds of governments, others encourage additional activities, such as international partnerships, rather than supporting Covenant implementation in a narrower sense. German interviewees in particular expressed their disappointment in this regard.

In light of these mixed experiences, Covenant Coordinators reduced their efforts in Covenant promotion and shifted their activities in climate policy coordination to other schemes. This trend was often reinforced by changes in local political agendas. In cases where the commitment of a Covenant Coordinator had been actively supported by inter-municipal leaders (Greater Lyon, Val d'Ille, Rhine-Neckar Region) or mayors (Rennes Metropolis, Rhine-Neckar Region), their departure due to electoral outcomes necessarily marked a turning point for local climate policy coordination (Rennes Metropolis, Val d'Ille, Rhine-Neckar Region). Often new mayors and inter-municipal representatives do not seize the issue with the same enthusiasm as their predecessors did. The initial signer obtains a certain degree of visibility both locally and at the European level, tends to be personally committed to the Covenant and often vividly remembers the signature ceremony. This point is illustrated by the words of one interviewee recalling the signature ceremony:

> "At the time I was very much impressed by the event in Brussels where I talked a lot with colleagues [...] from Sweden and Italy, just to learn from each other: What are good examples and what works where, how to win people over to participate, where are funding opportunities and schemes" (interview with S. Dallinger, 29 January 2015).[237]

After an electoral change, the glamour of the signature ceremony is long gone, and only obligations remain. The Covenant's focus on engaging high-level politicians and the prerequisite of deliberation by the local council have the virtue of ensuring a certain level of leadership and commitment that administrative staff

---

237 Author's translation; original quote: "*Ich war damals sehr beeindruckt von einer Veranstaltung [signature ceremony 2009] in Brüssel, wo ich mich ganz viel dann auch mit Kollegen [...] aus Schweden und Italien unterhalten habe, einfach voneinander zu lernen: Was sind gute Beispiele und was funktioniert wo wie, wie gewinnt man Menschen mittun [sic], wo sind Fördermöglichkeiten und Fördertöpfe.*"

alone could not achieve. But this approach also personalizes the Covenant commitment. It is the political leader rather than the city who commits to climate action. Once she or he is voted out of office, the continuity of the commitment is challenged.

Elections of course are not the only possible factor which can cause a shift in political priorities. Climate policy coordination was part of an overarching strategy for shaping territorial governance in Greater Lyon and Rennes Metropolis. Consequently, institutional change through territorial reform impacted inter-municipal priorities. When Greater Lyon and Rennes Metropolis obtained metropolitan status, not only was there no more need for lobbying in this regard, inter-municipal administrations were also busy adapting to the new situation for a certain period of time. Both factors reduced the attention paid to climate policy coordination by political leaders. While the latter were particularly important during the starting phase of inter-municipal climate policy coordination, it is up to technical personnel to stay in touch with the Covenant of Mayors Office. Thus, a change in staff also entails the risk of weakening commitment to the Covenant. New staff is not necessarily informed about the commitment, reporting obligations, and next steps to undertake. Comparatively anonymous e-mails from the CoMO do not suffice to fill this gap.

Interestingly, Covenant Coordinators do not rely on the Covenant in view of coordinating or synchronizing their activities. Learning and exchange of experiences figured among the motivations for joining the Covenant in some cases, but interviewees consistently reported that few networking occasions had emerged and no new contacts had been forged. Also, they did not rely on Covenant publications for identifying innovations and best practices when searching for ways to implement their commitment. They preferred well-established personal contacts and pre-existing networks. It is also worth noting that the revision of membership categories was not communicated to supporting structures, at least not successfully. Several interviewees from inter-municipal associations insisted on having the status of a supporting structure even after this category was abandoned, and associated the role of Coordinators with climate managers within their member municipalities. This information gap points to communication problems between Covenant Coordinators and the CoMO.

Case studies have found that Covenant Coordinators often have weak relationships with the CoMO and Covenant signatories. Interestingly, this applies to cases where Covenant Coordinators exhibited elevated levels of intermediate

Covenant promotion in the beginning that extenuated over time as well as to cases where their engagement remained weak in the first place. Apparently, different situations at the outset lead to similar outcomes. A closer look at the cases demonstrates that local governments' enthusiasm at the outset was even linked to particularly strong disappointments at later stages. This pattern can be explained by local governments' approach to Covenant support depending on their initial motivation to participate in the orchestration arrangement when differentiating between instrumentalist and normative orientations as defined by Benz et al. (2015) (see 3.1.1).

When local governments join the Covenant out of an instrumentalist orientation, they are more likely to actively rely on Covenant support. Here, the Covenant commitment is a tool for the realization of local governments' policy goals and serves to solve distinct policy problems. These policy goals include, but are not limited to, tangible contributions to climate mitigation. Accompanying policy goals typically relate to territorial competition and rescaling. In this situation, local governments have distinct expectations towards the Covenant. Successful orchestration then depends upon their ability to build upon Covenant support, i.e. the match between Covenant support and local needs. The opposite applies should a gap occur between the offer of the CoMO and the needs of a Covenant Coordinator or signatory, or between the offer of a Covenant Coordinator and a Covenant signatory. When the Covenant falls short of local governments' distinct expectations, they are likely to withdraw from the orchestration arrangement. In short, an instrumentalist motivation will lead to markedly high or low levels of Covenant engagement depending on the value added of Covenant support.

In contrast, local governments' Covenant engagement is likely to remain moderate when following a normative orientation. In these instances, the commitment is an expression of local governments' adherence to the Covenant's policy goals, but entails no distinct intentions with regard to local implementation. Local governments are then likely to approve of learning and exchange of best practices within municipal networks, citizen sensitizing, and exemplary local climate policies without actually relying on the Covenant to these ends. Hence, they will neither build upon nor be disappointed by Covenant support. As a result, orchestration will neither succeed nor fail in an obvious way. Rather, local governments will remain nominal Covenant participants. Their participation contributes to impressive membership numbers, but in practice only entails limited interactions and outcomes.

6.5 Summary of Discussion

Cross-case analysis of local climate policy coordination by Covenant Coordinators in Germany and France has shown why and how inter-municipal associations engage as intermediaries on behalf of the Covenant, how municipal target actors respond to the joint governance effort, and why the orchestration arrangement has limited impact. Inter-municipal associations commit to the Covenant in view of benefits for local leadership, territorial competition, and rescaling. Their approaches to implementation can be characterized as political, technical, concerted, or practical, depending on the primary target group. Municipalities participate in the orchestration arrangement in the hope of benefits with regard to internal politics and external relations, but struggle to realize these because of elevated costs of program participation and adverse domestic frameworks. In response to critical municipal feedback and their own disappointments about the level of Covenant support, Covenant Coordinators reduced their orchestration activities over time. This does not necessarily imply a decline of municipal climate action or inter-municipal climate policy coordination more generally, but it does signal that the Covenant failed to impact local climate governance effectively.

The typologies presented in this chapter consist of analytical categories which are not mutually exclusive in practice. Local actors frequently pursue several motivations and approaches to local climate policy at a time. In particular, priorities diverge across administrative departments, or between staff and political leaders.

These results imply careful adjustments in understanding the Covenant and local climate governance, and for the O-I-T-model, that will be discussed conclusively in the following chapter.

7 Conclusions

The O-I-T-model has been applied to the study of the Covenant of Mayors in order to better understand local contributions to global climate governance within a multi-level governance system. Five case studies on inter-municipal climate policy coordination in Germany and France have exemplified the co-benefits pursued by local governments in voluntary climate action. Their participation within the orchestration arrangement of the Covenant has been shown to be highly selective, and subject to local and national conditions that significantly complicate sustained coordination, namely competing policies and volatile agendas with regard to territorial governance

This chapter takes stock of the main contributions of this book and discusses their implications with regard to the Covenant, climate governance research more broadly, and the theoretical framework of the O-I-T-model (7.1). It continues with an assessment of the scope of this research in light of methodological challenges in terms of normativity, data, comparability, and transferability (7.2). The chapter concludes with suggestions for further research (7.3) and selected policy recommendations (7.4).

7.1 Results and Implications

Conventional wisdom with regard to the role of networks and transnational initiatives in local climate governance has been challenged above with results that emphasize the importance of differentiating between mobilization and effective climate action, and to account for the genuinely political dimensions of climate governance and orchestration more broadly. This section subsequently discusses the main contributions for our understanding of the Covenant (7.1.1), local climate governance (7.1.2), and orchestration theorizing (7.1.3).

© Springer Fachmedien Wiesbaden GmbH, part of Springer Nature 2020
L. Bendlin, *Orchestrating Local Climate Policy in the European Union*,
Energiepolitik und Klimaschutz. Energy Policy and Climate Protection,
https://doi.org/10.1007/978-3-658-26506-9_7

7.1.1   The Covenant of Mayors: A Cautionary Tale

Climate governance is not only about mitigating global warming; it also is a battle over competences and resources. Conflicts are not limited to cleavages between states in international negotiations. Within the EU, supranational institutions are known to strategically expand their scopes of action, and local governments lobby for recognition. The Covenant, a highly-acclaimed EU-initiative, constitutes an exemplary governance arrangement for exposing the interplay of different levels of government in climate governance. When the Committee of the Regions and other interested actors praise the Covenant as a model case of multi-level governance, it is important to keep in mind that they refer to multi-level governance as a community method of coordination and not to scholarly theorizing. Investigating Covenant Coordinators in Germany and France based on the O-I-T-model has allowed for a better understanding of patterns in interactions between Covenant participants – especially the strategic use made of the Covenant by municipalities and inter-municipal associations – and in Covenant participation, namely limited numbers of Covenant signatories in Germany and France, and reduced engagement of Covenant Coordinators over time. The importance of domestic frameworks in shaping incentives for Covenant participation, the limited overlap of policy goals among Covenant participants, and the restricted regulatory impact of the Covenant amount to a rather cautionary tale (Steinberg 2015, 169).

Counterintuitively, domestic frameworks in seeming harmony with the Covenant's goals have been found to represent a significant hindrance for orchestration. German municipalities are reluctant to commit to Covenant targets not despite, but because of their record in local climate policy dating back to the 1990s. On the one hand, there is no need to kick off local action that is already under way. Thus, Covenant support in the form of convening relevant actors, agenda setting, and endorsement is superfluous. Covenant Coordinators are partly redundant, too, because assistance and coordination for local climate policy are already taken care of by other actors, such as the Climate Alliance, regional energy agencies, or programs at the state and the federal level. German municipalities know from experience that GHG emission reductions in sectors other than their own energy consumption are hard to achieve, and shy away from making venturous commitments (Alber 2009, 8). In contrast, French municipalities welcomed Covenant support when they engaged in local climate policy during the 2000s, but were in need of methodological guidance. Since this time, French climate policy seems to have experienced a rise in local level activities. PCETs have become mandatory, and the decentralized National Energy Transition Debate explicitly recognized the role of subnational governments in climate and energy policy. The mainstreaming of low-carbon development at the local level entailed a distancing from the kind of bottom-up approaches developed in Local Agenda 21 processes during the 1990s (Béal and Pinson 2013). Since the mid-2000s, the climate and energy policy objectives prescribed by the state and national schemes offered to local governments, such as a common ADEME methodology, started state dominated development. Consequently, Covenant methodology has become redundant. In parallel, local budgets were adversely afflicted by the financial crisis in France (Kuhlmann and Bouckaert 2016), which further weakened minuscule municipalities in particular, a challenge that did not emerge to the same degree from case studies in Germany. In both countries, Covenant participation now provides municipalities with limited benefits, which explains the modest share of German and French Covenant signatories. Covenant participation has become an extra that municipalities afford when they are already well positioned for a strong local climate policy performance. Driven by normative orientations, they comply with Covenant reporting obligations, but do not rely much on the Covenant as a source of support. This raises the question of whether the Covenant's limited effectiveness in Germany and France point to deficiencies in orchestration in a practical sense. The Covenant might work best where it is most needed – in countries where local climate policy and its coordination have lagged behind (Collier 1997).

Chapter 6 has shown that Covenant participants' goals overlap, but their interests are not identical. If governments at multiple levels share climate policy goals, this does not yet entail fruitful and sustained cooperation and coordination. Municipalities question Covenant Coordinators' focality in local climate governance, and inter-municipal associations struggle to design helpful programs for operational assistance to municipal climate policy that would divert their self-will. This divergence highlights competing interests within the Covenant. Motivations of Covenant stakeholders from the orchestrator, intermediary and target level overlap to a certain extent – a basic prerequisite for orchestration in the first place. But policy goals do not fully coincide, and participants have to carefully manage their interdependencies.

Internal benefits of participation as perceived by Covenant signatories largely coincide with the interests of the European Commission to incentivize and steer local climate and energy policy. Perceived external benefits tend to go beyond the European Commission's concept of the Covenant. Those realized at an early stage of policy – visibility, legitimacy, and funding – still overlap with the vision of the European Commission, but those realized at later stages of local climate policy development – windfall auditing and advocacy at the European level – risk conflicting with the interests of the European Commission. At the same time, Covenant signatories with an established record in local climate and energy policy are indispensable participants, as they provide the Covenant with best practices to guide other signatories and presentable outcomes to document success. The interests of inter-municipal associations are more difficult to assess in terms of their match with those of the European Commission. Individual leadership, territorial competition and rescaling do not interfere directly with EU goals, but inter-municipal associations' focus on instrumental orientations calls for more tangible benefits from program participation than the Covenant can provide on a sustained basis. Where the interests of local governments and the European Commission diverge, it is up to the CoMO consortium to bridge that gap and moderate conflict. At best, new suggestions emerge and are taken up in the next contract period, in accompanying projects or additional programs.

Last but not least, a tangible regulatory impact of the Covenant could hardly be observed. The case studies in chapter 5 have shown that the Covenant does not permanently influence local climate governance to a significant extent, with the notable exception of providing tools and orientation for early stages of local climate planning when no prior experience or superordinate legislation are available. But inter-municipal associations are not in regular contact with the CoMO, do not rely on the Covenant for networking, and use their own or national methodologies for local climate planning instead of relying solely on Covenant reporting tools. The Covenant becomes instrumental for the initiation, or possibly revitalization, of local climate action. Case studies entailed similar accounts for Covenant signatories. Municipalities tend to use and refer to their membership rather selectively. Elevated levels of delayed submission of reporting documents and suspension of members show that local governments maintain autonomy and only comply with Covenant rules when opportune. The Covenant's voluntary basis naturally limits the European Commission's assertiveness and obliges it to design programs in ways to attract local governments' participation, e.g. by allowing extensions for overdue submissions. In addition, Covenant participants do not rely on the initiative for learning. This might be linked to its overlap with the genuine TMCNs from the CoMO consortium, and to the emphasis put on SMART targets, which induces participants to focus on low-hanging fruit instead of particularly ambitious commitments, experimentation, or innovation (Abbott 2017). Rather than promoting innovation, the Covenant mainstreams the state of the art of local climate policy.

Driven by co-benefits, used selectively, struggling to sustain participation over time, and hardly groundbreaking – these results refute enthusiastic accounts of the Covenant that seem to be inspired by its numerical success rather than its actual performance in coordinating local climate policy in the EU. The Covenant demonstrates that multiple levels of government have an interest in engaging in climate action and try to do so, but it also illustrates their difficulties in coping with a lack of competences and resources, and in finding common ground for coordinating their actions.

7.1.2   Advances for Understanding Multi-level Games in Local Climate Governance

In the EU, local governments pursue European climate and energy targets within their respective national frameworks based on regional planning and inter-municipal coordination. These interlinkages contradict claims in the literature on multi-level governance according to which "more than three levels are never relevant in political practice" (Benz 2007, 27) because of prohibitive complexities. Subnational coordination requirements have attracted limited scholarly interest to date and cannot be captured by studies focusing exclusively on transnational dimensions of climate governance. A more differentiated assessment of the respective roles of different kinds of local governments is necessary for better understanding local climate governance and its role within the global climate regime.

When inter-municipal associations engage in local climate policy coordination, they can make significant contributions to coherent climate governance in terms of mobilization, planning, and municipal capacity building. In this way, the local level of government is able to self-coordinate to a certain extent. But inter-municipal associations encounter numerous difficulties with regard to overlaps with other governance actors, their own level of capacities, and matching their offer to municipal needs. As will be summarized next, their ability to effec-

tively coordinate local climate policy depends largely on their domestically determined endowment with resources and competences, and on opportunity structures for rescaling. The section concludes with a critical assessment of local climate policy initiatives' potential to make significant contributions to global climate governance.

Firstly, the resources and competences of an inter-municipal association
predefine the modes of governance it can choose from for local climate policy
coordination. A major difference between German and French inter-municipal
associations is the operational level of support they can provide for member municipalities. Interestingly, this is hardly linked to in-house human resources. Although French intercommunalities have significantly more staff in total, the number of staff in charge of climate and energy is not necessarily more restricted at
German regional associations that have additional departments for energy-related economic development. What enables French intercommunalities to provide more detailed assistance for municipal climate policy making are not in-
house capacities, but their local energy agencies – an asset not available to all
intercommunalities. These are mandated with energy monitoring and other services to support member municipalities in terms of data, knowledge, and person
hours. In Germany, too, the landscape of energy agencies is patchy. While some
urban and rural districts have their own local energy agency, others share with
neighboring local governments or rely on regional offices of a state-level agency.
None of these situations enables regional associations to appoint a particular local or regional energy agency to assist their member municipalities on their behalf.

In contrast, German regional associations are better positioned to govern by
authority with regard to land-use and energy planning. Their competence for establishing mandatory regional plans enables them to integrate climate and energy
considerations with regard to mitigation and adaptation. French intercommunalities can only indirectly influence SCoTs that are established by superordinate
associations of neighboring intercommunalities. These differences trace back to
the different levels of the inter-municipal associations under study. In short,
French intercommunalities are smaller and have more operative responsibilities,
while German regional associations are larger and are more in charge of planning. As a result, French intercommunalities are closer to their member municipalities and more operative in their interpretation of the Covenant commitment,

while German inter-municipal associations provide more general information and networking opportunities.

Secondly, the domestic agenda for territorial reform determines opportunities for inter-municipal associations to engage in rescaling and thereby influences the national framework for local climate policy coordination. Struggles over territorial governance can have an important impact on local climate policy coordination. Inter-municipal climate policy coordination involves inter-municipal associations and member municipalities that share resources in collaborative governance arrangements (Hooghe 1996, 18). Inter-municipal associations provide guidance, expertise, and support, while municipalities contribute their own programs, services, and staff. Both levels of government rely on each other in the attempt to spread voluntary climate action and to improve planning, implementation, and monitoring. These arrangements are not carved in stone, nor are they exogenous to climate governance. The respective responsibilities and resources of the participating governance actors are continuously negotiated. Internally, inter-municipal associations and member municipalities negotiate about voluntary transfers of tasks from the municipal to the inter-municipal levels.

But inter-municipal associations and municipalities also compete externally for additional competences and resources and lobby the superordinate levels of government that determine territorial organization for rescaling in their favor. In this sense, inter-municipal climate policy coordination involves many levels of government. In Germany, recent territorial reforms have not substantially questioned traditional levels of government.[238] The competences and resources of regional associations are determined at the *Länder*-level and have increased repeatedly since the turn of the millennium, but not primarily to the detriment of municipalities. Cooperation in fields such as public transport is ruled by contracts between municipalities, inter-municipal associations and additional public agencies or service providers. Regional associations seek to legitimize their increased role. For municipalities, legitimacy-seeking behavior is of limited importance in local climate governance because they are not put into question. In contrast, French territorial governance has been under constant construction during the past decades. Even the abolition of the departmental level of government and landslide consolidations of the fragmented municipal landscape has been repeat-

---

238 An exception applies to governmental districts, the lower level of regional government Hooghe, Marks, and Schakel (2010, 119).

edly under discussion (West 2007). The role of intercommunalities in particular has evolved importantly in law and in practice (Epstein 2008). While inter-municipal cooperation was voluntary in the beginning, membership has now become mandatory for all municipalities. Beyond the mere intensification of inter-municipal cooperation, metropolitan governance is a field of constant experimentation and reform, as seen in the case of Greater Lyon (see 5.2.2). The process of metropolitan integration has been described as a set of three subsequent phases, the first one being most politicized: territorial composition, regime establishment, and public policy implementation (Négrier 2007, 41). Case studies show that these phases tend to overlap and repeat in practice.[239] The second and third phase can be highly politicized, too, even if only because of their overlap with the first one. In this context, inter-municipal climate policy coordination becomes an arena for legitimizing capacity gains and repelling competing actors (Ben Mabrouk 2007). In both countries, though, climate governance is an arena for inter-municipal associations to assert their collective actorness by successful cooperation and improved governance performance (Le Galès 2011, 417–19).

Local climate policy coordination comes with conflicting objectives. Sponsors of voluntary non-state action programs such as the Covenant are naturally inclined to aim for mobilizing local resources with the least possible investment, which can have adverse effects on program design. Examples include a preference for measurable over effective action, and for program promotion over implementation. SMART objectives hold the promise to ensure transparency and compliance in voluntary commitment systems, but tend to discourage experimentation and ambition. Increasing numbers of participants seem likely indicators of program success, but this focus can lead to an imbalance when benefits from program participation occur largely at the adherence stage rather than with sustained participation and implementation.

In light of these results, hopes expressed in the literature that local climate policy can substitute for the lack of ambitious international treaties and national programs or the absence of adequate national support seem overly optimistic.[240] The contribution of local governments to achieving the European climate and energy targets depends on member states' preferences, too. Local climate policy relies on co-benefits to a large extent, and on favorable frameworks in terms of

---

239   This parallels criticism of the so-called policy cycle as a model of the political process as put
         forward e.g. by Jann and Wegrich (2006); Sabatier and Weible (2014b).
240   See chapter 1.

responsibilities, capacities, and resources. These are determined at superordinate levels. Ambitious climate policy cannot rely on the local level only; there is no way to circumvent the national level. It is up to the latter to pave the way from singular projects to systemic change. Crucial questions for climate governance research and practice consist in how superordinate levels of government can best activate and enable local governments for climate action, and to facilitate and support actual innovation and diffusion. This is all the more important as an overlap of policies has been shown to lead to program cannibalization.

7.1.3   Implications for Orchestration Theorizing

An extension of the O-I-T-model from the study of international relations to the realm of European politics was possible because the underlying assumptions of the O-I-T-model refer to instances of multi-level governance, which is at the core of governing in the EU. The O-I-T-model constitutes a useful lens for studying European multi-level governance, as it provides suitable categories for analyzing non-hierarchical interactions between the European Commission and its agents on the one hand, and non-state actors on the other. In particular, the supranational level relies on intermediary governors and uses ideational and material forms of support to address target actors. The O-I-T-model represents a relatively simple concept of governance; it cannot pretend to be a holistic theory of European multi-level governance, but it can be used gainfully in a modular approach to the study of European policy processes (Scharpf 2010, 75). The application of the O-I-T-model for the study of subnational governments was fruitful, for it highlighted their agency in climate governance and shed light on their motivations and co-benefits. These insights improved our understanding of local climate governance, and pointed out how superordinate levels can stimulate local climate action on a voluntary basis.

At the same time, the above results suggest caution when applying the O-I-T-model. The researcher should not assume that governance actors that participate in a given governance arrangement necessarily adopt the associated normative framework or policies (Forsyth and Levidow 2015; Steinberg 2015). Two issues have repeatedly occurred throughout this book deserving more attention in orchestration theorizing: effectiveness, and the importance of timing.

The selective recourse of inter-municipal associations to the Covenant raises several questions with regard to the effectiveness of orchestration arrangements in terms of the durability of participation, the intensity of interactions between the actors involved, and the resulting causal effects. Firstly, the threshold of benefits required for active participation points to the fact that the overlap of policy goals between the orchestrator and the intermediary is likely to be only a partial one. More precisely, both sides pursue several goals at a time within the same orchestration arrangement. Some overlap, others do not; some might even conflict with each other. Hence, shared policy goals remain a prerequisite for orchestration, but the importance of resource exchange in forging the actual alliance can hardly be overestimated. This tempers overly optimistic accounts of orchestration arrangements as coalitions of reason in the pursuit of common goods. Secondly, even instances of active participation of intermediaries cannot be expected to yield predictable results in accordance with orchestrators' preferences. Inter-municipal Covenant Coordinator status does not imply predetermined outcomes or even policies but creates opportunities for agency, including, but not limited to, the institutionalization of local climate policy. Thirdly, limited records of Covenant Coordinators in terms of Covenant promotion and sustained support to Covenant signatories raise the question how impacts of orchestration are to be assessed within the O-I-T-model. To date, it provides no clear guidance for measuring impacts. In extreme cases, the question arises of how intense and durable an interaction of governance actors with overlapping interests has to be to be justifiably identified as a case of orchestration. The O-I-T-model also contributes to understanding unsuccessful instances of orchestration; however, future orchestration theorizing should differentiate more carefully between orchestration attempts, and effects.

A second important issue for orchestration theorizing consists in the significance of timing. Timing has proven to be an important factor for explaining participation in orchestration. Local governments' motivations to join the Covenant depend largely on the political agenda at the time of signature. In contrast, the implementation approach, i.e. the means and strategies chosen, are typically determined by the organizational setup and the larger context of territorial and climate governance. This is because they are subject to institutional provisions to a larger extent than a mere council resolution. When challenges for Covenant implementation occur, changes in the political agenda easily affect the level of engagement and activity. When applying the O-I-T-model, it is important to distinguish between actors' motivations to begin and to continue to participate, and between policy goals as determined by deeply anchored belief systems (Sabatier 1988; Sabatier and Jenkins-Smith 1993; Sartori 1969) as opposed to more volatile priorities on the institutional and decision agenda (Alexandrova, Carammia, and Timmermans 2012; Baumgartner and Jones 2009; Birkland 2006; Jones, Sulkin, and Larsen 2003). This finding supports scholarly criticism of overly variable-centered social research as put forward by Pierson (2004), who argues that positive feedback and path dependence, timing and sequence, as well as long-term processes, intervene in causal relations in the social world. Institutional design alone thus has limited impact, and institutional development over time also needs to be taken into account; he further calls for social science inquiry to address the temporal context of its study object, related results, and their transferability.

7.2 Scope of this Research

The black box of local climate policy making has been opened by differentiating between different kinds of government within the local bloc: municipalities and inter-municipal associations.[241] The inter-municipal level of government was at the core of this undertaking; it consists of so-called regional associations in Germany and so-called intercommunalities in France. The equivalence is not a perfect one because levels of government in Germany and France do not fully correspond (see section 4.2.2), which has certain implications for the motivations and possibilities of German and French inter-municipal associations that act as Covenant Coordinators. A lack of predictions in the literature on the impacts of differences in territorial organization on orchestration prohibited a fully systematic comparison, i.e. an unrestrictedly variable-driven research design. Nevertheless, cross-country comparison delivered insightful lessons. The incomplete match of governmental levels does not put to question the legitimacy of comparatively studying inter-municipal associations, but a careful consideration of findings and conclusions as summarized above necessarily raises the question of their transferability.

Building upon the concept of resonance groups of cases (see section 4.1.3), conclusions can be extended in two ways. Firstly, conclusions on Covenant participation and local climate policy coordination can inform our understanding of cases that are subject to similar domestic factors, such as national-level low-carbon policies, local experience in climate policy making, and comparable territo-

---

241   As opposed to distinctions based on settlements' rural, urban, or metropolitan character.

rial organization. Secondly, conclusions on the orchestration of local climate policy are also relevant to similar governance arrangements.

Since the case study research design did not include any provisions for representativeness, conclusions cannot pretend to allow for predictions about numerical distribution. Apart from that, the transferability of generalizations differs
between conclusions about the Covenant, on climate and territorial governance,
and on orchestration arrangements more generally. Regarding the behavior of
Covenant participants from countries other than Germany and France, it is important to differentiate between the typologies for analysis as established within
this exploratory study, and the specific dynamics as observed and discussed
based on these categories. The dynamics of local climate policy coordination
with help from the Covenant in different EU member states develop within a
shared multi-level system, but territorial governance has been shown to determine local governments' priorities and scope of action. The Covenant's regulatory impact depends significantly on participants' level of prior climate policy
development, and competing national legislation. Local governments' sincerity
with regard to their Covenant commitment might also be affected by political
culture in a broader sense. Thus, the *predominant* motivations and performances
of Covenant Coordinators and signatories are likely to differ across countries,
and the findings from this study should not be mistaken for representative.

Caution should apply to the study of Covenant participants from Eastern
Europe. By design, cases studied here stem from Western Europe.[242] Urban governance research has shown that an integrative understanding across Europe is
difficult to develop (so far) because Eastern European agglomerations and cities
operate in a distinct situation coined by experiences of shrinkage and other specific impacts of post-socialist transition (Le Galès 2011). In addition, inner-Eu-
ropean comparison has to account for a specific structure of political cleavages
(De Waele 2004). Differences in GHG emission profiles and EU burden sharing
contributions (Jordan et al. 2010) further indicate that findings from research on
Covenant participation in the old EU member states should only very cautiously
be transferred to cases from Eastern Europe. This applies all the more to Covenant participants from outside the EU. Outside the European multi-level system,
their resources for and benefits from local climate policy coordination are likely

---

242   This choice was made because of Germany and France's roles in terms of GHG emissions,
      European integration, and for the sake of comparison (see 4.2.1).

to differ significantly from those of EU municipalities. The Global Covenant of Mayors, launched in January 2017, might thus be subject to different dynamics.

In a similar vein, distinct conclusions about the behavior of municipalities and inter-municipal associations respectively should only be cautiously transferred to other levels of subnational government. That differentiation is among the primordial contributions of this book and points to specific dynamics between local and superordinate governments depending on their specific institutional form. This yields a critique of the frequent practice in local climate policy research of not differentiating between different kinds of local governments and settlements. Examples include the indistinctive study of cities (i.e. urban settlements) and local governments, cities and metropolises of largely different size, and of metropolitan areas that span across several cities and local governments. The dynamics of territorial governance – which have been shown to matter significantly for local climate policy and its coordination – are more than likely to differ importantly under these circumstances.

In contrast, there is no indication that conclusions about orchestration in general could not be transferrable to governance arrangements from outside the EU or from other policy fields. Although a transfer of conclusions about Covenant participation to actors from outside the EU multi-level system is questionable, as emphasized above, this is due to case specificities, not to limitations of the O-I-T-model per se. Other policy fields certainly entail different patterns in how policy goals overlap, and how support attracts and actually empowers intermediaries and target actors, but the underlying challenges relate to intergovernmental relations as determined by institutional frameworks, notably multi-level governance. Developed for the study of governance arrangements within this realm, the O-I-T-model and related conclusions will enable illuminating analyses.

Beyond the Covenant, conclusions can inform research on voluntary nonstate action programs more broadly. Although the Covenant has been frequently presented and perceived as a unique, particularly innovative initiative (Covenant of Mayors 2014a, 14; Heyvaert 2013, 78) or as a policy experiment (Hoffmann 2011), its properties are not new to climate governance. Other voluntary nonstate action programs rely on similar proceedings, although in varying combinations, with different sponsors and resource levels. These similarities – just as the successful application of the O-I-T-model, which was developed for an even broader universe of cases – give reason to believe that conclusions about local governments' motivations and uses of participation hold true for other voluntary climate action programs. This applies in particular to voluntary commitment systems sponsored by superordinate governors, and only to a lesser extent to TMCNs. Most TMCNs have been founded by the local level, are funded by membership contributions, and only secondarily involve superordinate levels or other, e.g. non-public actors. However, a diversity of actors, membership categories, and corresponding interests can be found in other TMCNs as well. The increased recognition of local climate policy at the global level goes hand in hand with growing cooperation between TMCNs and international organizations for the purpose of documenting the groundswell of climate action. Umbrella initiatives such as the newly established Global Covenant of Mayors for Climate and Energy share many characteristics with the Covenant and represent the most obvious resonance group.

7.3 Suggestions for Further Research

Some starting points for future study have been identified above within the discussion of the implications of this research. Needless to say, limitations to the transferability of conclusions as discussed above also offer natural inspiration for future research. For instance, a complete picture of Covenant participation cannot be drawn without more systematic research into the landscape of local governments' response to the Covenant in different EU member states. Beyond a rather mechanical deduction of follow-up research, this book identified two primordial gaps in research that lend themselves to further study, namely limits to our current understanding of agency and impacts with regard to the Covenant and other transnational climate governance initiatives, including TMCNs. Filling these gaps will require empirically grounded, comparative research rather than purely conceptual work.

The distribution of agency within the Covenant requires a very differentiated assessment. Established and funded by the European Commission, the Covenant cannot be accurately described as a horizontal, bottom-up initiative. The analysis has shown that a conceptualization as a voluntary commitment system is better suited to providing an understanding of the Covenant than treating it as a TMCN. The consortium running the CoMO depends on the European Commission within a principal-agent-relationship. But agents tend to try to extend their scope of action beyond the mere implementation of their principal's goals.

Common instances of agency slack consist of "shirking, when an agent minimizes the effort it exerts on its principal's behalf, and slippage, when an agent shifts policy away from its principal's preferred outcome and toward its own preferences" (Hawkins et al. 2006, 8). The interests of the European Commission, the CoMO, and the participating local governments do not fully coincide. Examples of tensions between different Covenant stakeholders include the levels of ambition and control, and standard documentation of local climate action vs. municipalities' workload of reporting. Non-active participation in the Covenant with accordingly minimal impacts on local climate policy is relatively common.

Moreover, local governments also influence the architecture of the governance arrangement in the first place. Representatives of the networks within the CoMO consortium referred to the Covenant as a bottom-up movement and to the creation of the Covenant as a success of their own lobbying efforts (Giest and Howlett 2013, 345), i.e. they claim that they as the agents initiated and, to a certain extent, also shaped the relationship with the principal. Since the Covenant's creation, several changes to the initial design have been implemented. Examples of local governments' agency in this process include the introduction of requested adaptations in methodology, such as joint SEAP submission, the creation of national Covenant clubs as institutionalized channels to the European level, and a consultation process before the Covenant's relaunch in 2015.

Pondering intended discretion by the EU and unavoidable autonomy of the CoMO has not been at the core of this research project (Hawkins et al. 2006, 8). As Jörgens et al. (2016, 71) point out, orchestration studies are not suited to establish a detailed distinction between the agency of the principal and the agent(s), especially if orchestrators "mostly act in line with their principals' preferences, that is, if their IO's plenary or council back their efforts to orchestrate the individual actions of a wide range of transnational actors". Future research on the Covenant could focus on the relation between the CoMO and its principal and assess in more detail how agency is distributed within the Covenant. It would also be interesting to investigate the extent to which different Directorates-General within the European Commission participate in Covenant development, and whether they agree about local governments' role in European climate governance.

A future task for research on transnational climate governance more generally consists in a more differentiated typology of networks and the kinds of governance at work, rather than merely assuming a purely horizontal, bottom-up arrangement. Other TMCNs, too, are not funded exclusively based on membership contributions. For example, C40 and ICLEI receive support from the World Bank (Hale and Roger 2014, 64), and Energy Cities and the Climate Alliance rely on project-based funding to a large extent. Systematic examination in a comparative perspective could unveil possible impacts of funding schemes on the activities and agency of TMCNs and contribute to an improved understanding of, and distinction between, the agency of member municipalities and network secretariats within TMCNs.

The above results raise suspicion about the impact of external interventions on local climate policy making. Local governments make strategic use of schemes offered at the national and European level, but balk at accepting additional tasks or rules if these are not in their own perceived interest. Empirically grounded research is needed in order not to overestimate the impact of transnational governance arrangements in terms of local climate mitigation. Instead, it can be reasonably assumed that transnational governance arrangements produce a whole range of impacts depending on local circumstances and deliberate use by local actors. Climate change as well as climate policies entail significant distributive effects among generations, states, and groups with serious justice implications (Bendlin 2014; Großmann, Schaffrin, and Smigiel 2017; Meyer and Roser 2006; Page 2008; Ruth and Ibarrarán 2009), and this not only at the global level. In other words, local climate policy research needs to consider how its objective is not just solving climate change – it is political.

7.4 Policy Recommendations

If continuous growth in terms of participating actors was proof enough of the Covenant's success, any recommendations regarding its future development would be superfluous. But the above analysis has pointed to several instances of failure to effectively orchestrate local climate policy. Two recommendations can be derived: Firstly, orchestration arrangements have to sustain benefit provision over time so as to ensure effective implementation rather than nominal or temporal participation. Secondly, successful experimentation in local climate policy has to leave the laboratory stage behind.

Covenant participants do not necessarily make active use of the program over time. Contacts with the CoMO, use of Covenant support, and reference to the Covenant in local climate policy often remain strictly limited. Covenant Coordinators even stopped Covenant promotion at all. Due to nominal participation, numerical success does not necessarily imply tangible impacts on the ground. Also, considerable numbers of Covenant participants have been excluded for lack of compliance. Although exact numbers are not publicly available, there is reason to suspect that a considerable share of Covenant signatories has virtually been replaced, their exit being overshadowed by new signatures. Yet, it is over time that local climate planning can make a verifiable difference in terms of GHG emissions. It would be in the interest of the European Commission to carefully assess the reasons for nominal and temporary participation. An important deficiency of the program consists in the fact that it favors the commitment maker over the supplier. Most benefits occur at the time of Covenant signature; continuous participation provides few additional benefits. To date, program revisions have either tackled this problem selectively within accompanying projects that are necessarily limited in scope and time, or through changes that approximate the Covenant to a TMCN. Not only have attempts from the latter approach, such as a best practices database and national Covenant clubs, partly remained empty shells. They also risk increasing the perceived overlap, from a municipal perspective, between the Covenant and the TMCNs from the CoMO consortium. After all, no program revision will make up for a lack of municipal capacities without enhancing the availability of resources. Dedicated funding schemes have to respond to municipal needs, and also must suit smaller municipalities' capacities for project management in terms of partnership requirements and financial dimensions. Once again, sustained action over time is crucial and requires reliable funding rather than selective allowances. This cannot be achieved with a project-based approach to strengthening local climate policy alone, but calls for reinforced advocacy efforts at the national and EU level. If national Covenant clubs are to become a veritable channel for lobbying in this regard, they need to obtain exclusive access to national ministries and to relevant departments within the European Commission.

A related challenge has stood out with regard to innovation in local climate policy. Today, tested policy instruments and available technologies for local climate mitigation are legion; still, local governments struggle to establish effective frameworks for transition beyond singular measures. But European and national climate mitigation programs for local action often cover initial funding only, such as for the development of a local climate plan, but not for implementing it over time. For additional projects, local governments are often required to submit innovative applications rather than replications of measures that have been successfully tested. Put differently, European and national governors focus on scaling down the global challenge of climate mitigation to local laboratories to the detriment of scaling up successful experiments. After more than two and a half decades of local climate policy experimentation, it is time for the next step.

References

100% RES Communities. 2015. *Steps towards 100% renewable energy at local level in Europe: Final report*. 100% RES Communities. http://www.100-ee.de/fileadmin/redaktion/100ee/PDFs/100ee_Europa/100-RES-COMMUNITIES_Final_Publishable_Report.pdf (Accessed July 8, 2016).

Abbott, Kenneth W. 2012. "The transnational regime complex for climate change." *Environment and Planning C: Government and Policy* 30 (4): 571–90.

———. 2014. "Strengthening the Transnational Regime Complex for Climate Change." *Transnational Environmental Law* 3 (1): 57–88.

———. 2015. "Orchestration." In *Encyclopedia of global environmental governance and politics*, eds. Philipp H. Pattberg and Fariborz Zelli. Cheltenham: Edward Elgar Publishing, 487–95.

———. 2016. "Orchestrating Experimentation in Non-State Environmental Commitments." *SSRN Electronic Journal*.

———. 2017. "Orchestrating experimentation in non-state environmental commitments." *Environmental Politics* 26 (4): 738–63.

———. 2018. "Orchestration: Strategic Ordering in Polycentric Governance." In *Governing Climate Change*, eds. Andrew Jordan, Dave Huitema, Harro van Asselt and Johanna Forster. Cambridge University Press, 188–209.

Abbott, Kenneth W., and Steven Bernstein. 2015. "The High-Level Political Forum on Sustainable Development: Orchestration by Default and Design." *Global Policy* 6 (3): 222–33.

Abbott, Kenneth W., Philipp Genschel, Duncan Snidal, and Bernhard Zangl. 2012. "Orchestration: Global Governance through Intermediaries." *SSRN Electronic Journal:* 3–36.

———. 2014. "Orchestrating global governance: from empirical findings to theoretical implications." http://ssrn.com/abstract=2452764.

———, eds. 2015. *International Organizations as Orchestrators*. Cambridge: Cambridge University Press.

———. 2016. "Two Logics of Indirect Governance: Delegation and Orchestration." *British Journal of Political Science* 46 (4): 719–29.

Abbott, Kenneth W., and Duncan Snidal. 2009. "Strengthening International Regulation through Transnational New Governance: Overcoming the Orchestration Deficit: Revised version." https://works.bepress.com/kenneth_abbott/2/ (June 16, 2016).

———. 2010. "International Regulation Without International Government: Improving International Organization Performance Through Orchestration." *SSRN Electronic Journal*.

© Springer Fachmedien Wiesbaden GmbH, part of Springer Nature 2020
L. Bendlin, *Orchestrating Local Climate Policy in the European Union*,
Energiepolitik und Klimaschutz. Energy Policy and Climate Protection,
https://doi.org/10.1007/978-3-658-26506-9

Acuto, Michèle. 2013. "The new climate leaders?" *Review of International Studies* 39 (4): 835–57.

AFP. 2011. "Primaire PS: Collomb soutient Hollande." *Le Figaro,* June 29. http://www. lefigaro.fr/flash-actu/2011/06/29/97001-20110629FILWWW00320-primaire-ps-collomb-soutient-hollande.php (Accessed July 11, 2016).

Agence de l'Environnement et de la Maîtrise de l'Energie (ADEME). 2012. "Ville de Rennes: Fiche d'identité PCET." Agence de l'Environnement et de la Maîtrise de l'Energie. September 21. http://observatoire.pcet-ademe.fr/pcet/fiche/40/ville-de-rennes (July 7, 2016).

———. 2014. "Communauté de communes du Val d'Ille: Fiche d'identité PCET." Agence de l'Environnement et de la Maîtrise de l'Energie. June 13. http://observatoire.pcet-ademe.fr/pcet/fiche/407/communaute-de-communes-du-val-d-ille (July 2, 2016).

———. 2015a. "Département du Rhône." Agence de l'Environnement et de la Maîtrise de l'Energie. December 10. http://observatoire.pcet-ademe.fr/pcet/fiche/454/departement-du-rhone?url_recherche=/ (January 7, 2017).

———. 2015b. "Grand Lyon: Fiche d'identité PCET." Agence de l'Environnement et de la Maîtrise de l'Energie. February 27. http://observatoire.pcet-ademe.fr/pcet/fiche/119/grand-lyon (July 9, 2016).

———. 2015c. "Région Rhône-Alpes." Agence de l'Environnement et de la Maîtrise de l'Energie. December 22. http://observatoire.pcet-ademe.fr/pcet/fiche/245/region-rhone-alpes (January 7, 2017).

———. 2015d. "Conseil départemental d'Ille-et-Vilaine: Fiche d'identité PCET." Agence de l'Environnement et de la Maîtrise de l'Energie. November 23. http://observatoire .pcet-ademe.fr/pcet/fiche/447/conseil-departemental-d-ille-et-vilaine (July 7, 2016).

———. 2015e. "Région Bretagne: Fiche d'identité PCET." Agence de l'Environnement et de la Maîtrise de l'Energie. December 22. http://observatoire.pcet-ademe.fr/pcet/fiche/135/region-bretagne (July 7, 2016).

———. 2015f. "Rennes Métropole: Fiche d'identité PCET." Agence de l'Environnement et de la Maîtrise de l'Energie. December 23. http://observatoire.pcet-ademe.fr/pcet/fiche/38/rennes-metropole (July 6, 2016).

Agence Locale de l'Energie du Pays de Rennes (ALEC). 2013. *Rapport d'activités 2012.* Agence Locale de l'Energie du Pays de Rennes. http://www.alec-rennes.org/PDF/RA_ALEC/ALEC-Rapport-Activites-2012-BD.pdf (Accessed August 26, 2016).

———. 2014. *Rapport d'activités 2013.* Agence Locale de l'Energie du Pays de Rennes. http://www.alec-rennes.org/PDF/RA_ALEC/ALEC-Rapport-Activites-2013-BD.pdf (Accessed August 26, 2016).

———. 2016. "Plans Energie et climat, Convention des Maires." Agence Locale de l'Energie du Pays de Rennes. www.alec-rennes.org/collectivites/plans-energie-et-climat-convention-des-maires/ (April 14, 2016).

Agranoff, Robert, and Beryl A. Radin. 1991. "The Comparative Case Study Approach in Public Administration." *Research in Public Administration* 1: 203–31.

Alber, Gotelind. 2009. "Reality and perspectives of climate policy in cities." In *Climate Change and the Cities of the Future: Art, Technology and Economics in the face of Climate Change. ECF Working Paper*, eds. Klaus Hasselmann, Antonio Ruiz de Elvira and Martin Welp. Trans. Aida Abdulah, 4–14.

Alber, Gotelind, and Kristine Kern. "Governing Climate Change in Cities: Modes of Urban Climate Governance in Multi-level Systems." In *Competitive Cities and Climate Change: Milan 9.-10.10.2008. OECD Conference Proceedings*, ed. Organisation for Economic Co-operation and Development, 171–96.

Alexandrova, Petya, Marcello Carammia, and Arco Timmermans. 2012. "Policy Punctuations and Issue Diversity on the European Council Agenda." *Policy Studies Journal* 40 (1): 69–88.

Andersson, Krister P., and Elinor Ostrom. 2008. "Analyzing decentralized resource regimes from a polycentric perspective." *Policy Sciences* 41 (1): 71–93.

Andonova, Liliana B., Michele M. Betsill, and Harriet Bulkeley. 2009. "Transnational Climate Governance." *Global Environmental Politics* 9 (2): 52–73.

Andonova, Liliana B., Thomas N. Hale, and Charles Roger. 2017. "National Policies and Transnational Governance of Climate Change: Substitutes or Complements?" *International Studies Quarterly*: 253–68.

Andresen, Steinar, and Shardul Agrawala. 2002. "Leaders, pushers and laggards in the making of the climate regime." *Global Environmental Change* 12 (1): 41–51.

Arnold, Annika, Gisela Böhm, Adam Corner, Claire Mays, Nick Pidgeon, Wouter Poortinga, Marc Poumadère, Dirk Scheer, Marco Sonnberger, Katharine Steentjes, and Endre Tvinnereim. 2016. *European Perceptions of Climate Change: Socio-political profiles to inform a cross-national survey in France, Germany, Norway and the UK*. Oxford: Climate Outreach.

Articus, Stephan. 2007. "Der Deutsche Städtetag." In *Handbuch der kommunalen Wissenschaft und Praxis*. 3rd ed., eds. Thomas Mann and Günter Püttner. Berlin, New York, NY: Springer, 937–44.

Arts, Bas, Arnoud Lagendijk, and Henk J. van Houtum, eds. 2009. *The disoriented state: Shifts in governmentality, territoriality and governance*. Vol. 49 of *Environment & policy*. Dordrecht, London: Springer.

Aylett, Alexander. 2015. "Relational Agency and the Local Governance of Climate Change: International Trends and an American Exemplar." In *The urban climate challenge: Rethinking the role of cities in the global climate regime*. Vol. 4 of *Cities and global governance*, eds. Craig Johnson, Noah Toly and Heike Schroeder. New York, NY: Routledge, 156–77.

Babbie, Earl. 2013. *The Practice of social research*. 13th ed. Belmont, CA: Wadsworth Cengage Learning.

Bache, Ian, and Matthew Flinders, eds. 2004. *Multi-level Governance*. Oxford University Press.

Bäckstrand, Karin, and Ole Elgström. 2013. "The EU's role in climate change negotiations: From leader to 'leadiator'." *Journal of European Public Policy* 20 (10): 1369–86.

Bäckstrand, Karin, Jamil Khan, Annica Kronsell, and Eva Lövbrand, eds. 2010. *Environmental politics and deliberative democracy: Examining the promise of new modes of governance*. Cheltenham, Northampton, MA: Edward Elgar Publishing.

Bäckstrand, Karin, and Jonathan W. Kuyper. 2017. "The democratic legitimacy of orchestration: The UNFCCC, non-state actors, and transnational climate governance." *Environmental Politics* 26 (4): 764–88.

Bäckstrand, Karin, Jonathan W. Kuyper, Björn-Ola Linnér, and Eva Lövbrand. 2017. "Non-state actors in global climate governance: From Copenhagen to Paris and beyond." *Environmental Politics* 26 (4): 561–79.

Bader, Nikolas, and Raimund Bleischwitz. 2009. "Measuring Urban Greenhouse Gas Emissions: The Challenge of Comparability." *S.A.P.I.EN.S Online* 2 (3).

Ballesteros Torres, Pedro, and Roman Doubrava. 2010. "The Covenant of Mayors: Cities Leading the Fight against the Climate Change." In *Local Governments and Climate Change: Sustainable Energy Planning and Implementation in Small and Medium Sized Communities*, eds. Maryke van Staden and Francesco Musco. Springer, 91–98.

Bansard, Jennifer S., Philipp H. Pattberg, and Oscar Widerberg. 2016. "Cities to the rescue?: Assessing the performance of transnational municipal networks in global climate governance." *International Environmental Agreements: Politics, Law and Economics*.

Barbey, Kristin. 2012. *Metropolregion im Klimawandel: Räumliche Strategien Klimaschutz und Klimaanpassung*. Zur Entwicklung gesamträumlicher Konzepte am Beispiel der Metropolregion Rhein-Neckar. Karlsruhe: KIT Scientific Publishing.

Barbier, Isabelle. 2012. "Le plan climat." Rennes Métropole. May 13. http://metropole .rennes.fr/politiques-publiques/transports-urbanisme-environnement/l-environne ment/le-plan-climat/ (August 15, 2016).

Barker, Terry, Tom Kram, Sebastian Oberthür, and Monique Voogt. 2001. "The Role of EU Internal Policies in Implementing Greenhouse Gas Mitigation Options to Achieve Kyoto Targets." *International Environmental Agreements: Politics, Law and Economics* 1 (2): 243–65.

Barnes, Pamela M. 2010. "The role of the Commission of the European Union: creating external coherence from internal diversity." In *The European Union as a Leader in International Climate Change Politics. UACES Contemporary European Studies*, eds. Rüdiger K. Wurzel and James Connelly. London, New York, NY: Routledge, 41–57.

Base nationale sur l'intercommunalité. 2014. "CU de Lyon (Grand Lyon): N° SIREN : 246900245." Base nationale sur l'intercommunalité. January 31. https://www.bana-tic.interieur.gouv.fr/V5/recherche-de-groupements/fiche-raison-sociale.php?siren= 246900245&arch=01/01/2014&dcou= (July 9, 2016).

———. 2016a. "CC du Val d'Ille: N° SIREN : 243500667." Base nationale sur l'inter-communalité. April 1. https://www.banatic.interieur.gouv.fr/V5/recherche-de-grou-pements/fiche-raison-sociale.php?siren=243500667 (June 30, 2016).

———. 2016b. "Métropole de Lyon: N° SIREN : 200046977." Base nationale sur l'inter-communalité. April 1. https://www.banatic.interieur.gouv.fr/V5/recherche-de-grou-pements/fiche-raison-sociale.php?siren=200046977&arch=01/04/2016&dcou= (July 9, 2016).

———. 2016c. "Rennes Métropole: N° SIREN : 243500139." Base nationale sur l'inter-communalité. April 1. https://www.banatic.interieur.gouv.fr/V5/recherche-de-grou-pements/fiche-raison-sociale.php?siren=243500139&arch=01/04/2016&dcou= (July 6, 2016).

Baumgartner, Frank R., and Bryan D. Jones. 2009. *Agendas and instability in American politics*. 2nd ed. *Chicago studies in American politics*. Chicago, IL: The University of Chicago Press.

Béal, Vincent, and Gilles Pinson. 2013. "Gouvernance et durabilité sont-elles (encore) les deux mamelles des politiques d'aménagement et d'urbanisme?" In *La gouvernance territoriale: Pratiques, discours et théories*. 2nd ed. *Droit et société*, eds. Romain Pasquier, Vincent Simoulin and Julien Weisbein. Paris: L.G.D.J., 247–68.

———. 2014. "From the Governance of Sustainability to the Management of Climate Change: Reshaping Urban Policies and Central–local Relations in France." *Journal of Environmental Policy & Planning* 17 (3): 402–19.

Beermann, Jan. 2014. "Urban partnerships in low-carbon development: Opportunities and challenges of an emerging trend in global climate politics." *URBE - Revista Brasileira de Gestão Urbana* 6 (541): 170.

Beisheim, Marianne, and Nils Simon. 2015. *Meta-Governance of Partnerships for Sustainable Development: Actors' Perspectives on How the UN Could Improve Partnerships' Governance Services in Areas of Limited Statehood*. Berlin: Collaborative Research Center (SFB) 700. SFB-Governance Working Paper Series, 68.

Ben Mabrouk, Taoufik. 2007. "Quand la métropole se fait territoire politique: Les ressorts politiques et symboliques du changement d'échelle de la coopération intercommunale." In *Action publique et changements d'échelles: Les nouvelles focales du politique* [Résumés bilingues français-anglais]. *Logiques politiques*, eds. Alain Faure, Jean-Philippe Leresche, Pierre Muller and Stéphane Nahrath. Paris: l'Harmattan, 105–19.

Bendlin, Lena. 2013. "Der Gipfel von Kopenhagen – Eine defensive deutsch-französische Partnerschaft." In *Die Konsenswerkstatt: Deutsch-französische Kommunikations- und Entscheidungsprozesse in der Europapolitik*, eds. Claire Demesmay, Martin Koopmann and Julien Thorel. Nomos, 159–72.

———. 2014. "Women's human rights in a changing climate: Highlighting the distributive effects of climate policies." *Cambridge Review of International Affairs* 27 (4): 680–98.

———. 2016. "Cities' Views and Ownership of the Covenant of Mayors." In *Städte und Energiepolitik im europäischen Mehrebenensystem: Zwischen Energiesicherheit, Nachhaltigkeit und Wettbewerb*. Vol. 95 of *Schriftenreihe des Arbeitskreises Europäische Integration e.V.*, eds. Jörg Kemmerzell, Michèle Knodt and Anne Tews. Baden-Baden: Nomos, 101–24.

Benner, Thorsten, Wolfgang H. Reinicke, and Jan M. Witte. 2004. "Multisectoral Networks in Global Governance: Towards a Pluralistic System of Accountability." *Government & Opposition* 39 (2): 191–210.

Benz, Arthur. 1998. "German regions in the European Union: From joint decision-making to multi-level governance." In *Regions in Europe* [Translated from the French]. *European public policy series*, eds. Patrick Le Galès and Christian Lequesne. London, New York, NY: Routledge, 111–29.

———. 2007. *Politik in Mehrebenensystemen*. Vol. 5 of *Lehrbuch*. Wiesbaden: VS Verlag für Sozialwissenschaften.

————. 2010. "The European Union as a loosely coupled multi-level system." In *Handbook on Multi-level Governance*, eds. Henrik Enderlein, Sonja Wälti and Michael Zürn. Cheltenham, Northampton, MA: Edward Elgar Publishing, 214–26.

Benz, Arthur, and Burkard Eberlein. 1998. *Regions in European Governance: The Logic of Multi-Level Interaction*. San Domenico: European University Institute. EUI Working Papers of the Robert Schuman Centre, 98/31.

Benz, Arthur, Jörg Kemmerzell, Michèle Knodt, and Anne Tews. 2015. "The trans-local dimension of local climate policy: Sustaining and transforming local knowledge orders through trans-local action in three German cities." *Urban Research & Practice:* 319–35.

Benz, Arthur, Susanne Lütz, Uwe Schimank, and Georg Simonis, eds. 2007. *Handbuch Governance: Theoretische Grundlagen und empirische Anwendungsfelder*. Wiesbaden: VS Verlag für Sozialwissenschaften.

Berny, Nathalie, ed. 2011. *L'intégration européenne par l'environnement: Le cas français*. Vol. 33 of *Politique européenne*. Paris: l'Harmattan.

Bertrana, Xavier, and Hubert Heinelt. 2013. "The Second Tier of Local Government in the Context of European Multi-level Government Systems: Institutional Setting and Prospects for Reform." *Revista catalana de dret públic* (46): 73–89. http://libros-revistas-derecho.vlex.es/vid/the-tier-context-european-prospects-473524750#section_6.

Bertrand, François, Elsa Richard, and Corinne Larrue. 2016. "Climate problem and territorial governance: an overview of adaptation initiatives at the French regional level." In *Climate adaptation governance in cities and regions: Theoretical fundamentals and practical evidence*, ed. Jörg Knieling. Chichester: Wiley Blackwell, 191-207.

Betsill, Michèle, Navroz K. Dubash, Matthew Paterson, Harro van Asselt, Antto Vihma, and Harald Winkler. 2015. "Building Productive Links between the UNFCCC and the Broader Global Climate Governance Landscape." *Global Environmental Politics* 15 (2): 1–10.

Betsill, Michele M., and Harriet Bulkeley. 2006. "Cities and the Multilevel Governance of Global Climate Change." *Global Governance* 12 (2): 141–59.

————. 2007. "Looking Back and Thinking Ahead: A Decade of Cities and Climate Change Research." *Local Environment* 12 (5): 447–56.

Betsill, Michele M., and Barry G. Rabe. 2009. "Climate Change and Multilevel Governance: The Evolving State and Local Roles." In *Toward sustainable communities: Transition and transformations in environmental policy*. 2nd ed. *American and Comparative Environmental Policy*, eds. Daniel A. Mazmanian and Michael E. Kraft. Cambridge, MA: MIT Press, 201–25.

Biela, Jan, Annika Hennl, and André Kaiser. 2013. *Policy making in multilevel systems: Federalism, decentralisation, and performance in the OECD countries. ECPR Monographs*. Colchester: ECPR Press.

Bieling, Hans-Jürgen, and Marika Lerch, eds. 2006. *Theorien der europäischen Integration*. 2nd ed. Wiesbaden: VS Verlag für Sozialwissenschaften.

Biermann, Frank, Philipp H. Pattberg, Harro van Asselt, and Fariborz Zelli. 2009. "The Fragmentation of Global Governance Architectures: A Framework for Analysis." *Global Environmental Politics* 9 (4): 14–40.

Birkland, Thomas A. 2006. "Agenda Setting in Public Policy." In *Handbook of public policy analysis: Theory, politics, and methods.* Vol. 125 of *Public administration and public policy,* eds. Frank Fischer, Gerald Miller and Mara S. Sidney. Boca Raton: CRC Press, 63–78.

Böcher, Michael, and Annette E. Töller. 2012. *Umweltpolitik in Deutschland: Eine politikfeldanalytische Einführung.* Vol. 50 of *Grundwissen Politik.* Wiesbaden: Springer VS.

Bocquillon, Pierre, and Aurélien Evrard. 2017. "French climate policy: diplomacy in the service of symbolic leadership?" In *The European Union in international climate change politics: Still taking a lead?* Vol. 1 of *Routledge studies in European foreign policy,* eds. Rüdiger K. Wurzel, James Connelly and Duncan Liefferink. New York, NY: Routledge.

Bodansky, Daniel. 2010. "The Copenhagen Climate Change Conference: A Postmortem." *The American Journal of International Law* 104 (2): 230.

Bogner, Walter. 2007. "Mehrstufige kommunale Organisationseinheiten." In *Handbuch der kommunalen Wissenschaft und Praxis.* 3rd ed., eds. Thomas Mann and Günter Püttner. Berlin, New York, NY: Springer, 245–67.

Boissieu, Christian d. 2006. *Division par quatre des émissions de gaz à effet de serre de la France à l'horizon 2050.* Paris: La Documentation française.

Bomberg, Elizabeth. 2012. "Mind the (Mobilization) Gap: Comparing Climate Activism in the United States and European Union." *Review of Policy Research* 29 (3): 408–30.

Bopp, Thomas S. 2008. "'Er hat die regionale Entwicklung unumkehrbar gemacht.': Nachruf des Regionalpräsidenten bei der Trauerfeier am 1. Oktober 2008." In *Region Stuttgart Aktuell,* 4–6.

Börzel, Tanja A. 2001. "Europeanization and Territorial Institutional Change: Toward Cooperative Regionalism?" In *Transforming Europe: Europeanization and domestic change. Cornell studies in political economy,* eds. Maria Green Cowles, James A. Caporaso and Thomas Risse-Kappen. Ithaca, NY: Cornell University Press, 137–58.

———. 2011. "Non-compliance in the European Union: Pathology or statistical artefact?" *Journal of European Public Policy* 8 (5): 803–24.

Brand, Stephan. 2015. "Paradigmenwechsel in der Kommunalfinanzierung — der lange Schatten der Finanzkrise." *Wirtschaftsdienst* 95 (1): 51–55.

Brenner, Neil. 2004. "Urban governance and the production of new state spaces in western Europe, 1960–2000." *Review of International Political Economy* 11 (3): 447–88.

———. 2009. "Open questions on state rescaling." *Cambridge Journal of Regions, Economy and Society* 2 (1): 123–39.

Brody, Samuel D., Sammy Zahran, Himanshu Grover, and Arnold Vedlitz. 2008. "A spatial analysis of local climate change policy in the United States: Risk, stress, and opportunity." *Landscape and Urban Planning* 87 (1): 33–41.

Brown, Douglas M. 2012. "Comparative Climate Change Policy and Federalism: An Overview." *Review of Policy Research* 29 (3): 322–33.

Bukowski, Jeanie J., Simona Piattoni, and Marc E. Smyrl. 2003. *Between Europeaniza-tion and local societies: The space for territorial governance. Governance in Eu-rope.* Lanham, MD: Rowman & Littlefield Publishers.

Bulkeley, Harriet. 2010. "Cities and the Governing of Climate Change." *Annual Review of Environment and Resources* 35 (1): 229–53.

———. 2013. *Cities and climate change. Routledge critical introductions to urbanism and the city.* Abingdon, Oxon, New York, NY: Routledge.

Bulkeley, Harriet, Liliana B. Andonova, Michele M. Betsill, Daniel Compagnon, Thomas N. Hale, Matthew J. Hoffmann, Peter Newell, Matthew Paterson, Charles Roger, and Stacy D. VanDeveer. 2014. *Transnational climate change governance.* New York, NY: Cambridge University Press.

Bulkeley, Harriet, and Michele M. Betsill. 2005. *Cities and climate change: Urban sus-tainability and global environmental governance.* Vol. 4 of *Routledge studies in physical geography and environment.* London: Routledge.

———. 2013. "Revisiting the urban politics of climate change." *Environmental Politics* 22 (1): 136–54.

Bulkeley, Harriet, Vanesa C. Broto, and Gareth Edwards. 2012. "Bringing climate change to the city: towards low carbon urbanism?" *Local Environment* 17 (5): 545–51.

Bulkeley, Harriet, Vanesa C. Broto, Mike Hodson, and Simon Marvin, eds. 2013. *Cities and low carbon transitions. Routledge studies in human geography.* Routledge.

Bulkeley, Harriet, Anna Davies, Bob Evans, David Gibbs, Kristine Kern, and Kate Theobald. 2003. "Environmental Governance and Transnational Municipal Net-works in Europe." *Journal of Environmental Policy & Planning* 5 (3): 235–54.

Bulkeley, Harriet, and Kristine Kern. 2006. "Local Government and the Governing of Climate Change in Germany and the UK." *Urban Studies* 43 (12): 2237–59.

Bulkeley, Harriet, and Peter Newell. 2010. *Governing Climate Change.* Vol. 41 of *Routledge global institutions.* London, New York, NY: Routledge.

Bulkeley, Harriet, Heike Schroeder, Katy Janda, Jimin Zhao, Andrea Armstrong, Shu Y. Chu, and Shibani Ghosh. 2011. "The Role of Institutions, Governance, and Urban Planning for Mitigation and Adaptation." In *Cities and Climate Change.* Vol. 1 of *urban Development Series*, eds. Daniel Hoornweg, Mila Freire, Marcus J. Lee, Perinaz Bhada-Tata and Belinda Yuen. Washington, D.C.: The World Bank, 125–59.

Bundesministerium für Umwelt. 2014. *Merkblatt Erstellung von Klimaschutzkonzepten: Hinweise zur Antragstellung.* Bundesministerium für Umwelt. https://www.klima schutz.de/sites/default/files/page/downloads/140912_MB_Konzepte.pdf (Accessed January 4, 2017).

———. 2016. "Erweiterte Fördermöglichkeiten in der Kommunalrichtlinie." Bundes-ministerium für Umwelt. https://www.klimaschutz.de/de/zielgruppen/kommunen/ foerderung/erweiterte-foerdermoeglichkeiten-der-kommunalrichtlinie (January 4, 2017).

Bundesministerium für Verkehr und digitale Infrastruktur (BMVI). 2016. "Schaufenster LivingLab BWe mobil." Bundesministerium für Verkehr und digitale Infrastruktur. http://www.bmvi.de/SharedDocs/DE/Artikel/G/modellregion-schaufenster-baden-wuerttemberg.html?nn=36210 (October 4, 2016).

Bundesministerium für Verkehr, Bau und Stadtentwicklung (BMVBS). 2011. *Umsetzungsbericht zum Förderprogramm "Elektromobilität in Modellregionen" des BMVBS.* Bundesministerium für Verkehr, Bau und Stadtentwicklung. https://www.bmvi.de/SharedDocs/DE/Anlage/VerkehrUndMobilitaet/modellregionen-elektromobilitaet-umsetzungsbericht-mai-2011.pdf?__blob=publicationFile (Accessed October 4, 2016).

―――. 2013. *Heute Zukunft gestalten: Raumentwicklungsstrategien zum Klimawandel.* Berlin: Bundesministerium für Verkehr, Bau und Stadtentwicklung. http://www.klimamoro.de/fileadmin/Dateien/Ver%C3%B6ffentlichungen/Publikatione_aus_dem_Modellvorhaben/Barrierefreie_Version_KlimaMORO_Broschuere_web.pdf (Accessed September 25, 2016).

―――. 2014. *Regionale Fragestellungen – regionale Lösungsansätze: Ergebnisbericht der Vertiefungsphase des Modellvorhabens der Raumordnung „Raumentwicklungsstrategien zum Klimawandel" (KlimaMORO).* Berlin: Bundesministerium für Verkehr, Bau und Stadtentwicklung. BMVBS, 01/2014.

Bürgin, Alexander. 2014. "National binding renewable energy targets for 2020, but not for 2030 anymore: Why the European Commission developed from a supporter to a brakeman." *Journal of European Public Policy* 22 (5): 690–707.

Busch, Henner. 2015. "Linked for action?: An analysis of transnational municipal climate networks in Germany." *International Journal of Urban Sustainable Development* 7 (2): 213–31.

Busch, Henner, and Stefan Anderberg. 2015. "Green Attraction — Transnational Municipal Climate Networks and Green City Branding." *Journal of Management and Sustainability* 5 (4): 1–16.

Busch, Henner, Lena Bendlin, and Paul Fenton. 2018. "Shaping local response – The influence of transnational municipal climate networks on urban climate governance." *Urban Climate* 24: 221–30.

Cao, Xun, and Hugh Ward. 2016. "Transnational Climate Governance Networks and Domestic Regulatory Action." *International Interactions:* 1–27.

Carlarne, Cinnamon P. 2010. *Climate Change Law and Policy: EU and US Approaches.* Oxford, New York, NY: Oxford University Press.

Castán Broto, Vanesa, and Harriet Bulkeley. 2013. "A survey of urban climate change experiments in 100 cities." [ENG]. *Global environmental change : human and policy dimensions* 23 (1): 92–102.

Chabason, Lucien, and Corinne Larrue. 1998. "France: Fragmented policy and consensual implementation." In *Governance and environment in Western Europe: Politics, policy and administration,* eds. Kenneth Hanf and Alf-Inge Jansen. Harlow, Essex, New York, NY: Longman, 59–80.

Chan, Sander, Clara Brandi, and Steffen Bauer. 2016. "Aligning Transnational Climate Action with International Climate Governance: The Road from Paris." *Review of European, Comparative & International Environmental Law* 25 (2): 238–47.

Chan, Sander, Robert Falkner, Harro van Asselt, and Matthew Goldberg. 2015a. *Strengthening non-state climate action: a progress assessment of commitments launched at the 2014 UN Climate Summit.* Centre for Climate Change Economics and Policy. Working Paper, 242.

Chan, Sander, Harro van Asselt, Thomas N. Hale, Kenneth W. Abbott, Marianne Beisheim, Matthew J. Hoffmann, Brendan Guy, Niklas Höhne, Angel Hsu, Philipp H. Pattberg, Pieter Pauw, Céline Ramstein, and Oscar Widerberg. 2015b. "Reinvigorating International Climate Policy: A Comprehensive Framework for Effective Nonstate Action." *Global Policy* 6 (4): 466–73.

Clark, Sue. 2015. "The Sustainable Mobility Challenge: GreenFit – Re-energising business parks." http://www.greenfitproject.eu/resources/Mobility_complete_uk.pdf (Accessed July 8, 2016).

Cole, Alistair. 2013. *Beyond Devolution and Decentralisation: Building Regional Capacity in Wales and Brittany*. Devolution Series. Manchester: Manchester University Press.

Collet, Philippe. 2014. "Loi de transition énergétique : vers une refonte de la gouvernance énergétique française." [FR;fr]. Actu-Environnement. March 21. http://www.actu-environnement.com/ae/news/gouvernance-energie-france-loi-transition-budgets-carbone-srcae-pcet-21152.php4 (January 8, 2017).

Collier, David, and James Mahoney. 1996. "Insights and Pitfalls: Selection Bias in Qualitative Research." *World Politics* 49 (01): 56–91.

Collier, Ute. 1996. "The European Union's climate change policy: Limiting emissions or limiting powers?" *Journal of European Public Policy* 3 (1): 122–38.

———. 1997. "Local authorities and climate protection in the European union: Putting subsidiarity into practice?" *Local Environment* 2 (1): 39–57.

———. 2013. "The EU and Climate Change Policy: The Struggle over Policy Competences." In *Cases in Climate Change Policy: Political Reality in the European Union*. 2nd ed., eds. Ragnar E. Löfstedt and Ute Collier. New York, NY: Earthscan, 43–64.

Collier, Ute, and Ragnar E. Löfstedt. 2013. "The Climate Change Challenge." In *Cases in Climate Change Policy: Political Reality in the European Union*. 2nd ed., eds. Ragnar E. Löfstedt and Ute Collier. New York, NY: Earthscan, 1–13.

Comité 21. 2017a. "Agenda 21 de Territoires." Comité 21. http://www.agenda21france .org/agenda-21-de-territoire/index.html (January 8, 2017).

———. 2017b. "History and process of french local Agenda 21." Comité 21. http://www. agenda21france.org/agenda-21-de-territoire/history-of-local-agenda-21.html (January 8, 2017).

———. 2017c. "Le Comité 21." Comité 21. http://www.comite21.org/comite21/index .html (January 8, 2017).

Committee of the Regions. 2009. *The White Paper on multi-level governance*. Brussels: Committee of the Regions. http://cor.europa.eu/en/activities/governance/Documents/mlg-white-paper/0387_inside-en-last.pdf (Accessed January 8, 2017).

———. 2014. "Europe's regions and cities: global climate deal success rides on recognition of local government." December 9. http://cor.europa.eu/en/news/Pages/europes-regions-and-cities-global-climate.aspx (September 28, 2015).

Communauté de Communes du Val d'Ille. 2011. *Extrait du registre des délibérations du conseil communautaire: Séance du 1er Mars 2011*. Vignoc: Communauté de Communes du Val d'Ille.

———. 2014. "Le Mot du Président." Communauté de Communes du Val d'Ille. http://www.valdille.fr/component/content/article/29-non-categorise/2-legi-patres-colendissimi.html (June 29, 2016).

———. 2016. "La performance énergétique." Communauté de Communes du Val d'Ille. http://www.valdille.fr/environnement/la-performance-energetique.html (April 12, 2016).

Communauté urbaine de Lyon. 2008. *Signature de la Convention des Maires (Covenant of Mayors) et de la déclaration Eurocities sur le changement climatique: Extrait du registre des délibérations du conseil de communauté. Délibération n° 2008-0236.* Communauté urbaine de Lyon.

———. 2009. "Budget 2009: Le budget du Grand Lyon en grandes lignes." Communauté urbaine de Lyon. http%3A%2F%2Fwww.grandlyon.com%2FBudget-2009.4030.0.html (November 20, 2009).

Communautés urbaines de France, Assemblée des communautés de France, Association des Maires de Grandes Villes de France, Association des Régions de France, AMORCE, Fédération des agences locales de maîtrise de l'énergie et du climat, Services publics locaux de l'énergie, de l'eau, de l'environnement et des e-communications, and Réseau des Agences Régionales de l'Energie et de l'Environnement. 2013. *Proposition de schéma d'organisation et de mise en œuvre de la transition énergétique territoriale.* http://www.adcf.org/files/THEME-Environnement/ContributionnInterAssociativeEnergie_v10avril2013.pdf (Accessed May 25, 2015).

Conseil municipal de Thorigné-Fouillard. 2009. *Séance du 10 Décembre 2009: N° 162/2009 – Adhésion à la convention des Maires.* Thorigné-Fouillard: Conseil municipal de Thorigné-Fouillard.

———. 2010. *Séance du 15 Décembre 2010: N° 144/2010 – Convention des Maires - Approbation du Plan d'Action en faveur de l'Energie Durable.* Thorigné-Fouillard: Conseil municipal de Thorigné-Fouillard.

Conzelmann, Thomas. 2008. "A New Mode of Governing? Multi-level Governance between Cooperation and Conflict." [en]. In *Multi-Level Governance in the European Union· Taking Stock and Looking Ahead*, eds. Thomas Conzelmann and Randall Smith. Baden-Baden: Nomos, 9–31.

Conzelmann, Thomas, and Randall Smith, eds. 2008. *Multi-Level Governance in the European Union: Taking Stock and Looking Ahead.* Nomos.

COOPENERGY. 2014. *Rhine-Neckar Metropolitan Region, DE - Promotion of the Covenant of Mayors.* COOPENERGY. http://www.coopenergy.eu/sites/default/files/good_practice_files/27_Rhine-Neckar%20Metropolitan%20Region%2C%20DE%20-%20Promotion%20of%20the%20Covenant%20of%20Mayors.pdf (Accessed November 3, 2016).

———. 2016a. "About COOPENERGY." COOPENERGY. http://www.coopenergy.eu/content/about-coopenergy (October 27, 2016).

———. 2016b. "Welcome to COOPENERGY: Cooperating in sustainable energy planning." COOPENERGY. http://www.coopenergy.eu/ (October 27, 2016).

Corfee-Morlot, Jan, Lamia Kamal-Chaoui, Michael G. Donovan, Ian Cochran, Alexis Robert, and Jonathan-Pierre Teasdale. 2009. *Cities, Climate Change and Multilevel Governance.* Organisation for Economic Co-operation and Development. OECD

Environmental Working Paper, 14. http://www.oecd.org/governance/regional-policy/44232263.pdf (Accessed November 7, 2014).

Couzigou, Irène. 2013. "France: Territorial decentralisation in France: Towards autonomy and democracy." In *Local Government in Europe: The 'Fourth Level' in the EU Multi-Layered System of Governance. Routledge Research in European Union Law*, eds. Carlo Panara and Michael R. Varney. Hoboken, NJ: Taylor and Francis.

Covenant Club Deutschland. 2012. *Offener Brief zur Umsetzung der Konvent-Ziele in Deutschland.* Covenant Club Deutschland. http://www.klimabuendnis.org/fileadmin/inhalte/dokumente/2012/covclub-offener-Brief_de-web.pdf (Accessed April 15, 2016).

Covenant of Mayors. 2009. *Covenant Signatories and Supporting Structures begin sharing their best practices via the 'Benchmarks of Excellence catalogue'.* Brussels.

————. 2010. "Jeannette Wopperer Covenant of Mayors Ceremony 2010: Jeannette Wopperer, Regional Director of Verband Region Stuttgart (DE), Covenant of Mayors Ceremony 2010, statement II." Covenant of Mayors. http://www.eumayors.eu/media/videos_en.html?videoid=64 (September 26, 2016).

————. 2012. "SEAP monitoring and follow-up evolves in Val d'Ille Joint Community: RURENER Benchmarks." Covenant of Mayors. December 7. http://www.covenantofmayors.eu/about/covenant-supporters_en?structure_id=238&benchmarks=1569 (April 11, 2016).

————. 2013a. "Arnold Schwarzenegger joins Covenant Signatories in flexing climate muscles." June 24. http://www.eumayors.eu/news_en.html?id_news=474 (September 28, 2015).

————. 2013b. "Covenant Supporters." Covenant of Mayors. http://www.eumayors.eu/about/covenant-supporters_en.html (June 6, 2013).

————. 2013c. *Coordinators: Crucial contributors to the Covenant of Mayors success: Report on the activities of Covenant Coordinators 2011 – 2012.* Covenant of Mayors. http://www.eumayors.eu/IMG/pdf/Monitoring_report_CTCs.pdf (Accessed February 11, 2015).

————. 2014a. *Covenant Coordinators: Crucial contributors to the Covenant of Mayors success: Report on the activities of Covenant Territorial and National Coordinators (CTCs, CNCs) 2013.* Covenant of Mayors. http://www.eumayors.eu/IMG/pdf/Monitoring_report_CTCs_2013-final.pdf (Accessed February 9, 2015).

————. 2014b. "Esslingen am Neckar: Profile." Covenant of Mayors. June 25. http://www.eumayors.eu/about/signatories_en.html?city_id=2393 (September 27, 2016).

————. 2014c. "Quick Reference Guide: Joint Sustainable Energy Action Plan." August 1. http://www.covenantofmayors.eu/IMG/pdf/Joint_SEAP_guide-2.pdf (September 28, 2015).

————. 2014d. *Report on the activities of Covenant Supporters for the year 2013: Covenant Supporters demonstrate their potential to strengthen the implementation of the Covenant of Mayors.* Covenant of Mayors. http://www.eumayors.eu/IMG/pdf/Monitoring_report_CSs_2013-final-2.pdf (Accessed February 9, 2015).

————. 2014e. "Signatories." Covenant of Mayors. http://www.eumayors.eu/about/signatories_en.html (March 14, 2014).

————. 2014f. "Communauté de Communes du VAL D'ILLE: Signatory." Covenant of Mayors. February 28. http://www.covenantofmayors.eu/about/signatories_en.html?city_id=226 (April 11, 2016).

————. 2015a. "Covenant of Mayors: Committed to local sustainable energy." http://www.covenantofmayors.eu/ (September 28, 2015).

————. 2015b. "Covenant of Mayors for Climate & Energy: About." Covenant of Mayors. http://www.covenantofmayors.eu/about/covenant-of-mayors_en.html (September 28, 2015).

————. 2015c. "Stuttgart: Profile." Covenant of Mayors. February 10. http://www.covenantofmayors.eu/about/signatories_en.html?city_id=38&overview (September 27, 2016).

————. 2015d. "Stuttgart Region: Benchmarks." Covenant of Mayors. January 21. http://www.eumayors.eu/about/covenant-coordinators_en.html?structure_id=43&benchmarks (September 19, 2016).

————. 2015e. "Stuttgart Region: Profile." Covenant of Mayors. January 21. http://www.eumayors.eu/about/covenant-coordinators_en.html?structure_id=43 (September 16, 2016).

————. 2016a. "Climate Alliance: Profile." Covenant of Mayors. February 12. http://www.eumayors.eu/about/covenant-supporters_en.html?structure_id=33 (September 29, 2016).

————. 2016b. "Contact." Covenant of Mayors. http://www.eumayors.eu/about/contact_en.html (November 29, 2016).

————. 2016c. "Covenant Coordinators." Covenant of Mayors. http://www.eumayors.eu/about/covenant-coordinators_en.html (November 29, 2016).

————. 2016d. "Covenant of Mayors for Climate and Energy." Covenant of Mayors. http://www.covenantofmayors.eu/index_en.html (December 15, 2016).

————. 2016e. "Covenant Supporters." Covenant of Mayors. http://www.eumayors.eu/about/covenant-supporters_en.html (December 31, 2016).

————. 2016f. "Monitoring Action Plans." Covenant of Mayors. http://www.eumayors.eu/actions/monitoring-action-plans_en.html (December 4, 2016).

————. 2016g. "Rennes Metropolis: Covenant Coordinators." Covenant of Mayors. http://www.eumayors.eu/about/covenant-coordinators_en.html?structure_id=46&signatories (December 15, 2016).

————. 2016h. "Signatories." Covenant of Mayors. http://www.eumayors.eu/about/signatories_en.html?q=Search%20for%20a%20Signatory…&country_search=de&population=&date_of_adhesion=&status=&commitments1=&commitments2=&commitments3=&start=2 (December 31, 2016).

————. 2016i. "Signatories." Covenant of Mayors. http://www.eumayors.eu/about/signatories_en.html (June 8, 2016).

————. 2017a. "Signatories." Covenant of Mayors. http://www.eumayors.eu/about/signatories_en.html (January 5, 2017).

————. 2017b. "The Covenant of Mayors going global." Covenant of Mayors. January 8. http://www.covenantofmayors.eu/The-Covenant-of-Mayors-going,2332.html (January 8, 2017).

Covenant of Mayors Office. 2009. *Supporting structures make the Covenant a reality for all*. Brussels: Covenant of Mayors Office. http://www.konventderbuergermeister.eu/IMG/pdf/Thematic_Leaflet_Supporting_Structures.pdf (Accessed April 3, 2016).

Creswell, John W. 2013a. *Qualitative Inquiry & Research Design: Choosing Among Five Approaches*. 3rd ed. Los Angeles: Sage Publications.

———. 2013b. *Research design: Qualitative, quantitative, and mixed method approaches*. 4th ed.

Damro, Chad, and Pilar L. Méndez. 2003. "Emissions trading at Kyoto: From EU resistance to Union innovation." *Environmental Politics* 12 (2): 71–94.

De Waele, Jean-Michel, ed. 2004. *Les clivages politiques en Europe centrale et orientale. Sociologie politique*. Brussels: Editions de l'Université de Bruxelles.

Delcamp, Alain. 1995. "La démocratie municipale chez nos voisins : une typologie." *Pouvoirs. Revue française d'études constitutionnelles et politiques* (73): 125–39.

Demesmay, Claire, Martin Koopmann, and Julien Thorel, eds. 2013. *Die Konsenswerkstatt: Deutsch-französische Kommunikations- und Entscheidungsprozesse in der Europapolitik*. Nomos.

Département Ille-et-Vilaine. 2016. *Recueil des actes administratifs: Session du 1er trimestre 2016*. Rennes: Département Ille-et-Vilaine. http://www.ille-et-vilaine.fr/sites/default/files/asset/document/raa_493_-_session_janvier_2016.pdf (Accessed January 7, 2017).

Deroubaix, José-Frédéric, and François Lévèque. 2006. "The rise and fall of French Ecological Tax Reform: Social acceptability versus political feasibility in the energy tax implementation process." *Energy Policy* 34 (8): 940–49.

Desage, Fabien. 2011. "Intercommunalité." In *Dictionnaire des politiques territoriales. Références Gouvernances*, eds. Romain Pasquier, Sébastien Guigner and Alistair Cole. Paris: Presses de Sciences Po, 283–89.

Deutsche Gesellschaft für Internationale Zusammenarbeit. 2013. *Konvent der Bürgermeister/innen: Eine Klimaschutzinitiative europäischer Kommunen – Erfahrungen, Praxisbeispiele und erfolgversprechende Ansätze*. Trans. Andrea Hahn. Eschborn: Deutsche Gesellschaft für Internationale Zusammenarbeit.

Deutscher Bundestag. 2005. *Nationales Klimaschutzprogramm Sechster Bericht der Interministeriellen Arbeitsgruppe CO2-Reduktion: Drucksache 15/5931*. Berlin. Drucksachen, 773.

Deutscher Städte- und Gemeindebund. 2014. *Energiewende: Statement zur Energie- und Umweltpolitik*. Trans. Gerd Landsberg. Berlin: Deutscher Städte- und Gemeindebund. Position. http://www.dstgb.de/dstgb/Homepage/Publikationen/Positionspapiere/Energiewende%3A%20Neue%20Marktmechanismen%20etablieren%20-%20Netzausbau%20beschleunigen%20-%20Versorgungssicherheit%20gew%C3%A4hrleisten%20-%20Kosten%20begrenzen/PP%20Energiewende%2001_08_14.pdf (Accessed December 31, 2016).

Deutscher Städtetag. 2014. *Klimaschutz und Energiepolitik aktiv gestalten*. Berlin, Köln, Brussels: Deutscher Städtetag. http://www.staedtetag.de/imperia/md/content/dst/internet/presse/2013/positionspapier_klimaschutz_und_energiepolitik.pdf (Accessed December 31, 2016).

Dey, Ian. 1993. *Qualitative data analysis: A user-friendly guide for social scientists*. London: Routledge.

Di Martino, Luigi Alberto. 2013. "The Covenant of Mayors: Multi-level Governance of Energy Policy Implementation in the European Union." Paper presented at the 17th annual meeting of the Japan Association for Evolutionary Economics. March 17. http://www.setsunan.ac.jp/~k-yagi/2013intrrep9DiMartino.pdf (April 25, 2016).

———. 2015. "The Covenant of Mayors: Multi-level Governance of Energy Transition in the European Union." Paper presented at the 19th annual meeting of the Japan Association for Evolutionary Economics. March 21. http://www.setsunan.ac.jp/~k-yagi/chap10dimartino.pdf (April 25, 2016).

Diefenbacher, Hans, Benjamin Held, Dorothee Rodenhäuser, and Roland Zieschank. 2016. *Wohlfahrtsmessung "beyond GDP": Der Nationale Wohlfahrtsindex (NWI2016)*. Institut für Makroökonomie und Konjunkturforschung. IMK Study, 48.

Dimitrov, Radoslav S. 2010. "Inside Copenhagen: The State of Climate Governance." *Global Environmental Politics* 10 (2): 18–24.

———. 2016. "The Paris Agreement on Climate Change: Behind Closed Doors." *Global Environmental Politics* 16 (3): 1–11.

Dinnie, Keith, ed. 2011. *City Branding: Theory and Cases*. London: Palgrave Macmillan.

Dolšak, Nives, and Aseem Prakash. 2016. "Join the Club: How the Domestic NGO Sector Induces Participation in the Covenant of Mayors Program." *International Interactions*: 1–22.

Doran, George T. 1981. "There's a S.M.A.R.T. way to write management's goals and objectives." *Management Review* 70 (11): 35–36.

Dupont, Claire, and Sebastian Oberthür. 2015. "Decarbonizing the EU: Setting the Scene." In *Decarbonization in the European Union: Internal policies and external strategies. Energy, climate and the environment*, eds. Claire Dupont and Sebastian Oberthür. Basingstoke: Palgrave Macmillan, 1–24.

Eberlein, Burkard, and Dieter Kerwer. 2004. "New Governance in the European Union: A Theoretical Perspective." *Journal of Common Market Studies* 42 (1): 121–42.

Eckstein, Harry. 1975. "Case Study and Theory in Political Science." In *Strategies of Inquiry: Handbook of political science*. Vol. 2607 ot *Addison-Wesley series in political science*. Volume 7, eds. Fred I. Greenstein and Nelson W. Polsby. Reading, MA: Addison-Wesley, 79–137.

Eichener, Volker. 1997. "Effective European problem-solving: Lessons from the regulation of occupational safety and environmental protection." *Journal of European Public Policy* 4 (4): 591–608.

Eisermann, Miriam. 2009. "Växjö and Heidelberg formulate respective energy strategies." Energy Cities. October 14. http://www.energy-cities.eu/Vaxjo-and-Heidelberg-formulate (November 4, 2016).

Elgström, Ole, and Christer Jönsson. 2000. "Negotiation in the European Union: Bargaining or problem-solving?" *Journal of European Public Policy* 7 (5): 684–704.

———, eds. 2004. *European Union Negotiations: Processes, Networks and Institutions. Routledge Advances in European Politics*. London, New York, NY: Routledge.

Eltrop, Ludger. 2011. "Führung ist gefragt." In *Region Stuttgart Aktuell*, 28–29.

Emmerich-Fritsche, Angelika. 2007. "Importance of the Subsidiarity Principle in the Multi-Level System – Protection for Local Authorities?" *German Journal of Urban Studies* 46 (1).

e-mobil BW GmbH. 2015. *NETZ-E-2-R: Vernetzte E-Bike-Anschlussmobilität an Bahnhaltepunkten in der Region Stuttgart.* e-mobil BW GmbH. http://www.e-mobilbw.de/files/e-mobil/content/DE/Publikationen/PDF%20Schaufenster%20Projekte/1_Intermodalitaet/NETZE2R.pdf (Accessed October 4, 2016).

Enderlein, Henrik, Sonja Wälti, and Michael Zürn, eds. 2010. *Handbook on Multi-level Governance.* Cheltenham, Northampton, MA: Edward Elgar Publishing.

Energy Cities. 2017. "All members." Energy Cities. http://www.energy-cities.eu/cities/members_in_europe_en.php (January 6, 2017).

Epstein, Renaud. 2005. "Gouverner à distance: Quand l'État se retire des territoires." *Esprit* (319): 96–111.

———. 2008. "L'éphémère retour des villes.: L'autonomie locale à l'épreuve des recompositions de l'État." *Esprit* (342): 136–49.

Eurocities. 2008. *EUROCITIES Declaration on Climate Change.* http://www.dublincity.ie/sites/default/files/content//YourCouncil/LordMayorDublin/Documents/Published_Declaration.pdf (Accessed May 23, 2015).

———. 2015. "Rennes Metropole signs Green Digital Charter." Eurocities. October 2. http://www.eurocities.eu/eurocities/news/Rennes-Metropole-signs-Green-Digital-Charter-WSPO-A2WD9L (November 11, 2016).

European Climate Adaptation Platform (Climate-ADAPT). 2016. "Stuttgart: combating the heat island effect and poor air quality with green ventilation corridors (2014)." European Climate Adaptation Platform. http://climate-adapt.eea.europa.eu/metadata/case-studies/stuttgart-combating-the-heat-island-effect-and-poor-air-quality-with-green-ventilation-corridors (September 28, 2016).

European Commission. 2001. "European governance - A white paper: Communication from the Commission of 25 July 2001." *Official Journal of the European Union* (C 287). COM(2001) 428 final.

———. 2006a. *Analysis of the Action Plan for Energy Efficiency: Realising the Potential. Commission staff working document.*

———. 2006b. *Commission Decision of 14 December 2006 determining the respective emission levels allocated to the Community and each of its Member States under the Kyoto Protocol pursuant to Council Decision 2002/358/EC (notified under document number C(2006) 6468): (2006/944/EC).*

———. 2008. "The NUTS classification: Eurostat." European Commission. http://ec.europa.eu/eurostat/web/regions/nuts-classification (May 6, 2016).

———. 2014a. "Sharing urban sustainable energy strategies - promoting the Covenant of Mayors (COME2COM)." July 17. https://ec.europa.eu/energy/intelligent/projects/en/projects/come2com (September 28, 2015).

———. 2014b. "Communication from the Commission to the European Parliament and the Council: European Energy Security Strategy." (COM(2014) 330 final). Brussels. May 28.

———. 2015. "NUTS - Nomenclature of territorial units for statistics: Overview." European Commission. http://ec.europa.eu/eurostat/web/nuts/overview (May 6, 2016).

————. 2016a. "Statistics on environmental infringements." [en]. European Commission. June 8. http://ec.europa.eu/environment/legal/law/statistics.htm (January 7, 2017).

————. 2016b. "Intelligent Energy Europe: Project Database - Results for 'Covenant'." European Commission. April 4. http://ec.europa.eu/energy/intelligent/projects/en/project-search?search_api_views_fulltext=covenant (April 4, 2016).

European Energy Award. 2016a. *France*. European Energy Award. http://www.european-energy-award.org/fileadmin/Documents/Fact_sheets_Countries/FS_France.pdf (Accessed January 8, 2017).

————. 2016b. *Germany*. European Energy Award. http://www.european-energy-award.org/fileadmin/Documents/Fact_sheets_Countries/FS_Germany.pdf (Accessed January 8, 2017).

European Environmental Agency. 2007. *Greenhouse gas emission trends and projections in Europe 2007: Tracking progress towards Kyoto targets*. Copenhagen: European Environmental Agency. EEA Report, 5/2007.

————. 2012. *Greenhouse gas emission trends and projections in Europe 2012: Tracking progress towards Kyoto and 2020 targets*. European Environmental Agency. EEA Report, 6/2012.

————. 2015. *Trends and projections in Europe 2015: Tracking progress towards Europe's climate and energy targets*. European Environmental Agency. EEA Report, 4/2015.

European Sustainable Cities Platform. 2015. "The Aalborg Charter." European Sustainable Cities Platform. http://www.sustainablecities.eu/aalborg-process/charter (July 21, 2016).

European Union. 2015. *Regions in the European Union: Nomenclature of territorial units for statistics NUTS 2013/EU-28. Manuals and guidelines*. Luxembourg: Publications Office of the European Union.

Eurostat. 2016a. "Greenhouse gas emissions by source sector (source: EEA)." Eurostat. http://appsso.eurostat.ec.europa.eu/nui/show.do?dataset=env_air_gge&lang=en (December 29, 2016).

————. 2016b "Gross domestic product at market prices: At current prices." Eurostat. http://ec.europa.eu/eurostat/tgm/table.do?tab=table&plugin=1&language=en&pcode=tec00001 (December 29, 2016).

Fahl, Ulrich, Maike Sippel, Markus Blesl, Christoph Kruck, Marlies Härdtlein, Ludger Eltrop, Ralph Schelle, Jochen Lambauer, Erik Heyden, Christina Benighaus, Annika Arnold, and Ortwin Renn. 2011. *Integriertes Klimaschutz- und Energiekonzept für Ludwigsburg*. Stadt Ludwigsburg. http://www.ludwigsburg.de/site/Ludwigsburg-Internet/get/params_E-729216728/1426612/GEK_Abschlussbericht.pdf (Accessed September 28, 2016).

Falkner, Gerda, ed. 2011. *The EU's decision traps: Comparing policies*. Oxford, New York, NY: Oxford University Press.

Faltin, Thomas. 2015. "Nicola Schelling: Mit 385 PS zur nachhaltigen Mobilität." *Stuttgarter Zeitung*, January 7. http://www.stuttgarter-zeitung.de/inhalt.nicola-schelling-mit-385-ps-zur-nachhaltigen-mobilitaet.75bec78d-194c-4e14-8c8a-9776b8a02119.html (Accessed October 5, 2016).

Faure, Alain, and Pierre Muller. 2007. "Introduction générale: Objet classique, équations nouvelles." In *Action publique et changements d'échelles: Les nouvelles focales du politique* [Résumés bilingues français-anglais]. *Logiques politiques*, eds. Alain Faure, Jean-Philippe Leresche, Pierre Muller and Stéphane Nahrath. Paris: l'Harmattan, 9–19.

Fédération des agences locales de maîtrise de l'énergie et du climat. 2017. "La fédération: Présentation." Fédération des agences locales de maîtrise de l'énergie et du climat. http://www.federation-flame.org/la-federation/presentation/ (January 8, 2017).

Finck, Michèle. 2014. "Above and Below the Surface: The Status of Sub-National Authorities in EU Climate Change Regulation." *Journal of Environmental Law* 26 (3): 443–72.

Fischer, Severin. 2012. "Die letzte Runde in der Atomdebatte? Der Parteienwettbewerb nach Fukushima." In *Superwahljahr 2011 und die Folgen*. Vol. 2 of *Parteien und Wahlen*, ed. Eckhard Jesse. Baden-Baden: Nomos, 365–85.

———, ed. 2017. *Die Energiewende und Europa: Europäisierungsprozesse in der deutschen Energie- und Klimapolitik*. Wiesbaden: Springer VS.

Forsyth, Tim, and Les Levidow. 2015. "An Ontological Politics of Comparative Environmental Analysis: The Green Economy and Local Diversity." *Global Environmental Politics* 15 (3): 140–51.

Fougère, Louis, Jean-Pierre Machelon, and François Monnier, eds. 2002. *Les communes et le pouvoir: Histoire politique des communes françaises de 1789 à nos jours*. Paris: Presses universitaires de France.

Fournier, Fabien. 2012. "Pôle métropolitain: un nouveau mammouth administratif?" *Lyon Capitale*, June 6. https://www.lyoncapitale.fr/Journal/Lyon/Politique/Grand-lyon/Pole-metropolitain-un-nouveau-mammouth-administratif (Accessed July 10, 2016).

François, Jean-Luc. 2013a. "Brittany: The Val d'Ille's action plan in the CoM framework: A long long procedure." Territoires à énergie positive. January 18. http://www.territoires-energie-positive.fr/eng/accompagnement/bretagne-le-plan-d-actions-du-val-d-ille-et-la-convention-des-maires-une-longue-procedure (July 4, 2016).

———. 2013b. "Brittany: The "Val d'Ille" Territory and the Covenant of Mayors." Territoires à énergie positive. March 8. http://www.territoires-energie-positive.fr/pol/methods/brittany-the-val-d-ille-territory-and-the-covenant-of-mayors (April 11, 2016).

Frenzel, Eike M. 2013. "Germany: Local government in Germany: An indispensable level of EU governance." In *Local Government in Europe: The 'Fourth Level' in the EU Multi-Layered System of Governance. Routledge Research in European Union Law*, eds. Carlo Panara and Michael R. Varney. Hoboken, NJ: Taylor and Francis, 97–127.

Fünfgeld, Hartmut. 2015. "Facilitating local climate change adaptation through transnational municipal networks." *Current Opinion in Environmental Sustainability* 12: 67–73.

Fürst, Dietrich. 2004. "Metropolitan governance in Germany." In *Metropolitan Governance in the 21st Century: Capacity, Democracy and the Dynamics of Place*, eds. Hubert Heinelt and Daniel Kübler. Abingdon, Oxon, New York, NY: Routledge, 151–68.

Galimberti, Deborah, Sylvaine Lobry, Gilles Pinson, and Nicolas Rio. 2014. "La métropole de Lyon. Splendeurs et fragilités d'une machine intercommunale." *Hérodote*
154 (3): 191–209.

Gandon, Jean-Baptiste. 2015. "Devenir une véritable éco-métropole." Rennes Métropole.
November 25. http://metropole.rennes.fr/actualites/institutions-citoyennete/institu-
tion/devenir-une-veritable-eco-metropole/ (August 23, 2016).

Garello, Pierre. 2016. "French Subnational Public Finances: On the Difficulty of Being a
Decentralized Unitary State." In *Multi-level finance and the Euro crisis: Causes and
effects. Studies in fiscal federalism and state-local finance*, eds. Ehtisham Ahmad,
Massimo Bordignon and Giorgio Brosio. Cheltenham: Edward Elgar Publishing,
103–31.

Geden, Oliver, and Severin Fischer. 2008. *Die Energie- und Klimapolitik der Europäischen Union: Bestandsaufnahme und Perspektiven*. Baden-Baden: Nomos.

Geels, Frank. 2013. "The role of cities in technological transitions: analytical clarifications and historical examples." In *Cities and low carbon transitions. Routledge studies in human geography*, eds. Harriet Bulkeley, Vanesa C. Broto, Mike Hodson and
Simon Marvin. Routledge, 13–28.

Gemeinde Böhl-Iggelheim. 2011. *Aktionsplan zum Klimaschutz: Erstellt im Rahmen des
Projektes der Europäischen Union „Konvent der Bürgermeister/innen für lokale
nachhaltige Energie"*. Gemeinde Böhl-Iggelheim. http://mycovenant.eumayors.eu/
seap-monitoring/index.php?page=iframe_graphs&process=download_docu-
ments&item_id=8730&report_id=7830 (Accessed October 14, 2016).

George, Alexander L., and Andrew Bennett. 2005. *Case Studies and Theory Development
in the Social Sciences. BCSIA studies in international security*. Cambridge, MA:
MIT Press.

George, Stephen. 2004. "Multi-level Governance and the European Union." In *Multi-level
Governance*, eds. Ian Bache and Matthew Flinders. Oxford University Press,
107–26.

Gerring, John. 2004. "What Is a Case Study and What Is It Good for?" *The American
Political Science Review* 98 (2): 341–54.

———. 2007. *Case Study Research: Principles and Practices*. New York: Cambridge
University Press.

Ghorra-Gobin, Cynthia. 2015. *La métropolisation en question. La Ville en débat*. Paris:
Presses universitaires de France.

Giddens, Anthony. 2011. *The politics of climate change*. 2nd ed. Cambridge, Malden, MA:
Polity Press.

Giest, Sarah, and Michael Howlett. 2013. "Comparative Climate Change Governance:
Lessons from European Transnational Municipal Network Management Efforts."
Environmental Policy and Governance 23 (6): 341–53.

Glastra, Kathrin, and Andreas Rüdinger. 2016. "In guter Gesellschaft. Die deutsche Energiewende und ihr Echo in Frankreich." In *Frankreich und Deutschland - Bilder, Ste-
reotype, Spiegelungen*. 1st ed. *DGAP-Schriften zur Internationalen Politik*, eds.
Claire Demesmay, Christine Pütz and Hans Stark. Baden-Baden: Nomos, 191–202.

Gorden, Raymond L. 1987. *Interviewing: Strategy, techniques, and tactics*. 4th ed. Chicago, IL: Dorsey Press.

Gordon, David J. 2013. "Between local innovation and global impact: cities, networks, and the governance of climate change." *Canadian Foreign Policy Journal* 19 (3): 288–307.

———. 2016. "The Politics of Accountability in Networked Urban Climate Governance." *Global Environmental Politics* 16 (2): 82–100.

Gordon, David J., and Craig A. Johnson. 2017. "The orchestration of global urban climate governance: Conducting power in the post-Paris climate regime." *Environmental Politics* 26 (4): 694–714.

Gordon, John. 1993. "Letting the genie out: Local government and UNCED." *Environmental Politics* 2 (4): 137–55.

Gore, Christopher D. 2010. "The Limits and Opportunities of Networks: Municipalities and Canadian Climate Change Policy." *Review of Policy Research* 27 (1): 27–46.

Götze, Susanne, and Sandra Kirchner. 2016. *Die Umweltpolitik der Alternative für Deutschland (AfD): Eine politische Analyse.* Dresden: Heinrich Böll Stiftung.

Grande, Edgar. 2012. "Governance-Forschung in der Governance-Falle? – Eine kritische Bestandsaufnahme." *Politische Vierteljahresschrift* 53 (4): 565–92.

Green, Jessica F., Thomas Sterner, and Gernot Wagner. 2014. "A balance of bottom-up and top-down in linking climate policies." *Nature Climate Change* 4 (12): 1064–67.

Groen, Lisanne, and Arne Niemann. 2013. "The European Union at the Copenhagen climate negotiations: A case of contested EU actorness and effectiveness." *International Relations* 27 (3): 308–24.

Groen, Lisanne, Arne Niemann, and Sebastian Oberthür. 2012. "The EU as a global leader? The Copenhagen and Cancún UN climate change negotiations." *Journal of Contemporary European Research* 8 (2): 173–91.

Große Hüttmann, Martin. 2010. "Bundesrepublik Deutschland - Der unitarische Föderalismus wird vielfältiger." In *Regional Governance in EU-Staaten*, eds. Roland Sturm and Jürgen Dieringer. Opladen, Farmington Hills, MI: Budrich, 37–62.

Großmann, Katrin, André Schaffrin, and Christian Smigiel, eds. 2017. *Energie und soziale Ungleichheit: Zur gesellschaftlichen Dimension der Energiewende in Deutschland und Europa.* Wiesbaden: Springer.

Gupta, Joyeeta, and Michael J. Grubb, eds. 2000. *Climate change and European leadership: A sustainable role for Europe?* Vol. 27 of *Environment & policy.* Dordrecht, Boston, MA: Kluwer Academic.

Gupta, Joyeeta, and Lasse Ringius. 2001. "The EU's Climate Leadership: Reconciling Ambition and Reality." *International Environmental Agreements: Politics, Law and Economics* 1 (2): 281–99.

Haahr, Jens H., and William Walters. 2004. *Governing Europe: Discourse, Governmentality and European Integration. Routledge Advances in European Politics.* London, New York, NY: Routledge.

Hakelberg, Lukas. 2014. "Governance by Diffusion: Transnational Municipal Networks and the Spread of Local Climate Strategies in Europe." *Global Environmental Politics* 14 (1): 107–29.

Hale, Thomas. 2016. ""All Hands on Deck": The Paris Agreement and Nonstate Climate Action." *Global Environmental Politics* 16 (3): 12–22.

Hale, Thomas N., and Charles Roger. 2014. "Orchestration and transnational climate governance." *The Review of International Organizations* 9 (1): 59–82.

Hall, David, Emanuele Lobina, and Philipp Terhorst. 2013. "Re-municipalisation in the early twenty-first century: Water in France and energy in Germany." *International Review of Applied Economics* 27 (2): 193–214.

Halpern, Charlotte. 2011. "L'Union européenne, vecteur d'innovation instrumentale?: Les logiques d'instrumentation de la politique française de l'environnement (1971–2006)." *Politique européenne* 33 (1): 89.

Hamedinger, Alexander, and Alexander Wolffhardt, eds. 2010. *The Europeanization of cities: Policies, urban change, & urban networks.* Amsterdam: Techne Press.

Harrison, Kathryn. 2015. "International Carbon Trade and Domestic Climate Politics." *Global Environmental Politics* 15 (3): 27–48.

Hasselmann, Klaus, Antonio Ruiz de Elvira, and Martin Welp. 2009. *Climate Change and the Cities of the Future: Art, Technology and Economics in the face of Climate Change.* Trans. Aida Abdulah. European Climate Forum e.V. ECF Working Paper, 2/2009.

Hawkins, Darren G., David A. Lake, Daniel L. Nielson, and Michael J. Tierney, eds. 2006. *Delegation and agency in international organizations. Political economy of institutions and decisions.* Cambridge: Cambridge University Press.

Hegele, Yvonne, and Nathalie Behnke. 2013. "Die Landesministerkonferenzen und der Bund – Kooperativer Föderalismus im Schatten der Politikverflechtung." *Politische Vierteljahresschrift* 54 (1): 21–49.

Heidbreder, Eva G. 2011. "Structuring the European administrative space: Policy instruments of multi-level administration." *Journal of European Public Policy* 18 (5): 709–27.

———. 2014. "Kehrtwende in der Koordinierung europäischer Politikumsetzung: Horizontale Kapazitätsbündelung statt vertikaler Kompetenzverlagerung." *der moderne staat (dms) - Zeitschrift für Public Policy, Recht und Management* 7 (1): 37–54.

———. 2015. "Horizontal Capacity Pooling: Direct, Decentralized, Joint Policy Execution." In *The Palgrave Handbook of the European Administrative System. European administrative governance*, eds. Michael W. Bauer and Jarle Trondal. Houndmills, Basingstoke, Hampshire, New York, NY: Palgrave Macmillan, 369–82.

Heinelt, Hubert. 2008. "How to Achieve Governability in Multi-Level Policymaking: Lessons from the EU Structural Funds and EU Environment Policy." In *Multi-Level Governance in the European Union: Taking Stock and Looking Ahead*, eds. Thomas Conzelmann and Randall Smith. Baden-Baden: Nomos, 53–71.

Heinelt, Hubert, and Michèle Knodt. 2011. *Policies within the EU multi-level system: Instruments and strategies of European governance.* Baden-Baden: Nomos.

Heinelt, Hubert, and Wolfram Lamping. 2014. "Städte im Klimawandel: Zwischen Problembetroffenheit und Innovationserwartung." *Forschungsjournal Soziale Bewegungen* 27 (2): 79–89.

Heinz, Werner. 2007. "Inter-Municipal Cooperation in Germany: The Mismatch Between Existing Necessities and Suboptimal Solutions." In *Inter-municipal cooperation in Europe*, eds. Rudie Hulst and André van Montfort. Dordrecht: Springer, 91-1'15.

Hendriks, Gisela, and Annette Morgan. 2001. *The Franco-German Axis in European Integration*. Cheltenham, Northampton, MA: Edward Elgar Publishing.

Henneke, Hans-Günter. 2007a. "Der Deutsche Landkreistag." In *Handbuch der kommunalen Wissenschaft und Praxis*. 3rd ed., eds. Thomas Mann and Günter Püttner. Berlin, New York, NY: Springer, 945–61.

———. 2007b. "Funktionen und Aufgaben der kommunalen Spitzenverbände im europäisierten Bundesstaat." In *Handbuch der kommunalen Wissenschaft und Praxis*. 3rd ed., eds. Thomas Mann and Günter Püttner. Berlin, New York, NY: Springer, 981–1011.

Hey, Christian. 2010a. "Die Defizite der Umweltverwaltungen." In *Verwaltung und Raum: Zur Diskussion um Leistungsfähigkeit und Integrationsfunktion von Verwaltungseinheiten*. Vol. 34 of *Schriften der Deutschen Sektion des Internationalen Instituts für Verwaltungswissenschaften*, ed. Dieter Schimanke. Baden-Baden: Nomos, 89–100.

———. 2010b. "The German Paradox: Climate Leader and Green Car Laggard." In *The new climate policies of the European Union: Internal legislation and climate diplomacy*. Vol. 15 of *Institute for European Studies publication series*, eds. Sebastian Oberthür and Marc Pallemaerts. Trans. Claire Roche Kelly. Brussels: VUB PRESS Brussels University Press, 211–30.

———. 2012. "Low-carbon and Energy Strategies for the EU: The European Commission's Roadmaps: A Sound Agenda for Green Economy?" *GAIA - Ecological Perspectives for Science and Society* 21 (1): 43–47.

Hey, Christian, and Christian Calliess. 2013. "Multilevel Energy Policy in the EU: Paving the Way for Renewables?" *Journal for European Environmental & Planning Law* 10 (2): 87–131.

Heyvaert, Veerle. 2013. "What's in a Name?: The Covenant of Mayors as Transnational Environmental Regulation." *Review of European, Comparative & International Environmental Law* 22 (1): 78–90.

Hoffmann, Matthew J. 2011. *Climate governance at the crossroads: Experimenting with a global response after Kyoto*. Oxford, New York, NY: Oxford University Press.

Hoffmann-Martinot, Vincent, and Hellmut Wollmann, eds. 2006. *State and local government reforms in France and Germany: Divergence and convergence*. Vol. 7 of *Urban and regional research international*. Wiesbaden: VS Verlag für Sozialwissenschaften.

Holland, Detlef. 2016. "An der Doppelspitze des Regionalverbands rumort es." *Eßlinger Zeitung*, April 8. http://www.esslinger-zeitung.de/region/baden-wuerttemberg_artikel,-an-der-doppelspitze-des-regionalverbands-rumort-es-_arid,2037075.html (Accessed October 5, 2016).

Holzinger, Katharina, Helge Jörgens, and Christoph Knill, eds. 2007. *Transfer, Diffusion und Konvergenz von Politiken*. 38/2007 of *Politische Vierteljahresschrift Sonderheft*. Wiesbaden: VS Verlag für Sozialwissenschaften.

Hooghe, Liesbet. 1996. "Introduction: Reconciling EU-wide policy and national diversity." In *Cohesion policy and European integration: Building multi-level governance*, ed. Liesbet Hooghe. Oxford, New York: Oxford University Press, 1–24.

Hooghe, Liesbet, and Gary Marks. 2001a. *Multi-Level Governance and European Integration. Governance in Europe*. Lanham, MD: Rowman & Littlefield Publishers.

———. 2001b. "Types of Multi - Level Governance." *European Integration online Papers* 5 (11).

———. 2003. "Unraveling the Central State, but How?: Types of Multi-level Governance." *American Political Science Review* 97 (2): 233–43.

———. 2010. "Types of multi-level governance." In *Handbook on Multi-level Governance*, eds. Henrik Enderlein, Sonja Wälti and Michael Zürn. Cheltenham, Northampton, MA: Edward Elgar Publishing, 17–31.

———. 2016. *Community, Scale, and Regional Governance: A postfunctionalist theory of governance. Transformations in Governance*. Volume II. Oxford: Oxford University Press.

Hooghe, Liesbet, Gary Marks, and Arjan H. Schakel. 2010. *The rise of regional authority: A comparative study of 42 democracies*. London, New York, NY: Routledge.

Hooghe, Liesbet, Gary Marks, Arjan H. Schakel, Sara Niedzwiecki, Sandra Chapman Osterkatz, and Sarah Shair-Rosenfield. 2016. *Measuring Regional Authority: A postfunctionalist theory of governance. Transformations in Governance*. Volume I. Oxford: Oxford University Press.

Hoornweg, Daniel, Mila Freire, Marcus J. Lee, Perinaz Bhada-Tata, and Belinda Yuen, eds. 2011. *Cities and Climate Change*. Vol. 1 of *urban Development Series*. Washington, D.C.: The World Bank.

Hoppe, Thomas, Arjen van der Vegt, and Peter Stegmaier. 2016. "Presenting a Framework to Analyze Local Climate Policy and Action in Small and Medium-Sized Cities." *Sustainability* 8 (9).

Horga, Ioan. 2010. "The Multilevel Governance (MLG) and the respect of the subsidiarity principle." In *Cross-border partnership: With special regard to the Hungarian-Romanian-Ukrainian tripartite border*, eds. Ioan Horga and István Süli-Zakar. Oradea, Debrecen: University of Oradea Press, 169–74.

Hörster, Ansgar. 2007. "Höhere Kommunalverbände." In *Handbuch der kommunalen Wissenschaft und Praxis*. 3rd ed., eds. Thomas Mann and Günter Püttner. Berlin, New York, NY: Springer, 901–34.

Houle, David, Erick Lachapelle, and Mark Purdon. 2015. "Comparative Politics of Sub-Federal Cap-and-Trade: Implementing the Western Climate Initiative." *Global Environmental Politics* 15 (3): 49–73.

Hsueh, Lily, and Aseem Prakash. 2012. "Incentivizing self-regulation: Federal vs. state-level voluntary programs in US climate change policies." *Regulation & Governance* 6 (4): 445–73.

Huber, Michael. 2013. "Leadership and Unification: Climate Change Policies in Germany." In *Cases in Climate Change Policy: Political Reality in the European Union*. 2nd ed., eds. Ragnar E. Löfstedt and Ute Collier. New York, NY: Earthscan, 65–86.

Hueglin, Thomas O. 1999. "Government, governance, governmentality: understanding the EU as a project of universalism." In *The transformation of governance in the European Union*. Vol. 12 of *ECPR studies in European political science*, eds. Beate Kohler-Koch and Rainer Eising. London, New York, NY: Routledge, 249–65.

Hulst, Rudie, and André van Montfort, eds. 2007a. *Inter-municipal cooperation in Europe*. Dordrecht: Springer.

———. 2007b. "Inter-municipal Cooperation: A Widespread Phenomenon." In *Inter-municipal cooperation in Europe*, eds. Rudie Hulst and André van Montfort. Dordrecht: Springer, 1–21.

Ibrahim, Nadine, Lorraine Sugar, Daniel Hoornweg, and Christopher Kennedy. 2012. "Greenhouse gas emissions from cities: Comparison of international inventory frameworks." *Local Environment* 17 (2): 223–41.

ICLEI. 2017. "Who are our Members?" ICLEI. http://www.iclei-europe.org/members/who-are-our-members/ (January 4, 2017).

Ikrat, Alexander. 2012. "Fall Wopperer : Dürre Zeilen von umstrittener Direktorin." *Schwarzwälder Bote,* February 10. http://www.schwarzwaelder-bote.de/inhalt.fall-wopperer-duerre-zeilen-von-umstrittener-direktorin.ade2c4c7-72d3-4f9e-9d83-956b959fa377.html (Accessed October 5, 2016).

———. 2015. "Bilanz: Nicola Schelling: Regionaldirektorin, stets bemüht." *Stuttgarter Nachrichten,* March 31. http://www.stuttgarter-nachrichten.de/inhalt.bilanz-nicola-schelling-regionaldirektorin-stets-bemueht.a7bfbeb7-e348-484d-8489-95f391c7d449.html (Accessed October 5, 2016).

Initiativkreis Europäische Metropolregionen in Deutschland. 2016. "Stuttgart." Initiativkreis Europäische Metropolregionen in Deutschland. http://www.deutsche-metropolregionen.org/mitglieder/stuttgart/ (October 3, 2016).

Institut für Energie- und Umweltforschung (ifeu). 2014. *Konzept für den Masterplan 100% Klimaschutz für die Stadt Heidelberg: Endbericht im Auftrag der Stadt Heidelberg*. Institut für Energie- und Umweltforschung. https://www.heidelberg.de/site/Heidelberg_ROOT/get/documents_E-656386139/heidelberg/Objektdatenbank/31/PDF/Energie%20und%20Klimaschutz/31_pdf_Masterplan%20Bericht%20und%20Ma%C3%9Fnahmen.pdf (Accessed November 3, 2016).

Institut national de la statistique et des études économiques (INSEE). 2012. "EPCI de La CC du Val d'Ille (243500667): Résumé statistique." Institut national de la statistique et des études économiques. http://www.insee.fr/fr/themes/comparateur.asp?codgeo=epci-243500667 (June 28, 2016).

———. 2013. "Insee - Chiffres clés : Aire urbaine de Rennes (011)." Institut national de la statistique et des études économiques. http://www.insee.fr/fr/themes/comparateur.asp?codgeo=au2010-011 (August 14, 2016).

INTERREG IIIC. 2009. *AMICA - Adaptation and Mitigation – an Integrated Climate Policy Approach*. http://www.interreg4c.eu/uploads/media/pdf/AMICA_4W0167N.pdf (Accessed July 26, 2016).

Jacob, Klaus, and Hannah Kannen. 2015. *Integrated Strategies for Climate Policy Integration and Coherence: the Case of Germany*. Berlin: Forschungszentrum für Umweltpolitik. FFU-Report, 2/2015.

Jacob, Klaus, Franziska Wolff, Lisa Graaf, Dirk A. Heyen, and Anna-Lena Guske. 2016. *Dynamiken der Umweltpolitik in Deutschland: Rückschau und Perspektiven*. Dessau: Umweltbundesamt. Texte, 70/2016.

Jacobsson, Staffan, and Volkmar Lauber. 2006. "The politics and policy of energy system transformation—explaining the German diffusion of renewable energy technology." *Energy Policy* 34 (3): 256–76.

Jänicke, Martin. 1997. "Germany." In *National Environmental Policies: A Comparative Study of Capacity-Building*, eds. Martin Jänicke, Helge Jörgens and Helmut Weidner. Berlin, Heidelberg: Springer, 133–55.

———. 2005. "Voraussetzungen effizienter Klimastrategie: Anmerkungen aus Sicht der Politikanalyse." *Vierteljahrshefte zur Wirtschaftsforschung* 74 (2): 208–16.

———. 2006. "Umweltpolitik — auf dem Wege zur Querschnittspolitik." In *Regieren in der Bundesrepublik Deutschland: Innen- und Außenpolitik seit 1949*, ed. Manfred G. Schmidt. Wiesbaden: VS Verlag für Sozialwissenschaften, 405–18.

———. 2015. "Horizontal and Vertical Reinforcement in Global Climate Governance." *Energies* 8 (6): 5782–99.

———. 2017a. "Germany: Innovation and Climate Leadership." In *The European Union in international climate change politics: Still taking a lead?* Vol. 1 of *Routledge studies in European foreign policy*, eds. Rüdiger K. Wurzel, James Connelly and Duncan Liefferink. New York, NY: Routledge, 114–30.

———. 2017b. "The Multi-level System of Global Climate Governance – the Model and its Current State." *Environmental Policy and Governance* 27 (2): 108–21.

Jänicke, Martin, and Lutz Mez. 2000. "Umweltpolitik." In *Handwörterbuch des politischen Systems der Bundesrepublik Deutschland*, ed. Uwe Andersen. Wiesbaden: VS Verlag für Sozialwissenschaften, 596–608.

Jänicke, Martin, Miranda A. Schreurs, and Klaus Töpfer. 2015. *The Potential of Multi-Level Global Climate Governance*. Institute for Advanced Sustainability Studies. IASS Policy Brief, 2/2015. http://www.iass-potsdam.de/sites/default/files/files/potential_of_multi_level_governance.pdf (Accessed December 18, 2016).

Jann, Werner, and Kai Wegrich. 2006. "Theories of the Policy Cycle." In *Handbook of public policy analysis: Theory, politics, and methods*. Vol. 125 of *Public administration and public policy*, eds. Frank Fischer, Gerald Miller and Mara S. Sidney. Boca Raton: CRC Press, 43–62.

Janning, Josef. 2005. "Leadership coalitions and change: The role of states in the European Union." *International Affairs* 81 (4): 821–34.

Jeffery, Charlie. 2000. "Sub-National Mobilization and European Integration: Does it Make any Difference?" *Journal of Common Market Studies* 38 (1): 1–23.

———, ed. 2001. *The Regional Dimension of the European Union: Towards a Third Level in Europe?* 2nd ed. *Routledge Series in Federal Studies*. London, New York, NY: Routledge.

Jensen, Christian B., and Jonathan B. Slapin. 2015. "The politics of multispeed integration in the European Union." In *European Union: Power and policy-making*. 4th ed., eds. Jeremy Richardson and Sonia Mazey. New York, NY: Routledge, 63–80.

Johnson, Craig, Noah Toly, and Heike Schroeder, eds. 2015. *The urban climate challenge: Rethinking the role of cities in the global climate regime*. Vol. 4 of *Cities and global governance*. New York, NY: Routledge.

Joint Research Centre, and PBL Netherlands Environmental Assessment Agency. 2015. *Trends in Global CO2 Emissions: 2015 Report*. The Hague: Joint Research Centre and PBL Netherlands Environmental Assessment Agency. JRC Technical Note, RC98184. http://edgar.jrc.ec.europa.eu/news_docs/jrc-2015-trends-in-global-co2-emissions-2015-report-98184.pdf (Accessed December 16, 2016).

Jones, Bryan D., Tracey Sulkin, and Heather A. Larsen. 2003. "Policy Punctuations in American Political Institutions." *The American Political Science Review* 97 (1): 151–69.

Jordan, Andrew, Dave Huitema, Harro van Asselt, and Johanna Forster, eds. 2018. *Governing Climate Change*. Cambridge University Press.

Jordan, Andrew, Dave Huitema, Harro van Asselt, Tim J. Rayner, and Frans Berkhout, eds. 2010. *Climate change policy in the European Union: Confronting the dilemmas of mitigation and adaptation?* Cambridge, New York, NY: Cambridge University Press.

Jordan, Andrew, and Tim J. Rayner. 2010. "The evolution of climate policy in the European Union: an historical overview." In *Climate change policy in the European Union: Confronting the dilemmas of mitigation and adaptation?*, eds. Andrew Jordan, Dave Huitema, Harro van Asselt, Tim J. Rayner and Frans Berkhout. Cambridge, New York, NY: Cambridge University Press, 52–80.

Jordan, Andrew, and Adriaan Schout. 2008. *The coordination of the European Union: Exploring the capacities of networked governance*. Oxford: Oxford University Press.

Jordan, Andrew, Harro van Asselt, Frans Berkhout, Dave Huitema, and Tim J. Rayner. 2012. "Understanding the Paradoxes of Multilevel Governing: Climate Change Policy in the European Union." *Global Environmental Politics* 12 (2): 43–66.

Jordan, Andrew J., Dave Huitema, Mikael Hildén, Harro van Asselt, Tim J. Rayner, Jonas J. Schoenefeld, Jale Tosun, Johanna Forster, and Elin L. Boasson. 2015. "Emergence of polycentric climate governance and its future prospects." *Nature Climate Change* 5 (11): 977–82.

Jörgens, Helge. 2004. "Governance by Diffusion´ Implementing Global Norms Through Cross-National Imitation and Learning." In *Governance for sustainable development: The challenge of adapting form to function*, ed. William M. Lafferty. Cheltenham, Northampton, MA: Edward Elgar Publishing, 246–83.

Jörgens, Helge, Nina Kolleck, Barbara Saerbeck, and Mareike Well. 2016. "Orchestrating (Bio-)Diversity: The Secretariat of the Convention of Biological Diversity as an Attention-seeking Bureaucracy." In *International Bureaucracies: Challenges and Lessons for Public Administration Research*, eds. Christoph Knill, Michael W. Bauer and Mareike Well. Basingstoke: Palgrave Macmillan, 65–86.

Jörgensen, Kirsten. 2002. *Ökologisch nachhaltige Entwicklung im föderativen Staat: Das Beispiel der deutschen Bundesländer*. Berlin: Forschungsstelle für Umweltpolitik. FFU-Report, 4.

———. 2012. "Governance for Sustainable Development in the German Bundesländer." In *Sustainable development and subnational governments: Policy-making and multi-level interactions*, eds. Hans Bruyninckx, Sander Happaerts and Karoline van den Brande. Houndmills, Basingstoke, Hampshire, New York, NY: Palgrave Macmillan, 103–19.

Jörgensen, Kirsten, Anu Jogesh, and Arabinda Mishra. 2015. "Multi-level climate governance and the role of the subnational level." *Journal of Integrative Environmental Sciences* 12 (4): 235–45.

Jouve, Bernard. 1998. "Planification territoriale, dynamique métropolitaine et innovation institutionnelle: La Région Urbaine de Lyon." *Politiques et management public* 16 (1): 61–82 (Accessed April 11, 2016).

Karrenberg, Hanns, and Engelbert Münstermann. 1999. "Kommunale Finanzen." In *Kommunalpolitik: Politisches Handeln in den Gemeinden*. 2nd ed., eds. Hellmut Wollmann and Roland Roth. Opladen: Leske + Budrich, 437–60.

Keating, Michael. 2008. "Thirty Years of Territorial Politics." *West European Politics* 31 (1-2): 60–81.

———. 2011. "Capacité politique." In *Dictionnaire des politiques territoriales. Références Gouvernances*, eds. Romain Pasquier, Sébastien Guigner and Alistair Cole. Paris: Presses de Sciences Po, 52–57.

Kehrl, Claudia. 2009. *Gemeinsam für den Klimaschutz: Bürgermeister der Metropolregion trafen sich in Heidelberg und beschlossen einheitliche Klimaziele*. Heidelberg. Stadtblatt Amtsanzeiger der Stadt Heidelberg, 41. http://ww2.heidelberg.de/stadt blatt-online/index.php?artikel_id=5953&bf= (Accessed January 20, 2015).

Keohane, Robert O., and Michael Oppenheimer. 2016. "Paris: Beyond the Climate Dead End through Pledge and Review?" *Politics and Governance* 4 (3): 142.

Keohane, Robert O., and David G. Victor. 2011. "The Regime Complex for Climate Change." *Perspectives on Politics* 9 (01): 7–23.

Kern, Kristine. 2003. "Vereint macht stark: Noch ist der Einfluss der Städte und Gemeinden im EU-System nicht groß. Aber kommunale Kooperationen über die Grenzen hinweg sind die Basis für Veränderungen." PRO POLIS 21. http://www2000.wzb .eu/alt/ztn/pdf/kern_propolis03.pdf (November 7, 2014).

Kern, Kristine, and Harriet Bulkeley. 2009. "Cities, Europeanization and Multi-level Governance: Governing Climate Change through Transnational Municipal Networks." *Journal of Common Market Studies* 47 (2): 309–32. http://www.geos.ed.ac.uk/~sallen/dave/Kern%20and%20Bulkeley%20%282009%29.%20Cities,%20Europeanization%20and%20multi-level%20governance.pdf.

Kern, Kristine, Stefan Niederhafner, Sandra Rechlin, and Jost Wagner. 2005. *Kommunaler Klimaschutz in Deutschland – Handlungsoptionen, Entwicklung und Perspektiven*. Wissenschaftszentrum Berlin für Sozialforschung. Discussion Paper, SP IV 2005-101.

Kersting, Norbert, and Angelika Vetter. 2003. *Reforming local government in Europe: Closing the gap between democracy and efficiency*. Vol. 4 of *Urban and regional research international*. Wiesbaden: Springer.

Khan, Jamil. 2010. "Local climate mitigation and network governance: progressive policy innovation or status quo in disguise?" In *Environmental politics and deliberative democracy: Examining the promise of new modes of governance*, eds. Karin Bäckstrand, Jamil Khan, Annica Kronsell and Eva Lövbrand. Cheltenham, Northampton, MA: Edward Elgar Publishing, 197–214.

King, Gary, Robert O. Keohane, and Sidney Verba. 1994. *Designing social inquiry: Scientific inference in qualitative research*. *Princeton paperbacks*. Princeton, NJ: Princeton University Press.

Kinley, Richard. 2016. "Climate change after Paris: From turning point to transformation." *Climate Policy* 17 (1): 9–15.

Klima-Bündnis. 2017. "Das Netzwerk." Klima-Bündnis. http://www.klimabuendnis.org/kommunen/das-netzwerk.html (January 4, 2017).

Klima-Bündnis, Energy Cities, and Klimaschutz- und Energieagentur Baden-Württemberg. 2014. *Plan Climat-Energie Territorial und Klimaschutzkonzept: Praxis und Rahmenbedingungen in Frankreich und Deutschland*. Trans. Claire Mouchard, Enora Garreau, Christiane Maurer, Peter Schilken, Jenny-Claire Keilmann and Ulrike Janssen. Klima-Bündnis, Energy Cities and Klimaschutz- und Energieagentur Baden-Württemberg. http://www.klimabuendnis.org/fileadmin/inhalte/dokumente/2014/Tandem_Vergleichsanalyse.pdf (Accessed April 25, 2016).

Klimaschutz- und Energie-Beratungsagentur Heidelberg-Rhein-Neckar-Kreis gGmbH (KliBA). 2014. "Hand in Hand für den Klimaschutz. KliBA ist mit im Boot." Heidelberg: Klimaschutz- und Energie-Beratungsagentur Heidelberg-Rhein-Neckar-Kreis gGmbH. April 24. http://www.kliba-heidelberg.de/aktuelles_im_detail_kooperationsvertrag_klimaschutz_mrn_rnk.html (April 13, 2016).

Klingebiel, Stephan, and Sebastian Paulo. 2015. *Orchestration: An Instrument for Implementing the Sustainable Development Goals*. German Development Institute. Briefing Paper, 14.

Knill, Christoph. 2008a. "Emerging Patterns of Multi-Level Governance in EU Environmental Policy." In *Multi-Level Governance in the European Union: Taking Stock and Looking Ahead*, eds. Thomas Conzelmann and Randall Smith. Baden-Baden: Nomos, 143–63.

———. 2008b. *Europäische Umweltpolitik: Steuerungsprobleme und Regulierungsmuster im Mehrebenensystem*. 2nd ed. Vol. 4 of *Governance*. Wiesbaden: VS Verlag für Sozialwissenschaften.

———. 2015. "Implementation." In *European Union: Power and policy-making*. 4th ed., eds. Jeremy Richardson and Sonia Mazey. New York, NY: Routledge, 371–96.

Knodt, Michèle, and Nadine Piefer. 2014. "Energiesolidarität im Normdreieck aus Sicherheit, Wettbewerb und Nachhaltigkeit." In *Solidarität in der EU*. 1st ed. Vol. 81 of *Schriftenreihe des Arbeitskreises Europäische Integration e.V.*, eds. Jürgen Bast and Michèle Knodt. Baden-Baden: Nomos, 219–40.

Knodt, Michèle, and Anne Tews. 2014. "Städte im europäischen Mehrebenensystem – die lokale Generierung handlungsrelevanten Wissens im Klimawandel." In *Gemeinden im Europäischen Mehrebenensystem: Herausforderungen im 21. Jahrhundert*, eds. Elisabeth Alber and Carolin Zwilling. Nomos, 271–90.

Kohler-Koch, Beate, and Rainer Eising, eds. 1999. *The transformation of governance in the European Union*. Vol. 12 of *ECPR studies in European political science*. London, New York, NY: Routledge.

Kohler-Koch, Beate, and Fabrice Larat, eds. 2009. *European multi-level governance: Contrasting images in national research*. Cheltenham, Northampton, MA: Edward Elgar Publishing.

Kölliker, Alkuin. 2001. "Bringing together or driving apart the union?: Towards a theory of differentiated integration." *West European Politics* 24 (4): 125–51.

Kooiman, Jan. 2003. *Governing as governance*. Thousand Oaks, CA: Sage Publications.

Koontz, Tomas M., and Craig W. Thomas. 2006. "What Do We Know and Need to Know about the Environmental Outcomes of Collaborative Management?" *Public Administration Review* 66 (1): 111–21.

Kousky, Carolyn, and Stephen H. Schneider. 2003. "Global climate policy: will cities lead the way?" *Climate Policy* 3 (4): 359–72.

Krätke, Stefan. 2007. "Metropolisation of the European Economic Territory as a Consequence of Increasing Specialisation of Urban Agglomerations in the Knowledge Economy." *European Planning Studies* 15 (1): 1–27.

Kropp, Sabine. 2010. *Kooperativer Föderalismus und Politikverflechtung*. Vol. 7 of *Lehrbuch*. Wiesbaden: VS Verlag für Sozialwissenschaften.

Kropp, Sabine, and Nathalie Behnke. 2016. "Marble cake dreaming of layer cake: The merits and pitfalls of disentanglement in German federalism reform." *Regional & Federal Studies* 26 (5): 667–86.

Krotz, Ulrich, and Joachim Schild. 2015. "Embedded Bilateralism: Die deutsch-französischen Sonderbeziehungen in Europa." In *Sonderbeziehungen als Nexus zwischen Außenpolitik und internationalen Beziehungen. Außenpolitik und Internationale Ordnung*, eds. Sebastian Harnisch, Klaus Brummer and Kai Oppermann. Baden-Baden: Nomos, 287–312.

Kruse, Jan. 2015. *Qualitative Interviewforschung: Ein integrativer Ansatz*. 2nd ed. *Grundlagentexte Methoden*. Weinheim: Beltz Juventa.

Kuckartz, Udo. 2010. *Einführung in die computergestützte Analyse qualitativer Daten*. 3rd ed. *Lehrbuch*. Wiesbaden: VS Verlag für Sozialwissenschaften.

Kuhlmann, Sabine, and Geert Bouckaert, eds. 2016. *Local public sector reforms in times of crisis: National trajectories and international comparisons. Governance and public management*. London: Palgrave Macmillan.

Kulovesi, Kati, Elisa Morgera, and Miquel Muñoz. 2011. "Environmental Integration and Multi-Faceted International Dimensions of EU Law: Unpacking the EU's 2009 Climate and Energy Package." *Common Market Law Review* 48: 829–91.

Kuronen, Marjo, and Pascal Caillaud. 2015. "Vertical Governance, National Regulation and Autonomy of Local Policy Making." In *Local welfare policy making in European cities*. Vol. 59 of *Social indicators research series*, eds. Dagmar Kutsar and Marjo Kuronen. Cham: Springer, 71–85.

L'Agence Locale de l'Energie de l'agglomération lyonnaise (ALE). 2015. "Présentation." L'Agence Locale de l'Energie de l'agglomération lyonnaise. http://www.ale-lyon.org/qui-sommes-nous/presentation.html (April 3, 2016).

Laffan, Brigid. 1997. "From policy entrepreneur to policy manager: The challenge facing the European Commission." *Journal of European Public Policy* 4 (3): 422–38.

Landauer, Mia, Sirkku Juhola, and Maria Söderholm. 2015. "Inter-relationships between adaptation and mitigation: A systematic literature review." *Climatic Change* 131 (4): 505–17.

Landsberg, Gerd. 2007. "Der Deutsche Städte- und Gemeindebund." In *Handbuch der kommunalen Wissenschaft und Praxis*. 3rd ed., eds. Thomas Mann and Günter Püttner. Berlin, New York, NY: Springer, 963–80.

Lang, Dorothee. 2008. "Ein Europäer im Denken und Handeln: Der Versuch eines Rückblicks auf Arbeitsschwerpunkte von Dr. Bernd Steinacher." In *Region Stuttgart Aktuell*, 11–13.

———. 2011. "Für ein gutes Klima: Region ist beim Klimaschutz breit aufgestellt." In *Region Stuttgart Aktuell*, 22–23.

Larrue, Corinne. 2002. "Environmental Capacity Building in France." In *Capacity Building in National Environmental Policy: A Comparative Study of 17 Countries*, eds. Helmut Weidner and Martin Jänicke. Berlin, Heidelberg: Springer, 201–18.

Lascoumes, Pierre, and Patrick Le Galès. 2004. "Conclusion : De l'innovation instrumentale à la recomposition de l'Etat." In *Gouverner par les instruments. Références Gouvernances*, eds. Pierre Lascoumes and Patrick Le Galès. Paris: Presses de Sciences Po, 357–70.

Lauber, Volkmar, and Lutz Mez. 2004. "Three Decades of Renewable Electricity Policies in Germany." *Energy & Environment* 15 (4): 599–623.

Lauth, Hans-Joachim, Gert Pickel, and Susanne Pickel. 2009. *Methoden der vergleichenden Politikwissenschaft: Eine Einführung. Lehrbuch*. Wiesbaden: VS Verlag für Sozialwissenschaften.

Le Galès, Patrick. 2011. *Le retour des villes européennes: Sociétés urbaines, mondialisation, gouvernement et gouvernance*. 2nd ed. *Références Gouvernances*. Paris: Presses de Sciences Po.

———. 2013. "Postface: La gouvernance territoriale sous pression de la crise et de la restructuration de l'État." In *La gouvernance territoriale: Pratiques, discours et théories*. 2nd ed. *Droit et société*, eds. Romain Pasquier, Vincent Simoulin and Julien Weisbein. Paris: L.G.D.J., 289–300.

Le Guillou, Pascale. 2016. "Le maire André Crocq démissionne mais reste conseiller." *Ouest France*, April 2. http://www.ouest-france.fr/bretagne/chavagne-35310/le-maire-andre-crocq-demissionne-mais-reste-conseiller-4141058 (Accessed August 23, 2016).

Le Saout, Rémy. 2011. "Commune." In *Dictionnaire des politiques territoriales. Références Gouvernances*, eds. Romain Pasquier, Sébastien Guigner and Alistair Cole. Paris: Presses de Sciences Po, 80–85.

———. 2012. "Introduction." In *Réformer l'intercommunalité: Enjeux et controverses autour de la réforme des collectivités territoriales. Res publica*, ed. Rémy Le Saout. Rennes: Presses universitaires de Rennes, 11–17.

Lee, Taedong. 2013. "Global Cities and Transnational Climate Change Networks." *Global Environmental Politics* 13 (1): 108–27.

Lepenies, Philipp H. 2014. "Die Politik der messbaren Ziele: Die Millennium Development Goals aus gouvernementalitätstheoretischer Sicht." In *Entwicklungstheorien: Weltgesellschaftliche Transformationen, entwicklungspolitische Herausforderungen, theoretische Innovationen*. Vol. 48 of *Politische Vierteljahresschrift Sonderheft*, eds. Cord Jakobeit, Franziska Müller, Elena Sondermann, Ingrid Wehr and Aram Ziai. Baden-Baden: Nomos, 205–29.

Levin, Kelly, Benjamin Cashore, Steven Bernstein, and Graeme Auld. 2012. "Overcoming the tragedy of super wicked problems: Constraining our future selves to ameliorate global climate change." *Policy Sciences* 45 (2): 123–52.

Listera, Jane, René T. Poulsen, and Stefano Ponte. 2015. "Orchestrating transnational environmental governance in maritime shipping." *Global Environmental Change* 34: 185–95.

Löfstedt, Ragnar E., and Ute Collier, eds. 2013. *Cases in Climate Change Policy: Political Reality in the European Union*. 2nd ed. New York, NY: Earthscan.

Ludwig, Jürgen, ed. 2009. *Metropolregionen in Deutschland: 11 Beispiele für Regional Governance*. 2nd ed. Baden-Baden: Nomos.

Luterbacher, Urs, and Detlef F. Sprinz, eds. 2001. *International relations and global climate change*. Cambridge, MA, London: MIT Press.

Lutsey, Nicholas P., and Danniel Sperling. 2008. "America's Bottom-Up Climate Change Mitigation Policy." *Energy Policy* 36: 673–85.

Mahoney, James. 2010. "After KKV: The New Methodology of Qualitative Research." *World Politics* 62 (1): 120–47.

Majone, Giandomenico, ed. 1996. *Regulating Europe. European public policy series*. London, New York, NY: Routledge.

———. 2001. "Two Logics of Delegation: Agency and Fiduciary Relations in EU Governance." *European Union Politics* 2 (1): 103–22.

Mann, Thomas, and Günter Püttner, eds. 2007. *Handbuch der kommunalen Wissenschaft und Praxis*. 3rd ed. Berlin, New York, NY: Springer.

Marcou, Gérard. 2011. "Clause générale de compétence." In *Dictionnaire des politiques territoriales. Références Gouvernances*, eds. Romain Pasquier, Sébastien Guigner and Alistair Cole. Paris: Presses de Sciences Po, 57–62.

Marks, Gary. 1993. "Structural policy and multilevel governance in the EC." In *The State of the European Community: The Maastricht Debates and Beyond*. Vol. 2 of *European Community Studies Association Biennial series*, eds. Alan W. Cafruny and Glenda G. Rosenthal. Boulder, CO: Lynne Rienner, 391–410.

Marks, Gary, and Liesbet Hooghe. 2004. "Contrasting Visions of Multi-level Governance." In *Multi-level Governance*, eds. Ian Bache and Matthew Flinders. Oxford University Press, 15–30.

Marks, Gary, Liesbet Hooghe, and Kermit Blank. 1996. "European Integration from the 1980s: State-Centric v. Multi-level Governance." *Journal of Common Market Studies* 34 (3): 341–78.

Marrel, Guillaume. 2011. "Cumul des mandats." In *Dictionnaire des politiques territoriales. Références Gouvernances*, eds. Romain Pasquier, Sébastien Guigner and Alistair Cole. Paris: Presses de Sciences Po, 115–21.

Marsden, Greg, Antonio Ferreira, Ian Bache, Matthew Flinders, and Ian Bartle. 2013. "Muddling through with climate change targets: A multi-level governance perspective on the transport sector." *Climate Policy* 14 (5): 617–36.

Marshall, Adam. 2005. "Europeanization at the urban level: Local actors, institutions and the dynamics of multi-level interaction." *Journal of European Public Policy* 12 (4): 668–86.

Martínez Soria, José. 2007. "Kommunale Selbstverwaltung im europäischen Vergleich." In *Handbuch der kommunalen Wissenschaft und Praxis*. 3rd ed., eds. Thomas Mann and Günter Püttner. Berlin, New York, NY: Springer, 1015–43.

Mattli, Walter, and Jack Seddon. 2015. "Orchestration along the Pareto frontier." In *International Organizations as Orchestrators*, eds. Kenneth W. Abbott, Philipp Genschel, Duncan Snidal and Bernhard Zangl. Cambridge: Cambridge University Press, 315–48.

Mayer, Helmut, Hyunjung Lee, Annika Oertel, Robert Schulze-Dieckhoff, Matthias Schmid, Britta Steinerstauch, Thomas Lampen, Rainer Kapp, Ulrich Reuter, and Hermann Lambert Oediger. 2015. *KlippS - Klimaplanungspass Stuttgart*. Karlsruhe. KLIMOPASS-Berichte. https://www.stadtklima-stuttgart.de/stadtklima_filestorage/download/KlippS-Klimaplanungspass-Stuttgart.pdf (Accessed September 23, 2016).

Mayors Adapt. 2015. "Stuttgart: Germany." Mayors Adapt. http://mayors-adapt.eu/wp-content/uploads/2015/06/Stuttgart.pdf (September 28, 2016).

Meijers, Evert, Marloes Hoogerbrugge, and Koen Hollander. 2014. "Twin cities in the process of metropolisation." *Urban Research & Practice* 7 (1): 35–55.

Meijers, Evert J. 2007. *Synergy in polycentric urban regions: Complementarity, organising capacity and critical mass* [Summary in Dutch]. Vol. 13 of *Sustainable urban areas*. Delft: Delft University Press.

Mendonça, Miguel, David Jacobs, and Benjamin K. Sovacool. 2009. *Powering the green economy: The feed-in tariff handbook*. Sterling, VA: Earthscan.

Mercier, Geoffrey. 2014. *Lyon, comment Collomb a résisté: Sa stratégie face à la vague UMP et FN aux municipales. Les acteurs de la campagne 2014*. Lyon: Le Progrès.

Métropole de Lyon. 2015. "La loi MAPAM." Métropole de Lyon. http://www.grandlyon.com/metropole/la-loi-mapam.html (July 7, 2016).

———. 2016. "Budget." Métropole de Lyon. http://www.grandlyon.com/metropole/budget.html (July 10, 2016).

Metropolregion Rhein-Neckar GmbH. 2013. *Newsletter Metropolregion Rhein-Neckar Energie & Umwelt*. Metropolregion Rhein-Neckar GmbH. 1. https://www.m-r-n.com/fileadmin/user_upload/Image/05_Meta/Newsletter/Energie-Umwelt/2013/Newsletter_EuU_MRN_01_01.pdf (Accessed October 26, 2016).

Meuser, Michael, and Ulrike Nagel. 2009. "Das Experteninterview – konzeptionelle Grundlagen und methodische Anlage." In *Methoden der vergleichenden Politik- und Sozialwissenschaft: Neue Entwicklungen und Anwendungen*, eds. Susanne Pickel, Gert Pickel, Hans-Joachim Lauth and Detlef Jahn. Wiesbaden: VS Verlag für Sozialwissenschaften, 465–79.

Meyer, Lukas H., and Dominic Roser. 2006. "Distributive Justice and Climate Change. The Allocation of Emission Rights." *Analyse & Kritik* 28 (2).

Michaelowa, Axel. 2008. "German Climate Policy Between Global Leadership and Muddling Through." In *Turning down the heat: The politics of climate policy in affluent democracies*, eds. Hugh Compston and Ian Bailey. Basingstoke, New York, NY: Palgrave Macmillan, 144–63.

Michaelowa, Katharina, and Axel Michaelowa. 2016. "Transnational Climate Governance Initiatives: Designed for Effective Climate Change Mitigation?" *International Interactions:* 1–27.

Ministerkonferenz für Raumordnung. 2016. *Leitbilder und Handlungsstrategien für die Raumentwicklung in Deutschland.* Berlin. http://www.deutsche-metropolregionen.org/fileadmin/ikm/01_allgemein/2016_MKRO-Leitbildbroschuere_DE.pdf (Accessed January 1, 2017).

Monstadt, Jochen, and Stefan Scheiner. 2016. "Die Bundesländer in der nationalen Energie- und Klimapolitik: Räumliche Verteilungswirkungen und föderale Politikgestaltung der Energiewende." *Raumforschung und Raumordnung* 74 (3): 179–97.

Morgenstern, Richard D., and William A. Pizer, eds. 2010. *Reality Check: The Nature and Performance of Voluntary Environmental Programs in the United States, Europe, and Japan.* 2nd ed. Washington, D.C.: Resources for the Future.

Moser, Susanne C. 2012. "Adaptation, mitigation, and their disharmonious discontents: An essay." *Climatic Change* 111 (2): 165–75.

Müller, Patrick, and Peter Slominski. 2013. "Agree now – pay later: Escaping the joint decision trap in the evolution of the EU emission trading system." *Journal of European Public Policy* 20 (10): 1425–42.

MVV Energie AG. 2007. *Climate Protection Atlas: Climate Protection Projects in the Rhine-Neckar Metropolitan Region.* http://www.klimaschutz-rhein-neckar.de/pdf/Klimaatlas_abridged_english.pdf (Accessed November 2, 2016).

Nasiritousi, Naghmeh, Mattias Hjerpe, and Björn-Ola Linnér. 2014. "The roles of non-state actors in climate change governance: understanding agency through governance profiles." *International Environmental Agreements: Politics, Law and Economics.*

Naßmacher, Hiltrud. 2015. "Föderalismus aus kommunaler Perspektive." *Verwaltung & Management* 21 (4): 206–11.

Natural Resources Defense Council. 2013. *A New Architecture for a New Global Partnership for Sustainable Development: A proposal from the Natural Resources Defense Council to the UN Open Working Group on Sustainable Development Goals.* New York, NY: Natural Resources Defense Council. https://sustainabledevelopment.un.org/content/documents/1178A%20New%20Architecture%20for%20a%20New%20Global%20Partnership.pdf (Accessed December 2, 2016).

Négrier, Emmanuel. 2005. *La question métropolitaine: Les politiques à l'épreuve du changement d'échelle territoriale. Collection Symposium.* Grenoble: Presses universitaires de Grenoble.

———. 2007. "L'échelle métropolitaine pour repenser la politique." In *Action publique et changements d'échelles: Les nouvelles focales du politique* [Résumés bilingues français-anglais]. *Logiques politiques*, eds. Alain Faure, Jean-Philippe Leresche, Pierre Muller and Stéphane Nahrath. Paris: l'Harmattan, 29–44.

———. 2011. "Échelles d'action publique." In *Dictionnaire des politiques territoriales. Références Gouvernances*, eds. Romain Pasquier, Sébastien Guigner and Alistair Cole. Paris: Presses de Sciences Po, 195–200.

Négrier, Emmanuel, Philippe Teillet, and Julien Préau. 2008. *Intercommunalités: Le temps de la culture.* Grenoble: Observatoire des politiques culturelles.

Networking the Covenant of Mayors (NETCOM). 2017. "National Clubs." Networking the Covenant of Mayors. http://www.networkingcovenantofmayors.eu/National-Clubs,15.html (January 6, 2017).

Nierhaus, Michael. 1999. "Die kommunale Selbstverwaltung." In *Kommunale Selbstverwaltung*, eds. Christiane Büchner and Jochen Franzke. Berlin: Berliner Debatte Wissenschaftsverlag, 8–33.

Oberthür, Sebastian. 2016a. "Reflections on Global Climate Politics Post Paris: Power, Interests and Polycentricity." *The International Spectator* 51 (4): 80–94.

———. 2016b. "The Paris Agreement: Rebooting Climate Cooperation: Perspectives on EU Implementation of the Paris Outcome." *Carbon & Climate Law Review* 10 (1): 34–45.

Oberthür, Sebastian, and Lisanne Groen. 2016. "The European Union and the Paris Agreement: Leader, mediator, or bystander?" *Wiley Interdisciplinary Reviews: Climate Change*.

Oberthür, Sebastian, and Hermann Ott. 1999. *The Kyoto Protocol: International climate policy for the 21st century. International and European Environmental Policy Series*. New York, NY: Springer.

Oberthür, Sebastian, and Marc Pallemaerts. 2010a. "The EU's Internal and External Climate Policies: an Historical Overview." In *The new climate policies of the European Union: Internal legislation and climate diplomacy*. Vol. 15 of *Institute for European Studies publication series*, eds. Sebastian Oberthür and Marc Pallemaerts. Trans. Claire Roche Kelly. Brussels: VUB PRESS Brussels University Press, 27–63.

———, eds. 2010b. *The new climate policies of the European Union: Internal legislation and climate diplomacy*. Vol. 15 of *Institute for European Studies publication series*. Trans. Claire Roche Kelly. Brussels: VUB PRESS Brussels University Press.

Oberthür, Sebastian, and Claire Roche Kelly. 2008. "EU Leadership in International Climate Policy: Achievements and Challenges." *The International Spectator* 43 (3): 35–50.

Oekom Verlag. 2016. "Das Konzept des Klimasparbuchs." Oekom Verlag. http://www.klimasparbuch.net/konzept.html (October 1, 2016).

Ohlhorst, Dörte. 2015. "Germany's energy transition policy between national targets and decentralized responsibilities." *Journal of Integrative Environmental Sciences* 12 (4): 303–22.

Ohlhorst, Dörte, Kerstin Tews, and Miranda A. Schreurs. 2014. "Energiewende als Herausforderung der Koordination im Mehrebenensystem." In *Im Hürdenlauf zur Energiewende: Von Transformationen, Reformen und Innovationen*, eds. Achim Brunnengräber and Maria R. Di Nucci. Wiesbaden: Springer VS, 93–104.

Ollitrault, Sylvie. 2011. "Développement durable." In *Dictionnaire des politiques territoriales. Références Gouvernances*, eds. Romain Pasquier, Sébastien Guigner and Alistair Cole. Paris: Presses de Sciences Po, 171–76.

O'Neill, Kate. 2009. *The Environment and International Relations. Themes in international relations*. Cambridge, UK, New York: Cambridge University Press.

Organisation for Economic Co-operation and Development (OECD). *Competitive Cities and Climate Change: Milan 9.-10.10.2008*. OECD Conference Proceedings.

http://search.oecd.org/governance/regional-policy/50594939.pdf (Accessed April 25, 2016).

———. 2010. *Cities and Climate Change*. Paris: OECD Publishing.

———. 2016. *France: 2016*. OECD Environmental Performance Reviews. Paris: OECD Publishing.

Ostrom, Elinor. 2005. *Understanding institutional diversity*. *Princeton paperbacks*. Princeton, NJ: Princeton University Press.

———. 2009. *A Polycentric Approach For Coping With Climate Change: Background paper to the 2010 World Development Report*. World Bank. Policy Research Working Paper, 5095.

Pablo-Romero, María d. P., Antonio Sánchez-Braza, and José Manuel González-Limón. 2015. "Covenant of Mayors: Reasons for Being an Environmentally and Energy Friendly Municipality." *Review of Policy Research* 32 (5): 576–99.

Page, Edward A. 2008. "Distributing the burdens of climate change." *Environmental Politics* 17 (4): 556–75.

Panara, Carlo, and Michael R. Varney. 2013a. "Introduction: Local government in the EU multilayer system of governance." In *Local Government in Europe: The 'Fourth Level' in the EU Multi-Layered System of Governance*. *Routledge Research in European Union Law*, eds. Carlo Panara and Michael R. Varney. Hoboken, NJ: Taylor and Francis.

———, eds. 2013b. *Local Government in Europe: The 'Fourth Level' in the EU Multi-Layered System of Governance*. *Routledge Research in European Union Law*. Hoboken, NJ: Taylor and Francis.

Parker, Charles F., and Christer Karlsson. 2010. "Climate Change and the European Union's Leadership Moment: An Inconvenient Truth?" *Journal of Common Market Studies* 48 (4): 923–43.

Pasquier, Romain, and Gilles Pinson. 2004. "Politique européenne de la ville et gouvernement local en Espagne et en Italie." In *L'Europe au microscope du local*. Vol. 12 of *Politique européenne*, eds. Romain Pasquier and Julien Weisbein. Paris: l'Harmattan, 42–65.

Pasquier, Romain, Vincent Simoulin, and Julien Weisbein, eds. 2013. *La gouvernance territoriale: Pratiques, discours et théories*. 2nd ed. *Droit et société*. Paris: L.G.D.J.

Pasquier, Romain, and Julien Weisbein. 2013. "Conclusion: La 'gouvernance territoriale': une perspective toujours aussi plurielle." In *La gouvernance territoriale: Pratiques, discours et théories*. 2nd ed. *Droit et société*, eds. Romain Pasquier, Vincent Simoulin and Julien Weisbein. Paris: L.G.D.J., 269–87.

Pastille Consortium. 2002. *Indicators into action: Local sustainability indicator sets in their context*. London: Pastille Consortium.

Pegram, Tom. 2015. "Global human rights governance and orchestration: National human rights institutions as intermediaries." *European Journal of International Relations* 21 (3): 595–620.

Peters, B. G. 1997. "Escaping the joint-decision trap: Repetition and Sectoral politics in the European union." *West European Politics* 20 (2): 22–36.

———. 1998. "Managing Horizontal Government: The Politics of Co-Ordination." *Public Administration* 76 (2): 295–311.

Pettibone, Lisa. 2015. *Governing urban sustainability: Comparing cities in the USA and Germany. Urban planning and environment.* Farnham, Surrey, UK: Ashgate.

Piattoni, Simona. 2010. *The theory of multi-level governance: Conceptual, empirical, and normative challenges.* Oxford: Oxford University Press.

Pierson, Paul. 2004. *Politics in Time: History, Institutions, and Social Analysis.* Princeton, NJ: Princeton University Press.

Poppe, Stéphanie. 2013. "Rennes Métropole Case Study: OECD 9th Rural Development Policy Conference " Rural -Urban partnerships: an integrated approach to economic development"." SESSION IV: Rural and urban planning to promote greater co-operation 25.10.2013, Bologna. http://www.oecd.org/gov/regional-policy/Session%20IV%20presentations.pdf (Accessed August 15, 2016).

Potoski, Matthew, and Aseem Prakash, eds. 2009. *Voluntary Programs: A Club Theory Perspective.* Cambridge, MA, London: MIT Press.

2015. "Pour lutter contre le changement climatique, les villes s'engagent et l'Europe doit les appuyer." [fr]. *Le Monde,* March 25. http://www.lemonde.fr/idees/article/2015/03/25/pour-lutter-contre-le-changement-climatique-les-villes-s-engagent-et-l-europe-doit-les-appuyer_4600894_3232.html (Accessed July 21, 2016).

Prakash, Aseem, and Matthew Potoski. 2012. "Voluntary environmental programs: A comparative perspective." *Journal of Policy Analysis and Management* 31 (1): 123–38.

Presse- und Informationsstelle Viernheim. 2010. *Für mehr Klimaschutz: Elf Kommunen der Metropolregion treten in Heidelberg am 16.04. dem "Konvent der Bürgermeister" bei.*

Przeworski, Adam, and Henry Teune. 1970. *The Logic of Comparative Social Inquiry.* New York: Wiley-Interscience.

Purdon, Mark. 2015. "Advancing Comparative Climate Change Politics: Theory and Method." *Global Environmental Politics* 15 (3): 1–26.

Quitzow, Leslie, Weert Canzler, Philipp Grundmann, Markus Leibenath, Timothy Moss, and Tilmann Rave. 2016. "The German Energiewende – What's happening?: Introducing the special issue." *Utilities Policy* 41: 163–71.

Rabe, Barry G. 2004. *Statehouse and greenhouse: The emerging politics of American climate change policy.* Washington, D.C.: Brookings Institution Press.

———. 2007. "Beyond Kyoto: Climate Change Policy in Multilevel Governance Systems." *Governance* 20 (3): 423–44.

Rangeon, François. 2011. "Intérêt local." In *Dictionnaire des politiques territoriales. Références Gouvernances,* eds. Romain Pasquier, Sébastien Guigner and Alistair Cole. Paris: Presses de Sciences Po, 289–95.

Rayner, Tim J., and Andrew Jordan. 2013. "The European Union: The polycentric climate policy leader?" *Wiley Interdisciplinary Reviews: Climate Change* 4 (2): 75–90.

Reckien, Diana, Johannes Flacke, Richard J. Dawson, Oliver Heidrich, Marta Olazabal, Aoife Foley, Joël J.-P. Hamann, Hans Orru, Monica Salvia, Sonia d. Gregorio Hurtado, David Geneletti, and Filomena Pietrapertosa. 2014. "Climate change response in Europe: What's the reality? Analysis of adaptation and mitigation plans from 200 urban areas in 11 countries." *Climatic Change* 122 (1-2): 331–40.

Reed, Maureen G., and Shannon Bruyneel. 2010. "Rescaling environmental governance, rethinking the state: A three-dimensional review." *Progress in Human Geography* 34 (5): 646–53.

Région Bretagne. 2011. *Orientations pour le budget 2012: 8ème réunion – Décembre 2011*. Région Bretagne. http://www.bretagne.bzh/upload/docs/application/pdf/2011-12/orientations_budget_2012-_rapport_session_decembre_2011.pdf (Accessed August 26, 2016).

———. 2014a. *Plan climat-énergie territoire régional 2014-2019 du Conseil régional de Bretagne: Rapport adopté*. Région Bretagne. http://www.bretagne.bzh/upload/docs/application/pdf/2014-06/pcet_rb_09_04_14.pdf (Accessed August 26, 2016).

Region Stuttgart. 2009a. *Regionalplan*.

———. 2009b. *Klimaschutzaktivitäten des Verbands Region Stuttgart und der Wirtschaftsförderung Region Stuttgart GmbH: Anlage 1 zur Vorlage Nr. 10/2009 WIV am 25.11.2009*. Region Stuttgart (Accessed October 3, 2016).

Régnier, Yannick. 2012. "The community of communes of the Val d'Ille is aiming at becoming a positive energy territory." Territoires à énergie positive. February 1. http://www.tepos.be/eng/leurs-realisations/la-communaute-de-communes-du-val-d-ille-vise-a-devenir-un-territoire-a-energie-positive (July 1, 2016).

Renn, Ortwin, and Jonathan P. Marshall. 2016. "Coal, nuclear and renewable energy policies in Germany: From the 1950s to the "Energiewende"." *Energy Policy* 99: 224–32.

Rennes Métropole. 2012. *Rennes Métropole: Pour un développement durable et solidaire*. Rennes: Rennes Métropole. http://metropole.rennes.fr/politiques-publiques/elus-institution-citoyennete/institutions-et-competences/competences-metropolitaines/?no_cache=1&cid=2951&did=5081&sechash=83aa5028 (Accessed August 1, 2016).

Réseau action climat (RAC) France. *Quelle gouvernance territoriale pour la transition énergétique?* Réseau action climat (RAC) France. http://www.rac-f.org/IMG/pdf/Quelle_gouvernance_territoriale_pour_la_transition_e_nerge_tique_.pdf (Accessed April 25, 2016).

Rhein-Neckar-Kreis. 2016. "Koordination Klimaschutz." Rhein-Neckar-Kreis. http://www.rhein-neckar-kreis.de/,Lde/1873684.html (November 1, 2016).

RhônAlpEnergie Environnement. 2012. *Réunion des signataires de la Convention des Maires en Rhône-Alpes: Chambéry Métropole, 21 mars 2012*. RhônAlpEnergie Environnement. http://regions202020.eu/cms/assets/Uploads/Resources/Compte-rendu-Runion-21-mars-2012-CoM.pdf (Accessed June 8, 2016).

Risse-Kappen, Thomas. 2001. "Introduction." In *Transforming Europe: Europeanization and domestic change. Cornell studies in political economy*, eds. Maria Green Cowles, James A. Caporaso and Thomas Risse-Kappen. Ithaca, NY: Cornell University Press, 1–20.

Rocher, Laurence. 2016. "Governing metropolitan climate-energy transition: A study of Lyons strategic planning." *Urban Studies*: 1–16.

Roger, Charles, Thomas N. Hale, and Liliana B. Andonova. 2016. "The Comparative Politics of Transnational Climate Governance." *International Interactions*: 1–25.

Rohlfing, Ingo. 2009. "Vergleichende Fallanalysen." In *Methoden der vergleichenden Po-litik- und Sozialwissenschaft: Neue Entwicklungen und Anwendungen*, eds. Susanne Pickel, Gert Pickel, Hans-Joachim Lauth and Detlef Jahn. Wiesbaden: VS Verlag für Sozialwissenschaften, 133–51.

Rosenzweig, Cynthia, William D. Solecki, Stephen A. Hammer, and Shagun Mehrotra, eds. 2011. *Climate Change and Cities: First Assessment Report of the Urban Cli-mate Change Research Network*. Cambridge, MA: Cambridge University Press.

Roth, Dieter. 2013. "Baden-Württemberg 2011: Was entschied die Wahl?" In *Der histo-rische Machtwechsel: Grün-Rot in Baden-Württemberg*. 1st ed. Vol. 1 of *Verglei-chende Analyse politischer Systeme*, eds. Uwe Wagschal, Ulrich Eith and Michael Wehner. Baden-Baden: Nomos, 15–30.

Roussel, Florence. 2005. "Les collectivités reçoivent un guide pour la mise en place de leurs plans climats territoriaux." Actu-Environnement. November 28. http://www.actu-environnement.com/ae/news/1398.php4 (April 3, 2016).

Rowe, Carolyn, and Wade Jacoby, eds. 2013. *German Federalism in Transition: Reforms in a Consensual State*. Abingdon, Oxon, New York, NY: Routledge.

Ruth, Matthias, and María E. Ibarrarán, eds. 2009. *Distributional impacts of climate change and disasters: Concepts and cases. New horizons in environmental econom-ics*. Cheltenham, UK, Northampton, MA: Edward Elgar Publishing.

Ryan, Daniel. 2015. "From commitment to action: A literature review on climate policy implementation at city level." *Climatic Change* 131 (4): 519–29.

Sabatier, Paul A. 1988. "An advocacy coalition framework of policy change and the role of policy-oriented learning therein." *Policy Sciences* 21 (2-3): 129–68.

Sabatier, Paul A., and Hank C. Jenkins-Smith, eds. 1993. *Policy change and learning: An advocacy coalition approach. Theoretical lenses on public policy*. Boulder, CO: Westview Press.

Sabatier, Paul A., and Christopher M. Weible. 2014a. "The Advocacy Coalition Frame-work: Innovations and Clarifications." In *Theories of the policy process*. 3rd ed., eds. Paul A. Sabatier and Christopher M. Weible. Boulder, CO: Westview Press, 189–222.

———, eds. 2014b. *Theories of the policy process*. 3rd ed. Boulder, CO: Westview Press.

Sabel, Charles F., and Jonathan Zeitlin. 2008. "Learning from Difference: The New Ar-chitecture of Experimentalist Governance in the EU." *European Law Journal* 14 (3): 271–327.

Saerbeck, Barbara. 2014. *Unabhängige europäische Regulierungsagenturen: Ihr Ein-flusspotenzial am Beispiel der Europäischen Umweltagentur*. Wiesbaden: Springer VS.

Saldern, Adelheid von. 1999. "Rückblicke. Zur Geschichte der kommunalen Selbstver-waltung in Deutschland." In *Kommunalpolitik: Politisches Handeln in den Gemein-den*. 2nd ed., eds. Hellmut Wollmann and Roland Roth. Opladen: Leske + Budrich, 23–36.

Sartori, Giovanni. 1969. "Politics, Ideology, and Belief Systems." *The American Political Science Review* 63 (2): 398.

Sauviat, Agnès. 2016. "Decentralisation in France: A Principle in Permanent Evolution." In *The Palgrave Handbook of Decentralisation in Europe*, eds. José M. Ruano and Marius Profiroiu. Cham: Palgrave Macmillan, 157–200.

Schaffarzik, Bert. 2007. "Der Schutz der kommunalen Selbstverwaltung im europäischen Mehrebenensystem." In *Handbuch der kommunalen Wissenschaft und Praxis*. 3rd ed., eds. Thomas Mann and Günter Püttner. Berlin, New York, NY: Springer, 269–88.

Scharpf, Fritz W. 1985. "Die Politikverflechtungs-Falle: Europäische Integration und deutscher Föderalismus im Vergleich." *Politische Vierteljahresschrift* 26 (4): 323–56.

———. 1988. "The Joint-Decision Trap: Lessons From German Federalism And European Integration." *Public Administration* 66 (3): 239–78.

———. 1997. "Introduction: The problem-solving capacity of multi-level governance." *Journal of European Public Policy* 4 (4): 520–38.

———. 2006. "The Joint-Decision Trap Revisited." *Journal of Common Market Studies* 44 (4): 845–64.

———. 2010. "Multi-level Europe - the case for multiple concepts." In *Handbook on Multi-level Governance*, eds. Henrik Enderlein, Sonja Wälti and Michael Zürn. Cheltenham, Northampton, MA: Edward Elgar Publishing, 66–79.

Scharpf, Fritz W., Bernd Reissert, and Fritz Schnabel. 1976. *Theorie und Empirie des kooperativen Föderalismus in der Bundesrepublik*. Vol. 1 of *Ergebnisse der Sozialwissenschaften*. Kronberg: Scriptor-Verlag.

Schellnhuber, Hans J., Stefan Rahmstorf, and Ricarda Winkelmann. 2016. "Why the right climate target was agreed in Paris." *Nature Climate Change* 6 (7): 649–53.

Schilken, Peter, Christiane Maurer, Stéphane Dupas, Julia Wyssling, Claire Vasseur, and Jenny-Claire Kersting. 2013. ""Transition énergétique" vs "Energiewende": Eine Begegnung von Kommunen aus Frankreich und Deutschland Stuttgart, 19.-20. März 2013." http://www.kea-bw.de/fileadmin/user_upload/pdf/veranstaltungen/synthese_Stuttgart_de_final.pdf (Accessed August 14, 2016).

Schleich, Joachim, Wolfgang Eichhammer, Ulla Boede, Frank Gagelmann, Eberhard Jochem, Barbara Schlomann, and Hans-Joachim Ziesing. 2001. "Greenhouse gas reductions in Germany - lucky strike or hard work?" *Climate Policy* 1 (3): 363–80.

Schleifer, Philip. 2013. "Orchestrating sustainability: The case of European Union biofuel governance." *Regulation & Governance* 7 (4): 533–46.

Schneider, Diana, and Julia Eustachi. 2014. *Ergebnisse der Umfrage „Kommunaler Klimaschutz und Energiekonzepte"*. Verband Region Rhein-Neckar. http://www.energie-rhein-neckar.com/fileadmin/user_upload/Energie/06_Aktuelles/Auswertung_eUmfrage.pdf (Accessed November 1, 2016).

Schneidewind, Uwe, and Hanna Scheck. 2013. "Die Stadt als „Reallabor" für Systeminnovationen." In *Soziale Innovation und Nachhaltigkeit. Innovation und Gesellschaft*, ed. Jana Rückert-John. Wiesbaden: Springer, 229–48.

Schönberger, Philipp, and Daniel Reiche. 2016. "Why Subnational Actors Matter: The Role of Länder and Municipalities in the German Energy Transition." In *Germany's energy transition: A comparative perspective*, eds. Carol Hager and Christoph H. Stefes. New York, NY: Palgrave Macmillan, 27–61.

Schreurs, Miranda A. 2008. "From the bottom up: Local and subnational climate change politics." [eng]. *The journal of environment & development* 17 (4): 343–55.

———. 2010. "Multi-level Governance and Global Climate Change in East Asia." *Asian Economic Policy Review* 5 (1): 88–105.

———. 2016a. "Is Germany Really an Environmental Leader?" *Current History* 115 (779): 114–16.

———. 2016b. "The German Energiewende in a European Context." In *Germany's energy transition: A comparative perspective*, eds. Carol Hager and Christoph H. Stefes. New York, NY: Palgrave Macmillan, 91–110.

———. 2016c. "The Paris Climate Agreement and the Three Largest Emitters: China, the United States, and the European Union." *Politics and Governance* 4 (3): 219–23. 10.17645/pag.v4i3.666 (Accessed September 20, 2016).

Schreurs, Miranda A., and Sibyl Steuwer. 2015. "Der Koordinierungsbedarf zwischen Bund und Ländern bei der Umsetzung der Energiewende aus politikwissenschaftlicher Sicht." In *Energiewende im Föderalismus*. 1st ed. Vol. 18 of *Schriften zum Umweltenergierecht*, eds. Thorsten Müller and Hartmut Kahl. Baden-Baden: Nomos, 45–68.

Schreurs, Miranda A., and Yves Thiberghien. 2007. "Multi-Level Reinforcement: Explaining European Union Leadership in Climate Change Mitigation." *Global Environmental Politics* 7 (4): 19–46.

Schröder, Carolin, and Heike Walk. 2013. "Local Climate Governance and the Role of Cooperatives." In *Climate change governance. Climate Change Management*, eds. Jörg Knieling and Walter Leal Filho. Berlin, New York: Springer, 105–18.

Schröter, Fabian. 2017. "Ziele der deutschen Wirtschafts- und Energiepolitik." In *Industrielle Energiestrategie: Praxishandbuch für Entscheider des produzierenden Gewerbes*, eds. Frank J. Matzen and Ralf Tesch. Wiesbaden: Springer Gabler, 3–17.

Schwandt, Thomas A. 1997. *Qualitative inquiry: A dictionary of terms*. Thousand Oaks: Sage Publications.

Scott, Joanne, and David M. Trubek. 2002. "Mind the Gap: Law and New Approaches to Governance in the European Union." *European Law Journal* 8 (1): 1–18.

Seele, Günter. 2007. "Die übergemeindliche Kommunalverwaltung in Europa." In *Handbuch der kommunalen Wissenschaft und Praxis*. 3rd ed., eds. Thomas Mann and Günter Püttner. Berlin, New York, NY: Springer, 1045–75.

Selin, Henrik, and Stacy D. VanDeveer. 2007. "Political Science and Prediction: What's Next for U.S. Climate Change Policy?" *Review of Policy Research* 24 (1): 1–27.

———. 2011. "Climate Change Regionalism in North America." *Review of Policy Research* 28 (3): 295–304.

Setton, Daniela, and Sebastian Helgenberger. 2016. "Den Kohlekonsens befördern: Zum aktuellen Beitrag der transformativen Nachhaltigkeitsforschung." *GAIA - Ecological Perspectives for Science and Society* 25 (2): 142–44.

Setzer, Joana. 2015. "Testing the Boundaries of Subnational Diplomacy: The International Climate Action of Local and Regional Governments." *Transnational Environmental Law* 4 (02): 319–37.

Simoulin, Vincent. 2013. "Introduction - La gouvernance territoriale près d'une décennie plus tard: retour sur les discours, les stratégies et les cadres théoriques." In *La gou-*

vernance territoriale: Pratiques, discours et théories. 2nd ed. *Droit et société,* eds. Romain Pasquier, Vincent Simoulin and Julien Weisbein. Paris: L.G.D.J., 3–25.

Sippel, Maike, and Till Jenssen. 2009. *What about local climate governance? A review of promise and problems.* Munich Personal RePEc Archive. MPRA Paper, 20987. https://mpra.ub.uni-muenchen.de/20987/1/ (Accessed December 15, 2016).

Skjærseth, Jon B. 2014. "Linking EU climate and energy policies: Policy-making, implementation and reform." *International Environmental Agreements: Politics, Law and Economics.*

Small, Mario L. 2009. "'How many cases do I need?': On science and the logic of case selection in field-based research." *Ethnography* 10 (1): 5–38.

Sommerer, Thomas. 2011. *Können Staaten voneinander lernen?: Eine vergleichende Analyse der Umweltpolitik in 24 Ländern.* Wiesbaden: VS Verlag für Sozialwissenschaften.

Stadt Esslingen am Neckar. 2016. "Klimaschutz: 25 Prozent weniger Kohlendioxid bis 2020." Stadt Esslingen am Neckar. August 25. http://www.esslingen.de/,Lde/start/es_themen/klimaschutz.html (September 27, 2016).

Stadt Heidelberg. 2006. *Klimaschutzkonzept Heidelberg 2004: Fortschreibung des Handlungsorientierten kommunalen Konzepts zur Reduktion von klimarelevanten Spurengasen für die Stadt Heidelberg 1992.* Stadt Heidelberg. Schriftenreihe zur Umwelt, 1/2006. https://www.heidelberg.de/site/Heidelberg_ROOT/get/documents_E-105 9788406/heidelberg/Objektdatenbank/31/PDF/Energie%20und%20Klimaschutz/ 31_pdf_Klimaschutzkonzept_2004.pdf (Accessed November 3, 2016).

———. 2011. *Mehr Geld für den Klimaschutz in Sicht: Energy Cities freut sich über zusätzliche Fördermittel der EU für die Kommunen.* Heidelberg: Stadt Heidelberg. Stadtblatt Amtsanzeiger der Stadt Heidelberg, 7. http://ww2.heidelberg.de/ stadtblatt-online/index.php?artikel_id=7745&bf= (Accessed January 20, 2015).

———. 2016a. "Konvent der Bürgermeister." Stadt Heidelberg. www.heidelberg.de/hd,Lde/HD/Leben/Konvent+der+Buergermeister.html (April 12, 2016).

———. 2016b. *Konvent der Bürgermeister für Klima und Energie: Heidelberg beteiligt sich an neuen ehrgeizigen Zielen.* Heidelberg.

Stadt Ludwigsburg. 2011. *Europa- und Energieaktivitaten Referat Nachhaltige Stadtentwicklung: Sachstandsbericht Mai 2011.* Stadt Ludwigsburg. http://www.ludwigsburg.de/site/Ludwigsburg-Internet/get/params_E1915676208/1036776/SachstandsberichtEuE_2011_Mai.pdf (Accessed September 29, 2016).

———. 2014a. "Europäische Projekte." Stadt Ludwigsburg. http://www.ludwigsburg.de/ ,Lde/start/stadt_buerger/europa.html (September 29, 2016).

———. 2014b. "Klimaanpassung Region Stuttgart." Stadt Ludwigsburg. http://www.ludwigsburg.de/,Lde/start/stadt_buerger/Klimaanpassung+Region+Stuttgart.html (November 11, 2016).

———. 2016. *Strategisches Fachkonzept Klimaanpassung (KliK).* Stadt Ludwigsburg. http://www.ludwigsburg.de/site/Ludwigsburg-Internet/get/params_E-1985226817/ 13387018/Klimaanpassungskonzept_Ludwigsburg_onlineversion_160713.pdf (Accessed September 29, 2016).

Stadt Mannheim. 2010. *Amtsblatt.* https://www.mannheim.de/sites/default/files/page/ 1563/16_100422amtsblatt.pdf (Accessed April 11, 2016).

Stadt Stuttgart. 2013. *Klimawandel – Anpassungskonzept Stuttgart KLIMAKS*. Stuttgart. Schriftenreihe des Amtes für Umweltschutz, 1/2013. https://www.stadtklima-stuttgart.de/stadtklima_filestorage/download/kliks/KLIMAKS-Broschuere-2013.pdf (Accessed September 23, 2016).

———. 2014. *Energiekonzept für die Stadt Stuttgart - OB Kuhn: "Die Energiewende in Stuttgart geht alle an" - Bürgermeister Hahn: "Zentrale Energieleitplanung"*. Stuttgart.

———. 2016a. "300 Jahre Stadtklima in Stuttgart - ein historischer Rückblick." Stadt Stuttgart. https://www.stadtklima-stuttgart.de/index.php?luft_rueckblick_1698 (September 21, 2016).

———. 2016b. "Beteiligung an Netzwerken." Stadt Stuttgart. https://www.stadtklima-stuttgart.de/index.php?klima_kliks_netzwerke (October 4, 2016).

Statistische Ämter des Bundes und der Länder. 2016. *Verwaltungsgliederung am 31.12.2015: Daten aus dem Gemeindeverzeichnis* [DE]. Wiesbaden: Statistisches Bundesamt. https://www.destatis.de/DE/ZahlenFakten/LaenderRegionen/Regionales/Gemeindeverzeichnis/Administrativ/Archiv/Verwaltung-sgliederung/31122015_Jahr.html;jsessionid=5FA7C3FC4067C579D2571E3D9EF03050.cae1 (Accessed December 31, 2016).

Stecker, Christian. 2016. "The effects of federalism reform on the legislative process in Germany." *Regional & Federal Studies* 26 (5): 603–24.

Steinbacher, Karoline, and Michael Pahle. 2016. "Leadership and the Energiewende: German Leadership by Diffusion." *Global Environmental Politics* 16 (4): 70–89.

Steinberg, Paul F. 2015. "Can We Generalize from Case Studies?" *Global Environmental Politics* 15 (3): 152–75.

Sturm, Roland, and Michael W. Bauer. 2010. "Governance und Regionen – Die theoretische Debatte." In *Regional Governance in EU-Staaten*, eds. Roland Sturm and Jürgen Dieringer. Opladen, Farmington Hills, MI: Budrich, 11–34.

Sturm, Roland, and Jürgen Dieringer, eds. 2010. *Regional Governance in EU-Staaten*. Opladen, Farmington Hills, MI: Budrich.

2014b. "Suivi du SRCAE: Comité de pilotage 13 juin 2014." Rennes.

Szarka, Joseph. 2000. "Environmental policy and neo-corporatism in France." *Environmental Politics* 9 (3): 89–108.

———. 2011. "Climate policy in France: Between national interest and global solidarity?" *Politique européenne* 33 (1): 155–83.

Tallberg, Jonas. 2002. "Paths to Compliance: Enforcement, Management, and the European Union." *International Organization* 56 (3): 609–43.

Tansey, Oisín. 2009. "Process Tracing and Elite Interviewing: A Case for Non-probability Sampling." In *Methoden der vergleichenden Politik- und Sozialwissenschaft: Neue Entwicklungen und Anwendungen*, eds. Susanne Pickel, Gert Pickel, Hans-Joachim Lauth and Detlef Jahn. Wiesbaden: VS Verlag für Sozialwissenschaften, 481–96.

Tatham, Michaël. 2016. *With, Without, or Against the State?: How European Regions Play the Brussels Game. Transformations in Governance*. New York, NY: Oxford University Press.

Tews, Kerstin, and Martin Jänicke, eds. 2015. *Die Diffusion umweltpolitischer Innovationen im internationalen System*. Wiesbaden: VS Verlag für Sozialwissenschaften.

Thomas, Gary. 2011. *How to do your case study: A guide for students and researchers*. London: Sage Publications.

Tömmel, Ingeborg, ed. 2008. *Die Europäische Union*. Wiesbaden: VS Verlag für Sozialwissenschaften.

Touzard, Hubert. 2006. "Consultation, concertation, négociation." *Négociations* 5 (1): 67–74.

Treib, Oliver, Holger Bähr, and Gerda Falkner. 2007. "Modes of governance: Towards a conceptual clarification." *Journal of European Public Policy* 14 (1): 1–20.

Tsebelis, George. 2002. *Veto players: How political institutions work*. New York, NY, Princeton, NJ: Russell Sage Foundation; Princeton University Press.

United Nations Human Settlement Programme (UN-Habitat). 2011. *Cities and climate Change: Global report on human settlements 2011*. Washington, DC.

United Nations Human Settlements Programme (UN-HABITAT). 2008. *State of the World's Cities 2008/2009: Harmonious Cities*. London: Earthscan.

Urpelainen, Johannes. 2009. "Explaining the Schwarzenegger phenomenon: Local frontrunners in climate policy." *Global Environmental Politics* 9 (3): 82–105.

van Asselt, Harro, and Fariborz Zelli. 2014. "Connect the dots: Managing the fragmentation of global climate governance." *Environmental Economics and Policy Studies* 16 (2): 137–55.

van Bever, Eline, Herwig Reynaert, and Kristof Steyvers, eds. 2011. *The road to Europe: Main street or backward alley for local governments in Europe?* Brugge: Vanden Broele.

van der Heijden, Jeroen. 2018. "City and Subnational Governance: High Ambitions, Innovative Instruments and Polycentric Collaborations?" In *Governing Climate Change*, eds. Andrew Jordan, Dave Huitema, Harro van Asselt and Johanna Forster. Cambridge University Press, 81–96.

van Evera, Stephen. 1997. *Guide to methods for students of political science*. Ithaca: Cornell University Press.

van Schaik, Louise. 2010. "The Sustainability of the EU's Model for Climate Diplomacy." In *The new climate policies of the European Union: Internal legislation and climate diplomacy*. Vol. 15 of *Institute for European Studies publication series*, eds. Sebastian Oberthür and Marc Pallemaerts. Trans. Claire Roche Kelly. Brussels: VUB PRESS Brussels University Press, 251–80.

van Staden, Rian. 2014. *Climate Change: Implications for Cities: Key Findings from the Intergovernmental Panel on Climate Change Fifth Assessment Report*. http://www.cisl.cam.ac.uk/publications/publication-pdfs/ipcc-ar5-implications-for-cities-briefing-web-e.pdf/view (Accessed June 10, 2015).

Verband Region Rhein-Neckar. 2012a. *Regionales Energiekonzept Metropolregion Rhein-Neckar: Kurzfassung*. Verband Region Rhein-Neckar.

———. 2012b. *Regionales Energiekonzept Metropolregion Rhein-Neckar: Langfassung*. Verband Region Rhein-Neckar.

———. 2015. "Kooperationsvereinbarung MRN/Rhein-Pfalz-Kreis/Kommunen/Energieagentur RLP." Verband Region Rhein-Neckar. September 17. http://www.energie-rhein-neckar.com/aktuelles/detail/artikel/kooperationsvereinbarung-mrnrhein-pfalz-kreiskommunenenergieagentur-rlp.html?tx_ttnews%5BbackPid%5D=2381& cHash=24b79ee5e6c72760c6862c20517b2453 (November 1, 2016).

———. 2016a. "Energie Rhein-Neckar: Über uns." Verband Region Rhein-Neckar. March 2. http://www.energie-rhein-neckar.com/ueber-uns.html (October 25, 2016).

———. 2016b. "Kurzprofil der Rhein-Neckar-Region." Verband Region Rhein-Neckar. November 3. https://www.m-r-n.com/start/kurzprofil.html (November 11, 2016).

———. 2016c. "Umsetzungsbegleitung: Begleitung der Umsetzung des Regionalen Energiekonzepts der Metropolregion Rhein-Neckar." Verband Region Rhein-Neckar. September 29. http://www.energie-rhein-neckar.com/regionales-energie konzept/umsetzungsbegleitung.html (November 1, 2016).

———. 2016d. "Veranstaltungsreihe zum kommunalen Klimaschutz." Verband Region Rhein-Neckar. June 7. http://www.energie-rhein-neckar.com/veranstaltungen/kom-munaler-klimaschutz.html (October 26, 2016).

———. 2016e. "Regionalkonferenz Energie in Rhein-Neckar." Verband Region Rhein-Neckar. July 14. http://www.energie-rhein-neckar.com/veranstaltungen/regional-konferenz-energie-und-umwelt.html (October 25, 2016).

Verband Region Stuttgart. 2015. *Sitzungsvorlage Nr. 023/2015: Ausschuss für Wirtschaft, Infrastruktur und Verwaltung.* Verband Region Stuttgart.

———. 2016. "Netzwerkarbeit in Europa." Verband Region Stuttgart. https://www.re-gion-stuttgart.org/region-und-europa/netzwerke/ (September 26, 2016).

Ville de Rennes. 2014. *Rapport Annuel préalable au débat d'orientation budgétaire 2015 de la Ville de Rennes sur la situation en matière de Développement Durable: Bilan des activités et politiques publiques de l'année 2014.* Ville de Rennes. http://metro-pole.rennes.fr/fileadmin/rrm/documents/Politiques_publiques/Environn_Eco_Re-cherche/Environnement/Plan_climat/Docs/Rapport_Developpement_durable_2014 _-_Ville_de_Rennes.pdf (Accessed August 15, 2016).

Viola, Lora A. 2015. "The Governance Shift: From Multilateral IGOs to Orchestrated Networks." In *Negotiated Reform: The Multilevel Governance of Financial Regulation.* Vol. 85 of *Schriften des Max-Planck-Instituts für Gesellschaftsforschung Köln*, ed. Renate Mayntz. Frankfurt am Main: Campus-Verlag, 17–36.

Waldmann, Jörg. 2007. "Nie mehr low politics – oder: die EU auf dem Weg zum führen-den Akteur der internationalen Umweltpolitik?" In *Die Europäische Union im 21. Jahrhundert: Theorie und Praxis europäischer Außen-, Sicherheits- und Friedens-politik*, eds. Hans-Georg Ehrhart, Sabine Jaberg, Bernhard Rinke and Jörg Waldmann. Wiesbaden: VS Verlag für Sozialwissenschaften, 251–68.

Wallace, Helen, Mark A. Pollack, and Alasdair R. Young, eds. 2015. *Policy-making in the European Union.* 7th ed. *The new European Union series.* Oxford, UK: Oxford University Press.

Webber, Douglas, ed. 2005. *The Franco-German Relationship in the EU.* 2nd ed. *Rout-ledge Research in European Public Policy.* London, New York, NY: Routledge.

Weibust, Inger, and James Meadowcroft, eds. 2014. *Multilevel Environmental Govern-ance: Managing Water and Climate Change in Europe and North America.* Edward Elgar Publishing.

Weidner, Helmut, and Lutz Mez. 2008. "German Climate Change Policy: A Success Story With Some Flaws." *The journal of environment & development* 17 (4): 356–78.

West, Karen. 2007. "Inter-Municipal Cooperation in France: Incentives, Instrumentality and Empty Shells." In *Inter-municipal cooperation in Europe*, eds. Rudie Hulst and André van Montfort. Dordrecht: Springer, 67–90.

Wick, Martin. 2015. "Klimaschutz auf Länderebene." In *Energiewende im Föderalismus*. 1st ed. Vol. 18 of *Schriften zum Umweltenergierecht*, eds. Thorsten Müller and Hartmut Kahl. Baden-Baden: Nomos, 187–202.

Widerberg, Oscar. 2017. "The 'Black Box' problem of orchestration: How to evaluate the performance of the Lima-Paris Action Agenda." *Environmental Politics* 26 (4): 715–37.

Widerberg, Oscar, Philipp H. Pattberg, and Kristian Kristensen. 2016. *Mapping the Institutional Architecture of Global Climate Change Governance*. 2nd ed. IVM Institute for Environmental Studies. Technical Report, 2.

Wiener, Jonathan B. 2007. "Think Globally, Act Globally: The Limits of Local Climate Policies." *University of Pennsylvania Law Review* 155 (6): 1961–79.

Windhoff-Héritier, Adrienne. 1999. *Policy-making and diversity in Europe: Escaping deadlock. Theories of institutional design*. Cambridge, New York, NY: Cambridge University Press.

Windhoff-Héritier, Adrienne, and Martin Rhodes, eds. 2011. *New modes of governance in Europe: Governing in the shadow of hierarchy. Palgrave studies in European Union politics*. London, New York, NY: Palgrave Macmillan.

Wirtschaftsförderung Region Stuttgart GmbH. 2016a. "Modellregion für nachhaltige Mobilität: Regionalprogramm." Wirtschaftsförderung Region Stuttgart GmbH. http://nachhaltige-mobilitaet.region-stuttgart.de/regionalprogramm/ (October 4, 2016).

———. 2016b. "Our Office in Brussels." Wirtschaftsförderung Region Stuttgart GmbH. August 3. http://eu.region-stuttgart.de/en/office-in-brussels.html (September 26, 2016).

———. 2016c. "The Stuttgart Region's Work in Europe - Objectives and Focus Areas." Wirtschaftsförderung Region Stuttgart GmbH. August 3. http://eu.region-stuttgart.de/en/european-involvement.html (September 26, 2016).

Wolffhardt, Alexander, Herbert Bartik, Richard Meegan, Jens S. Dangschat, and Alexander Hamedinger. 2005. "The European engagement of cities: Experiences, motivations and effects on local governance in Liverpool, Manchester, Vienna, Graz, Dortmund & Hamburg." In *European Metropolitan Governance: Cities in Europe – Europe in the Cities*, eds. Eugen Antalovsky, Jens S. Dangschat and Michael Parkinson, 65–112.

Wollmann, Hellmut, and Geert Bouckaert. 2006. "State Organization in France and Germany between Territoriality and Functionality." In *State and local government reforms in France and Germany: Divergence and convergence*. Vol. 7 of *Urban and regional research international*, eds. Vincent Hoffmann-Martinot and Hellmut Wollmann. Wiesbaden: VS Verlag für Sozialwissenschaften, 11–38.

Woyke, Wichard. 2004. *Deutsch-französische Beziehungen seit der Wiedervereinigung: Das Tandem fasst wieder Tritt*. 2nd ed. Vol. 5 of *Reihe Grundlagen für Europa*. Wiesbaden: VS Verlag für Sozialwissenschaften.

Wurster, Stefan. 2016. "Energiewende in Baden-Württemberg: Ausmaß und Folgen." In *Das grün-rote Experiment in Baden-Württemberg: Eine Bilanz der Landesregierung Kretschmann 2011-2016*, eds. Felix Hörisch and Stefan Wurster. Wiesbaden: VS Verlag für Sozialwissenschaften, 251–78.

Wurzel, Rüdiger K. 2010. "Environmental, Climate and Energy Policies: Path-Dependent Incrementalism or Quantum Leap?" *German Politics* 19 (3-4): 460–78.

Wurzel, Rüdiger K., and James Connelly, eds. 2010. *The European Union as a Leader in International Climate Change Politics*. UACES Contemporary European Studies. London, New York, NY: Routledge.

Wurzel, Rüdiger K., James Connelly, and Duncan Liefferink, eds. 2017. *The European Union in international climate change politics: Still taking a lead?* Vol. 1 of *Routledge studies in European foreign policy*. New York, NY: Routledge.

Wurzel, Rüdiger K., Andrew Jordan, Anthony R. Zito, and Lars Brückner. 2003. "From High Regulatory State to Social and Ecological Market Economy?: New Environmental Policy Instruments in Germany." *Environmental Politics* 12 (1): 115–36.

Yin, Robert K. 2014. *Case study research: Design and methods*. 5th ed.

Zimmermann, Horst. 2016. *Kommunalfinanzen: Eine Einführung in die finanzwissenschaftliche Analyse der kommunalen Finanzwirtschaft*. 3rd ed. Vol. 3235 of *Schriften zur öffentlichen Verwaltung und öffentlichen Wirtschaft*. Berlin: Berliner Wissenschafts-Verlag.

Zürn, Michael, and Benjamin Faude. 2013. "Commentary: On Fragmentation, Differentiation, and Coordination." *Global Environmental Politics* 13 (3): 119–30.

Expert Interviews

Own material

The following list comprises all 47 expert interviews conducted within this research project with a total of 53 individuals including their affiliation, and the place and date of the interview, in chronological order.[243]

Interview with Ulrike Janssen; European Secretariat of the Climate Alliance; Frankfurt, 29.05.2012.

Interview with Frederic Boyer and Andrea Accorigi; Covenant of Mayors Office; Brussels, 30.05.2012.

Interview with Pedro Ballesteros Torres; Directorate-General for Energy of the European Commission; Brussels, 30.05.2012.

Interview with Markus Siehr; Stuttgart Region; Stuttgart, 29.11.2012.

Interview with Philippe Angotti; Association of Urban Communities of France (ACUF); Paris, 28.05.2013.

Interview with Luce Ponsar Greater Lyon; by phone, 13.06.2013.

Interview with Gérard Magnin; Energy Cities; by phone, 17.06.2013.

---

243 One staff at a municipality from Rhine-Neckar Region requested the conversation to remain confidential and objected to be mentioned by name or affiliation.

Interview with Guilhelm Isaac-Georges; Association of Regions of France (ARF); Paris, 21.06.2013.

Interview with Pierre Crépeaux; Greater Lyon; by phone, 04.07.2013.

Interview with Christian Cammal; General Council of Hérault; Montpellier, 05.07.2013.

Interview with Jérôme Barbaroux and Claire Revol-Buisson; Rhône-Alpes Region; Lyon, 09.07.2013.

Interview with Chiara Alice; Lyon City; Lyon, 10.07.2013.

Interview with Fabien Moudileno; Local Energy Agency of Lyon; Villeurbanne, 10.07.2013

Interview with Marion Athiel; Rillieux City; Rillieux, 11.07.2013.

Interview with Emmanuel Goy; AMORCE; Paris, 11.07.2013.

Interview with Benjamin Topper; Environment and Energy Management Agency (ADEME); Paris, 12.07.2013.

Interview with Geneviève Ancel; Greater Lyon; by phone, 16.07.2013.

Interview with Alexia Leseur; Caisse des Dépôts et Consignations (CDC); by phone, 19.07.2013.

Interview with Serge Miquel; General Council of Hérault; Montpellier, 23.07.2013.

Interview with Michel Pieyre; General Council of Hérault; Montpellier, 23.07.2013.

Interview with Jean Casteil; Montpellier City; Montpellier, 24.07.2013.

Interview with Pauline Delaere and Julia Barbier; Association of the Mayors of France (AMF); Paris, 25.07.2013.

Interview with Bernard Brillet; Ministry of the Environment, Energy and the Sea; Paris, 25.07.2013.

Interview with Jonathan Morice; Association of Mayors of French Large Cities (AMGVF); Paris, 25.07.2013.

Interview with François Panouille and Florent Yann Lardic; Association of Small Cities of France (APVF); Paris, 26.07.2013.

Interview with Brendan Catherine; Rennes Metropolis; Berlin, 03.11.2014.

Interview with Jean-Pierre De Nayer; Thorigné-Fouillard; Thorigné-Fouillard, 25.11.2014.

Interview with Steven Bobe; Brittany Region; Rennes, 26.11.2014.

Interview with Jean-Luc Daubaire; Rennes City; Rennes, 26.11.2014.

Interview with Laurent Polès; Local Energy Agency; Rennes, 26.11.2014.

Interview with Bernard Poirier; Rennes Metropolis, Mayor of Mordelles; 27.11.2014.

Interview with Nathalie Chargy; Regional Directorate of the Environment, Land-Use Planning, and Housing (DREAL) Brittany; Rennes, 28.11.2014.

Interview with Christelle Lefevre; Network of small RURal communities for ENERgetic neutrality (RURENER); by video call, 01.12.2014.

Interview with Daniel Guillotin; Local Energy Agency; by phone, 01.12.2014.

Interview with Michel Janssens; Val d'Ille; by phone, 11.12.2014.

Interview with Olivier Dehaese; Rennes Metropolis, Mayor of Acigné; by phone, 12.01.2015.

Interview with Claire Barais; ADEME Brittany; by phone, 14.01.2015.

Interview with Jörg Saalbach; Rhine-Neckar Region; by phone, 20.01.2015.

Interview with Thomas S. Bopp and Thomas Kiwitt; Stuttgart Region; Stuttgart, 28.01.2015.

Interview with Ulrich Reuter; Stuttgart City; Stuttgart, 28.01.2015.

Interview with Stefan Dallinger; Rhine-Neckar Region; Heidelberg, 29.01.2015.

Interview with Ralf Bermich; Heidelberg City; Heidelberg, 29.01.2015.

Interview with Volker Kienzlen; Climate and Energy Agency (KEA) Baden-Württemberg; Karlsruhe, 02.02.2015.

Interview with Julia Eustachi and Axel Finger; Rhine-Neckar Region; Mannheim, 03.02.2015.

Interview with Agnes Schönfelder; Mannheim City; Mannheim, 03.02.2015.

Interview with Svenja Schuchmann; European Secretariat of the Climate Alliance; by phone, 12.02.2015.

Additional Material

I gratefully acknowledge the possibility to include additional expert interview transcripts from a research project at Technical University Darmstadt:

Interview with Jürgen Görres and Stephanie Mehne. Energy Economy Department, Stuttgart City. Stuttgart, 28.01.2013. Conducted by Marina Hofmann, Christoph Stankiewicz, and Jasmin Boghrat.

Interview with Ulrich Reuter. Urban climatology service, Environmental department, Stuttgart City. Stuttgart, 28.01.2013. Conducted by Jasmin Boghrat, Christoph Stankiewicz, and Marina Hofmann.

Interview with Ulrich Reuter. Urban climatology service, Environmental department, Stuttgart City. Stuttgart, 13.02.2013. Conducted by Jörg Kemmerzell, Anne Tews, and Stefan Groer.

Interview with Robert Schulze-Dieckhoff, Britta Steinerstauch and Rainer Kapp. Urban planning department and urban climatology service, environmental department; Stuttgart City. Stuttgart, 19.07.2012. Conducted by Jasmin Boghrat and Anne Tews.

Interview with Wolfgang Schuster. Lord Mayor, Stuttgart City. Stuttgart, 16.05.2013. Conducted by Jörg Kemmerzell, Anne Tews, Wolfram Lamping, and Jasmin Boghrat.

FSC
www.fsc.org

MIX

Papier | Fördert
gute Waldnutzung

FSC® C083411

Zeitfracht Medien GmbH
Ferdinand-Jühlke-Straße 7
99095 Erfurt, Deutschland
produktsicherheit@kolibri360.de